ADVANCES IN CHEMICAL PHYSICS

VOLUME LXXIX

Advances in
CHEMICAL PHYSICS

EDITED BY

I. PRIGOGINE

University of Brussels
Brussels, Belgium
and
University of Texas
Austin, Texas

AND

STUART A. RICE

Department of Chemistry
and
The James Franck Institute
The University of Chicago
Chicago, Illinois

VOLUME LXXIX

AN INTERSCIENCE® PUBLICATION
John Wiley & Sons, Inc.
NEW YORK / CHICHESTER / BRISBANE / TORONTO / SINGAPORE

In recognition of the importance of preserving what has been written, it is a policy of John Wiley & Sons, Inc. to have books of enduring value published in the United States printed on acid-free paper, and we exert our best efforts to that end.

An Interscience® Publication

Library of Congress Catalog Number: 58-9935

ISBN 0-471-52768-8

Printed in the United States of America

10 9 8 7 6 5 4 3 2 1

CONTRIBUTORS TO VOLUME LXXIX

IRVING R. EPSTEIN, Department of Chemistry, Brandeis University, Waltham, Massachusetts

B. FAIN, School of Chemistry, Tel Aviv University, Tel Aviv, Israel

N. HAMER, Los Alamos National Laboratory, Los Alamos, New Mexico

S. H. LIN, Center for Study of Early Events in Photosynthesis, Arizona State University, Tempe, Arizona

T. P. LODGE, Department of Chemistry, University of Minnesota, Minneapolis, Minnesota

YIN LUO, Department of Chemistry, Brandeis University, Waltham, Massachusetts

S. PRAGER, Department of Chemistry, University of Minnesota, Minneapolis, Minnesota

N. A. ROTSTEIN, Department of Chemistry, University of Minnesota, Minneapolis, Minnesota

INTRODUCTION

Few of us can any longer keep up with the flood of scientific literature, even in specialized subfields. Any attempt to do more and be broadly educated with respect to a large domain of science has the appearance of tilting at windmills. Yet the synthesis of ideas drawn from different subjects into new, powerful, general concepts is as valuable as ever, and the desire to remain educated persists in all scientists. This series, *Advances in Chemical Physics*, is devoted to helping the reader obtain general information about a wide variety of topics in chemical physics, which field we interpret very broadly. Our intent is to have experts present comprehensive analyses of subjects of interest and to encourage the expression of individual points of view. We hope that this approach to the presentation of an overview of a subject will both stimulate new research and serve as a personalized learning text for beginners in a field.

ILYA PRIGOGINE
STUART A. RICE

CONTENTS

ADVANCES IN CHEMICAL PHYSICS

VOLUME LXXIX

DYNAMICS OF ENTANGLED POLYMER LIQUIDS: DO LINEAR CHAINS REPTATE?

T. P. LODGE, N. A. ROTSTEIN, and S. PRAGER

Department of Chemistry, University of Minnesota, Minneapolis, MN 55455

"... a long molecule is obliged to ... slalom around ... cross-lying molecules in order that its center of gravity can advance or retreat ..."

—"The Viscosity of High Polymers—The Random Walk of a Group of Connected Segments", H. Eyring, T. Ree, and N. Hirai, *Proc. Nat. Acad. Sci.* **44**, 1213 (1958).

"... [a] branched polymer may be more effectively trapped by its entanglements [than a linear polymer] because its various arms do not allow a simple snaking motion through the entanglements."

—"Solution and Bulk Properties of Branched Polyvinylacetate IV—Melt Viscosity," V. C. Long, G. C. Berry, and L. M. Hobbs, *Polymer* **5**, 517 (1964).

CONTENTS

Advances in Chemical Physics, Volume LXXIX, Edited by I. Prigogine and Stuart A. Rice.
ISBN 0-471-52768-8   © 1990 John Wiley & Sons, Inc.

I. INTRODUCTION

A. General Outline

The central problem in the dynamics of nondilute polymer liquids is to achieve a molecular-level description of the entanglement phenomenon. This goal is clearly of great fundamental interest as well as practical importance, and has been pursued vigorously for at least 50 years. In the last 15 years, attention has largely been focused on the applicability of the *reptation hypothesis*, in which individual chains, constrained by their neighbors, are postulated to move primarily along their own contours in a snakelike fashion.[1] Without doubt, reptation-based models have been very successful in reconciling a wide range of experimental observations, and have stimulated a great deal of novel experimental and theoretical work. Nevertheless, a number of important issues remain unresolved. For instance, the reptation idea remains a postulate, and has not yet been derived from a detailed consideration of intermolecular forces. Also, as will emerge in the subsequent discussion, while reptation does a remarkable job of describing certain phenomena, such as polymer diffusion, quantitative agreement is still lacking for viscoelastic properties. Thus it is appropriate to ask whether a better description will be obtained simply by modifications to the basic reptation picture, or whether a different approach is required. Underlying this question is a more direct one: do linear polymer chains undergo reptative motion in a nonvanishing region of molecular parameter space? The primary objective of this chapter is to address this issue.

This chapter is organized into four sections. In the Introduction the basic phenomenology of entanglement is recalled, and the classes of theoretical approach that have been explored are briefly reviewed. In addition, some fundamental properties of polymer molecules in the liquid state are presented as necessary background. The Introduction concludes with the framing of the questions that the chapter as a whole attempts to address. Section II begins by defining the reptation process and continues with a discussion of the development of the reptation hypothesis into predictions for diffusion, viscosity, and stress relaxation. Several modifications to the reptation hypothesis, which have been proposed in order to reduce discrepancies between theory and experiment, are also outlined. The chapter concludes with a discussion of some of the alternative treatments, emphasizing in particular fundamental differences with the reptation picture. Section III explores the ability of reptation-based models to interpret experimental data, in either a qualitative or a quantitative sense. The experiments are divided into three groups: those that sense chain dynamics in model systems over time scales longer than the longest single-chain relaxation time, those examining dynamics in the same systems over time scales shorter than this molecular time, and those that consider more complicated, but still closely related systems. In the fourth and final section the main evidence for and against the reptation hypothesis is summarized, and answers to the questions posed in the Introduction are proposed.

B.   Phenomenology of Entanglement

As suggested above, intermolecular entanglements dominate the dynamics of concentrated solutions and melts of high-molecular-weight, flexible polymers.[2] Four canonical experiments serve to illustrate the remarkable behavior of such liquids: elastic recovery after release of an imposed stress, stress relaxation after imposition of strain, the dynamic mechanical impedance in response to a sinusoidally time-varying strain, and the molecular-weight dependence of the limiting (zero shear rate) viscosity. Although this set is far from exhaustive, and its elements are not independent, any molecular theory that could describe these four properties in a reasonably quantitative way would represent a major achievement.

1.   Elastic Recovery

It is well known that cross-linked rubbers (i.e., materials comprising individual, flexible polymer molecules that have been covalently interconnected to form networks, and that are substantially above their glass transition temperatures at room temperature) can exhibit complete elastic recovery even after strains approaching 1000%.[3] In other words, even after stretching to 10 times their original length, such materials can recover to essentially their prestretch dimensions. What is more startling is that uncrosslinked samples of the same chemical composition can exhibit the same property, as illustrated in Fig. 1.1.[4] In the well-established theory of rubber elasticity, such recovery is (primarily) the result of an entropically driven return of each network strand to its unstretched, more probable random-coil conformation.[3] Recovery in uncross-linked sampled implies the presence of interchain entanglements which survive over the time scale of the experiment, and which therefore act as temporary cross-links.

2.   Stress Relaxation

After imposition of a step shear strain on an uncross-linked sample, the stress in the material relaxes to zero. The evolution of the relaxation can be described by the relaxation modulus $G(t)$, defined as the shear stress $\sigma(t)$ divided by the imposed (small) strain γ $(t = 0)$:

$$G(t) \equiv \frac{\sigma(t)}{\gamma} \qquad (1.1)$$

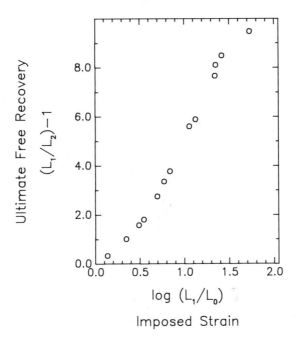

Figure 1.1.   Elastic recovery for molten polyethylene from Meissner.[4] L_0, initial length, L_1—maximum length; L_2—final length.

(Throughout the text we will use the symbols $\equiv$, $=$, $\approx$, and $\sim$ to mean "is defined as," "is equal to," "is approximately equal to," and "is proportional to," respectively.) This is illustrated in Fig. 1.2, where $G(t)$ is plotted logarithmically against time for three polymer samples of increasing molecular weight. At short times, the reduction in $G(t)$ reflects a distribution of relaxation times associated with the relaxation of stress in portions of the individual molecules which are shorter than a certain critical length. The "sub"-molecular nature of these processes is indicated by the molecular-weight independence of $G(t)$. At long times, and for the higher molecular weights, $G(t)$ exhibits a plateau, followed eventually by another relaxation to zero. This so-called rubbery plateau in $G(t)$ is attributable to interchain entanglements or temporary cross-links. Eventually the individual molecules are able to disentangle, and the liquid nature of the sample dominates. Note that the magnitude of $G(t)$ in the plateau is essentially independent of molecular weight, while the longest chain relaxation time, indicated by the onset of the final relaxation, is a strong function of chain length. For a cross-linked sample of the same chemical composition, the form of $G(t)$ would be identical to that for the highest molecular-weight sample in Fig. 1.2, except that the level of

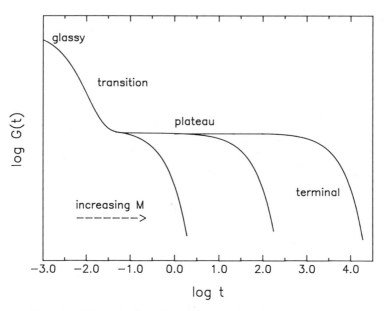

Figure 1.2. Schematic illustration of shear modulus for molten polymers.

the plateau would depend on the cross-link density, and the plateau would persist to infinite time.

3. Dynamic Shear Modulus

The viscoelastic properties of polymer materials are readily examined in an oscillatory shear (i.e., $\gamma(t) = \mathrm{Re}[\gamma_0 e^{i\omega t}]$). The dynamic shear modulus $G^*(\omega)$ is again defined as the ratio of (complex) stress to (complex) strain, with the in-phase (elastic or storage) component G' and the 90° out-of-phase (viscous or loss) component G'' directly related to the sine and cosine Fourier transforms of $G(t)$:

$$G^*(\omega) = \frac{\sigma(t)}{\gamma(t)} = G' + iG'' \qquad (1.2a)$$

$$G'(\omega) = \omega \int_0^\infty G(t)\cos(\omega t)\,dt \qquad (1.2b)$$

$$G''(\omega) = \omega \int_0^\infty G(t)\sin(\omega t)\,dt \qquad (1.2c)$$

Typical G' data for polystyrene samples of varying molecular weights are shown

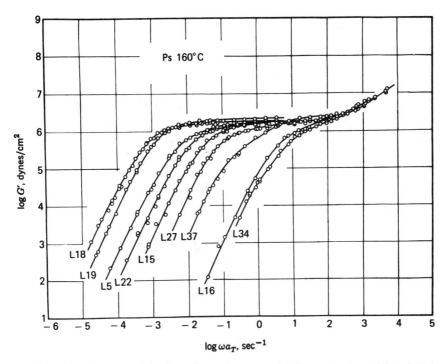

Figure 1.3. Storage modulus for molten polystyrenes of various molecular weights, plotted against reduced frequency. From Onogi et al.[5]

in Fig. 1.3.[5] Again, the crucial feature is the presence of a plateau over a substantial range in frequency, with magnitude independent of molecular weight. As in the stress relaxation experiment, the presence of the plateau is generally ascribed to intermolecular entanglements. [It should be noted that a plateau in either $G(t)$ or $G'(\omega)$ does not prove the existence of a temporary network; it is sufficient to have a wide separation in time scales between relaxation processes. However, the phenomenon of elastic recovery is not easily rationalized without such a network.]

4. Molecular-Weight Dependence of Viscosity

The zero-shear-rate limiting viscosity of a polymer melt, η, exhibits an interesting dependence on molecular weight: at low molecular weights η increases with approximately the first power of chain length, but above some critical molecular weight, designated M_c, the power-law exponent increases to 3.4 $\pm$ 0.2. The behavior is remarkably universal, as evidenced by the variety of monomer chemical structures included in Fig. 1.4.[6] As discussed in more detail

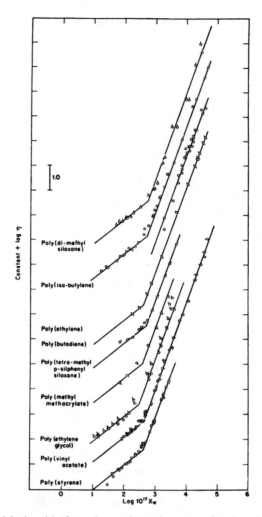

Figure 1.4. Melt viscosities for various molten polymers as a function of reduced molecular weight. An arbitrary vertical shift has been applied to each data set. From Berry and Fox.[6]

subsequently, the low-molecular-weight regime is reasonably well understood, while the high-molecular-weight behavior is also ascribed to intermolecular entanglement. However, both the value of 3.4 for the exponent and the location of the transition between the two regimes have proven to be rather recalcitrant problems for molecular theory.

Before proceeding, a few comments about the nature of entanglements are in order. First and foremost, it is undoubtedly the case that entanglements are topological in nature, reflecting the inability of two chains to pass through one another, and therefore are not dependent on any specific interchain interactions. This is evidenced primarily by the universality of phenomena such as those described. In other words, although many polymer systems feature specific interchain interactions, such as electrostatic forces in ionomers or microcrystallinity in some polymers of a particular tacticity, all flexible polymers of sufficiently high molecular weight exhibit the characteristics of entanglement. Furthermore, the magnitude of the modulus in the plateau region divided by kT, which may be taken as reflecting an entanglement density, is substantially independent of temperature. Second, given that entanglements are topological in origin, it is interesting to ask whether or not it is the presence of a particular interchain topological relationship, namely, a certain kind of knot, that is the main contributor. This issue can be clarified with the aid of a thought experiment. Suppose one had available a series of angstrom-resolution photographs of an entangled polymer liquid undergoing one of the experiments listed above, with the time increment between photographs chosen to reveal the chain motions over a long period of time. It is certainly conceivable that upon careful examination, the individual sites of long-lived entanglements could be identified and monitored, that is, small pieces of two distinct chains remaining in close proximity on a time scale comparable to the longest relaxation time defined by $G(t)$. The question posed earlier now becomes, is it possible to return to the first frame and learn to identify sites which are about to serve as long-lived entanglements, or not? If so, then there are particular local topological relationships that favor entanglement. This question has certainly not been answered definitively to date.

The entanglement phenomenon can be pictured in at least two ways. One is to assume the presence of particular sites where two chains meet which act as temporary cross-links, with or without an assumption about a particular topology. In other words, there are a subset of interchain contacts which can be called "entanglements." An alternative approach is to view entanglement as an effective field, or long-range interaction in the liquid, without designating any particular interchain contacts as entanglements. Both approaches have been used, and both raise puzzling questions. For example, if there are actual two-chain entanglements, why does such a low fraction of two-chain contacts serve as entanglements (the effective entanglement spacing in a melt is on the order of 100–200 monomers along a chain), and why does a chain have to be so long (150–300 monomers) before the effect of entanglement is seen? On the other hand, the entanglement interaction appears to have a correlation length, or entanglement spacing, of about 50 Å. If one chain end slips past another chain, a segment of the latter chain now has the opportunity to move laterally on the order of 5 Å. Thus on the entanglement length scale, nothing has changed; in other words, in the field picture it is hard to picture the release of entanglements. As will emerge in the subsequent discussion, the tube concept in the Doi–Edwards reptation model[7,8] for entangled polymer liquids is a representation of the entanglement field, yet subsequent attempts to modify the model to bring it into closer agreement with experiment (e.g., constraint release) are of the two-chain contact approach. It is not clear whether this is a fundamental inconsistency or two equivalent ways of treating the same phenomenon, but it serves to illustrate the difficulties that arise when one tries to imagine the behavior of actual polymer chains. As a cautionary note, use of the word entanglements in the remainder of this chapter is purely for convenience and should not be construed as endorsement of the particular site view.

C. Classes of Theoretical Treatment

The large majority of models for entangled polymer behavior may be partitioned into three groups, which are identified in chronological order. The first group comprises the so-called *temporary network models*, in which the theory of rubber elasticity is adapted to the uncross-linked case by the introduction of network junction points that have a finite lifetime. The second group may be termed *cooperative motion models*, where the focus is on the increased friction experienced by a test chain because it drags other chains or chain segments along with it over finite distances. The third group includes the *reptative motion models*, in which the primary result of entanglement is the constraining of a test chain to longitudinal motion along its own contour. Clearly, the major obstacle to developing a complete molecular theory is the complicated many-chain problem represented by entangled polymer liquids. The feature common to all three of these approaches is the adoption of a simplifying assumption about the major effect of interchain entanglement on the properties of the liquid; the assumptions themselves are quite different, however. Furthermore, these classes are not necessarily disjoint; for example, the aforementioned Doi–Edwards model is fundamentally a reptation model, but incorporates a major assumption of the temporary network models in the treatment of the response of the liquid to an applied stress. Nor are these classes mutually exclusive; it is conceivable that each captures some essential features of the problem which the others might ignore.

The temporary network models have been developed primarily by Green and Tobolsky,[9] Yamamoto,[10] and Lodge.[11] The starting point is the recognition that over certain time scales the polymer liquid behaves like a permanently cross-linked network, and that the (generally successful) theory of rubber elasticity may be applicable at short times. Therefore the latter serves as the basis for an extension of the model to polymeric liquids, with the primary step being the introduction of a function that describes the creation and loss of temporary junctions or entanglements. Different assumptions may be made about the form of this function, leading to different predictions for the dependence on flow history, but as yet no physical picture of individual molecular motions has been incorporated into this function. In a sense, then, the temporary network models approach the problem from a material, or continuum, perspective; predictions about the M dependence of the diffusion coefficient or the longest relaxation time have not yet been obtained.

The cooperative motion approach was pioneered by Bueche,[12] and a variety of related developments have more recently been presented.[13-17] In this type of approach, the common feature is the attempt to calculate the enhancement of the chain friction coefficient (relative to a Rouse chain, see later) due to entanglement with other chains. The resulting motion of the test chain is isotropic over all length scales, and the friction enhancement is in general a function of both test chain and surrounding chain molecular weight. A major difficulty in such an approach is the specification of the location and duration of entanglements, because the exact nature of an entanglement is not known. We have also included the generalized Langevin equation models of Ronca,[18] Fixman,[19] and Schweizer[20] in this category. In many cases these models do attempt to address the inherent multiple-chain nature of an entanglement, or at least the viscoelastic nature of the effective field felt by a test chain.

In reptation theories the fundamental postulate is that on time scales comparable to the longest relaxation time of a single chain, the primary effect of entanglement is to confine lateral motions of chain segments to distances less than the entanglement spacing, while longitudinal motions are not so restricted.[1,7,8,21-25] A test chain may thus be envisioned as being confined to a fictitious "tube" with diameter equal to the entanglement spacing,[7,8] but this tube is not, in fact, an essential part of reptation theories.[21-24] A direct result of this postulate is that the problem has been reduced to that of a single chain in an effective mean field, with a rather detailed picture of the dynamics down to the level of the polymer subchain. At the same time it is not yet clear to what extent such a reduction provides a reasonable representation of what is really a multichain process.

D.   Review of Basic Polymer Properties

1.   Coil Dimensions

A typical polymer molecule consists of hundreds or thousands of simple chemical units called monomers, which are covalently bonded together. For a linear homopolymer, all the monomers are identical and they are connected sequentially in one long chain. The backbone, most commonly formed of C–C bonds, has a great deal of conformational flexibility; in a simplified picture, if each bond has three rotational minima of comparable potential energy, an

n-bond chain has 3^n distinct conformations, which can be sampled at room temperature. Clearly, then, it only makes sense to consider appropriate average dimensions of a given molecule. The two size parameters in common usage are the mean square end-to-end distance $\langle h^2 \rangle$ and the mean square radius of gyration $\langle s^2 \rangle$. If the ith backbone bond is represented by a vector $\mathbf{l}_i$, then

$$\langle h^2 \rangle = \left\langle \sum_i \mathbf{l}_i \cdot \sum_j \mathbf{l}_j \right\rangle = \sum_i \sum_j \langle \mathbf{l}_i \cdot \mathbf{l}_j \rangle \tag{1.3}$$

and

$$\langle s^2 \rangle = n^{-1} \sum_i \langle \mathbf{s}_i \cdot \mathbf{s}_i \rangle \tag{1.4}$$

where the symbol $\langle \ \rangle$ denotes an ensemble average with respect to the appropriate equilibrium distribution function and $\mathbf{s}_i$ is the vector from the chain center of mass to the ith bond. The radius of gyration, often denoted by R_g, is the more useful quantity of the two, as it may be determined directly by light or neutron scattering experiments.

Each size parameter is determined by two kinds of interaction: short range and long range. The former refers to the details of potential barriers to rotation about backbone bonds, and is thus very sensitive to the chemical identity of the monomers, while the latter accounts for the excluded volume of different monomers on the same chain. Thus long range denotes the distance along the chain backbone, not the spatial separation; frequently the intermonomer excluded volume potential is approximated as a delta function. When the quantity $|i - j|$ exceeds a number of order 10, the orientations of the ith and jth backbone bonds become uncorrelated. The average end-to-end vector of this subchain can then be used as a renormalized backbone bond, which in the absence of excluded volume would be freely jointed with respect to its immediate neighbors. The subchain thus defined is also referred to as a persistence length, and the number of monomers in a subchain is a measure of flexibility. For a freely jointed chain of N links of average length b, the following relations apply:

$$\lim_{N \to \infty} \langle h^2 \rangle = Nb^2 = 6\langle s^2 \rangle \tag{1.5}$$

and the distribution function for h^2 (and s^2) is approximately Gaussian,

$$P(h, N)\, dh \approx \left(\frac{3}{2\pi Nb^2} \right)^{3/2} \exp\left(-\frac{3h^2}{2Nb^2} \right) 4\pi h^2\, dh \tag{1.6}$$

Thus in this case R_g scales with the 0.5 power of molecular weight, with only the prefactor dependent on the chemical details.

The attainment of a Gaussian monomer distribution function, and the concomitant scaling of R_g, requires two physically unachievable limits: infinite length and zero excluded volume. However, both conditions are well approximated under appropriate conditions. It turns out that if N is reasonably large but finite (e.g., $N > 100$), the difference between the Gaussian distribution and the exact freely jointed chain distribution function for h^2 is very small. The excluded volume can never truly vanish, but it can be counterbalanced almost completely. In dilute solution at one particular temperature, the so-called Flory theta temperature, the rather unfavorable monomer–solvent interaction energy favors compressed coil conformations to the extent that $R_g \sim M^{0.5}$. Thus in a theta solvent the chain is very nearly Gaussian. This behavior should be contrasted with the good solvent condition, in which the nearly equivalent monomer–monomer and monomer–solvent interaction energy allows the self-avoiding-walk nature of the chain to be revealed. In this instance,[26] $R_g \sim M^{0.588}$. In a melt of homopolymers, the excluded volume interaction is effectively screened out, as first predicted by Flory[27] and confirmed more recently by neutron scattering.[28] Physically this can be understood in the following manner. In dilute solution under good solvent conditions, if a chain coils back on itself, portions of the chain cannot occupy the same space, so the chain is effectively swollen. In the melt, however, any given monomer cannot distinguish whether another monomer belongs to the same chain or to another; there is therefore no tendency for a chain to swell beyond its ideal random-walk dimensions. Note that the absolute value of R_g need not be the same in different theta solvents (in fact, it is not), nor the same in the melt as in any particular theta solvent, but that the result $R_g \sim M^{0.5}$ is common to all. The fact that chains are very nearly Gaussian in the melt is a tremendously simplifying result, as the non-Markovian nature of the self-avoiding walk has so far made it impossible to obtain an exact distribution function for good solvents.

In the ensuing discussion it will be assumed that all chains are Gaussian, unless otherwise indicated. Furthermore, each polymer will be viewed as a freely jointed chain of subchains (or equivalently persistence lengths or Kuhn lengths). The term monomer will be reserved exclusively for the basic chemical building block of the chain. Note that this is far from universally followed; some authors have the unfortunate habit of using the term monomer when subchain is meant.

2. Dynamics of Unentangled Chains

The preceding discussion recalls all of the necessary background on *static* properties of molten polymers. It is also appropriate to review some well-established dynamic properties of *unentangled* chains in the melt, that is, those of sufficiently low molecular weight. In order to facilitate this development, it is actually most useful to begin with an approximate treatment of a different problem, namely, the Rouse model for an isolated chain in solution.[29] In this treatment a polymer is modeled as a freely jointed chain comprising $N + 1$ beads connected in a linear array by N springs, immersed in a Newtonian continuum with viscosity η_S. Each bead acts as a point source of friction, with friction coefficient ζ, while the frictionless springs are Hookean, with force constant $3kT/b^2$, where b is the root mean square separation of adjoining beads. Each bead–spring unit is intended to represent a subchain of the real molecule, not a monomer. The force balance on each bead in the chain involves three terms, representing Brownian, spring, and hydrodynamic forces. (Inertial terms are neglected, as is standard practice when considering the overdamped case of low-frequency motions in viscous media.) The first may be treated via any of the kinetic theory, Langevin, or Smoluchowski formalisms, with no difference in the results of interest here.[8, 24] The spring forces involve only the instantaneous difference in position of neighboring beads, and in the Rouse model they are the only way in which the motion of one bead can influence the motion of any other. The hydrodynamic drag force is a Stokes law friction experienced by a bead as it moves relative to the fluid continuum at that point. Externally imposed flow fields can be incorporated in this term. After a transformation to normal (intrachain) coordinates, the model may be solved to yield a spectrum of N relaxation times $\{\tau_i\}$, corresponding to the eigenvalues associated with each normal mode, as well as the eigenvectors themselves.

Some of the key results from this treatment are as follows. The relaxation times are expressible as

$$\tau_j = \frac{\zeta b^2}{24kT \sin^2(j\pi/2N)} \approx \frac{\zeta b^2 N^2}{6\pi^2 j^2 kT}, \qquad (j = 1, 2, \dots, N) \qquad (1.7)$$

with the last expression appropriate for the longer relaxation times ($j \ll N$) of chains with large N. The relaxation modulus is given by

$$G(t) = \frac{\rho RT}{M} \sum_j \exp(-t/\tau_j) \qquad (1.8)$$

which allows the viscosity to be calculated as

$$\eta = \int_0^\infty G(t)\, dt = \frac{\rho RT}{M} \sum_j \tau_j = \frac{\rho RT}{M} \tau_1 \frac{\pi^2}{6} \sim N\zeta \qquad (1.9)$$

The translational diffusion coefficient of the chain is given by

$$D = \frac{kT}{N\zeta} \tag{1.10}$$

Two characteristic results of the Rouse model are that the viscosity and the diffusion coefficient scale with the first power and the inverse first power of the molecular weight, respectively.

As a theory for the dynamics of an isolated chain in solution, the Rouse model has severe limitations. The primary difficulty is in the neglect of intramolecular hydrodynamic interactions, the process whereby the fluid velocity at bead j is perturbed by the wakes from the motions of all other beads. Approximate inclusion of this effect in the Zimm model[30] leads to the scaling laws $D \sim M^{-0.5}$ and $[\eta] \sim M^{0.5}$ (where $[\eta]$ is the intrinsic viscosity), in much better agreement with experiment, especially at large M. Furthermore, the frequency dependence of G' and G'' is also predicted much more successfully. A second major limitation of the Rouse model is the neglect of intramolecular excluded volume; the assumption of a freely jointed chain technically restricts the applicability of the model to theta solvents. (Approximate schemes for calculating the effect of excluded volume on $\{\tau_j\}$ are apparently quite successful in matching experiment, however.) Other limitations include the failure to predict the shear-rate dependence of viscosity (observed even in dilute solutions). The first two limitations, however, are suggestive in terms of the behavior of low M melts.

In a dense polymer system, the long-range hydrodynamic interactions are assumed to be screened out, much like the excluded volume, and the individual chains exhibit Gaussian statistics. Thus it has been suggested that a short chain in the melt can be viewed as a Rouse chain.[2] All properties expressible in terms of $\{\tau_i\}$ retain the same form, with the individual τ_i proportional to the subchain friction coefficient ζ, which is now much greater than in dilute solution. In particular, polymers obeying the scaling laws $\eta \sim M$ and $D \sim M^{-1}$ are said to be Rouse chains. (This nomenclature, although widespread, is somewhat unfortunate in our opinion, because its primary application is to polymer melts that were never the object of the original Rouse model. However, in the interests of conformity, we will adhere to the common usage.) For low M melts these scaling laws are at least approximately obeyed, although two cautionary points should be borne in mind. First, the range of M available to examine this regime is small, making the reliable extraction of power law exponents delicate, and second, the nonvanishing contribution from the extra mobility of the chain ends generally requires that the data be corrected for this effect before these exponents are observed; this correction has been applied to the data in Fig. 1.4. On the other hand, if the data are combined as $D\eta$, thus cancelling out the friction factor, they are generally in very good agreement with the magnitude predicted by the Rouse theory.

The main purpose of the preceding discussion was to introduce the concept of the Rouse chain, because it serves as an excellent reference state for entangled polymer behavior. In other words, the Rouse chain is thought to embody all the crucial features of polymer liquid behavior except for the effects of entanglement, and the goal of the theories to be discussed in Section II is to account for the difference between Rouse-like and entangled behavior. It is also important to note that the Rouse model (both as originally formulated and as applied to melts) is a single-chain approach, in which the behavior of one chain alone is considered in detail. In particular, the effect of the test chain on the surroundings is never considered. Clearly, the entanglement effect requires at least two chains; yet the reptation approach is also a single-chain theory, and it remains an interesting problem to establish what limits this places on the ability of reptation theory, or in fact any other mean-field model, to account for experimental observation.

E. Questions to Be Addressed

The primary question we wish to address has been introduced at the outset:

1. Do linear polymer chains execute primarily reptative motion over a nonvanishing region of parameter space?

If the answer to question 1 is affirmative, then the following questions arise:

What is the strongest evidence for the correctness of the reptation hypothesis?

What phenomena, if any, can reptation theory not describe, and what are its main weaknesses?

What sets the limits of applicability of reptation, and what other processes need to be considered?

How much further progress can be made by modifications to the basic reptation picture, or will a new approach ultimately be necessary?

If the answer to question 1 is negative, the following questions are pertinent:

What is fundamentally incorrect about the reptation model?

What is the strongest evidence against reptation?

What is the most promising alternative approach?

Why is the reptation picture in seeming agreement with so many observations?

2. What further experiments, if any, are required to resolve question 1 beyond any reasonable doubt?

3. In what direction should future theoretical efforts be aimed?

In addition to identifying the questions that we wish to address, it is appropriate to mention two issues that will not be discussed in any detail. The first is the issue of which models within a particular class are "better" than others, or which of various constitutive relations are more successful in describing different experiments; the thrust of this chapter is the process of reptation itself, which is distinct from any constitutive relation based upon it. Second, we will not discuss nonlinear rheological properties in any detail, as fascinating as they are. This choice is driven by the desire to keep the text to a manageable length, and because at this stage of development we do not feel that these experiments necessarily provide critical tests for the reptation process. However, reptation-based models do lead to powerful quantitative and qualitative predictions for such phenomena.[7,8,23–25]

To conclude this introductory section, a brief comment on perspective is offered. The subject of this chapter is of interest to a wide variety of scientists and engineers involved with polymers. It is not always clearly recognized that the natural priorities of the various practitioners can be quite distinct, which may cause unnecessary conflict. For example, a polymer processing engineer may be most concerned with the nonlinear rheological response of molten polymers at high shear rates or high strains. The ultimate goal would be a constitutive equation (relating stress to strain history) with a minimum number of parameters, which would yield reliable prediction of the optimum operating conditions for a given chemical system and processing operation. Arguments about the exact nature of the equilibrium monomer distribution function, or the reptative component of the chain diffusion coefficient, might be of little practical interest. On the other hand, polymer physicists and physical chemists have long emphasized the quasi-equilibrium domain of vanishing shear rates and strains, in an attempt to gain deeper molecular insight into this presumably more tractable regime. From this perspective, failure to make adequate predictions for phenomena such as stress overshoot or large-scale elastic recovery is not viewed necessarily as a serious limitation to a model. It is our opinion that, while both of these extreme perspectives are understandable, there is much to be gained from bridging the gap. In all cases, the same molecular system is involved, and it is surely not unreasonable to seek a theory that is firmly based on a molecular foundation, yet which accounts for the behavior observed in the rheological domain. Indeed, one definite benefit of the recent interest in reptation-based models has been a dramatic increase in the involvement of polymer scientists from a wide range of perspectives. However, it should be emphasized from the outset that our perspective is from the molecular end, and is based on the assumption that a widely applicable, predictive constitutive equation is unlikely to be developed without a basic understanding of macromolecular behavior in the near-equilibrium limit.

II. REPTATIVE AND NONREPTATIVE MODELS

A. Strict Reptation

In 1971 de Gennes published a theory of configurational relaxation and diffusion for an unattached chain in a network, featuring a chain dynamics, which he called reptation, very different from the Rouse-type motions associated with polymer chains in small-molecule solvents.[1] Applications of this model to polymer melts and solutions have since been studied extensively, with the first detailed treatment that of Doi and Edwards.[7,8]

1. The Basic Model

In molten polymers or concentrated polymer solutions the complex topological constraints imposed by entanglements are, in the strict reptation model, considered to restrict each chain to moving along its own contour, except at the chain ends. Often this view of motion in dense polymer systems is expressed by the statement that a chain is confined by its neighbors to a tube (Fig. 2.1). Excursions by interior chain segments at right angles to the tube axis are forbidden, in contrast to the isotropy inherent in the Rouse model. The tube picture is merely a convenient device for visualizing the local anisotropy of reptation, and some authors have preferred not to use it, on the grounds that the tube is not a real physical object.

Brownian motion of a chain takes on a quasi-unidimensional character under these restrictions. The motion within the tube can be represented as the random walk of a single particle: a step toward one end of the tube means that the entire chain slides along the tube in that direction, like a train moving on a track. (In fact, in de Gennes's original picture a long chain does not move as a single unit, but rather through propagation of chain defects, but this more physical picture does not alter the basic results.) The track in question here has two special features, however: it is itself a random walk in three-dimensional space, and it exists only under the train. As the train moves forward, the head car may move in any direction, since there is no existing track to guide it, and fresh track will be laid under it as it moves; the track abandoned by the tail car is torn up (i.e., the tube exists only because the chain is in it). If the train moves in reverse, the roles of the head and tail cars are simply interchanged.

The elemental step in reptation may be taken as the displacement of a polymer chain along its tube by one segment in either the forward or the

a)

b)

c)

d)

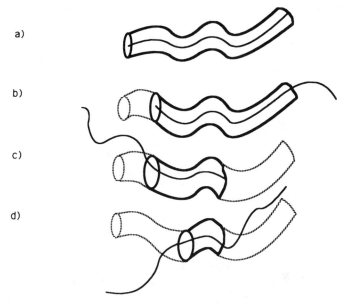

Figure 2.1. Evolution of a reptating chain, following Doi and Edwards.[8] (*a*) Initial conformation of primitive chain and original tube. (*b*), (*c*) Disappearance of original tube as chain diffuses to the right and to the left. (*d*) Chain conformation at a subsequent time prior to complete escape from original tube.

reverse direction. Such jumps occur with a frequency v, which should depend inversely on the chain length:

$$v = \frac{v_0}{N} \qquad (2.1)$$

where N is the number of segments and v_0 is independent of N. As a polymer chain executes this reptational Brownian motion, the tube in which it is initially confined is gradually destroyed from both ends, the lost sections being replaced by new ones whose conformations are completely independent of the initial conformation. The reptational relaxation time τ_{rep} is defined as the mean lifetime of a tube, that is, the average time over which there remains some correlation between current and initial configurations of the associated chain. The initial tube will have been completely destroyed when the head segment of the chain (or any other segment for that matter) has visited N different contiguous sites, that is, when its maximum excursions in the forward and reverse directions add up to N. For a random walk to cover a region of this width, on the order of N^2 steps are required; so τ_{rep} should be of order N^2/v or N^3/v_0.

In strict reptation the motion along the tube is a one-dimensional process, and for times long compared to $1/v$, it may be characterized by the diffusion equation and initial condition

$$\frac{\partial P}{\partial t} = \left(\frac{v}{2}\right)\frac{\partial^2 P}{\partial n^2}, \qquad P(n,0) = \delta(n) \tag{2.2}$$

where $P(n,t)\,dn$ is the probability that at time t the excess of forward over reverse jumps made by a particular chain is between n and $n + dn$. If we also impose the boundary conditions

$$P(-n_1,t) = 0 \qquad P(n_2,t) = 0 \tag{2.3}$$

we obtain the probability $P(n,t;n_1,n_2)\,dt$ that in addition the greatest forward excursion up to time t is below n_2 and the greatest reverse excursion below n_1. The probability of the first forward excursion of magnitude n_2 occurring before the greatest reverse excursion reaches n_1, and within the time interval t' to $t' + dt'$, is then

$$P_1(n_1,n_2,t')\,dt' = -\left(\frac{v}{2}\right)\frac{\partial P(n',t;n_1,n_2)}{\partial n'}\bigg|_{n'=n_2} dt' \tag{2.4}$$

From P_1 we can construct the probability that $n_1 + n_2$ first reaches N between t and $t + dt$ as a convolution of two first-passage probabilities:

$$P_R(t)\,dt = \left(\frac{2}{N}\right)\int_0^t\int_0^N\left[P_1(N - n_2,n_2,t')\frac{\partial P(N,n',t - t')}{\partial n'}\bigg|_{n'=0}\right]dn_2\,dt'\,dt \tag{2.5}$$

The mean tube relaxation time τ_{rep} is then just the average of t over the distribution P_R,

$$\tau_{rep} = \frac{N^3}{2v_0} \tag{2.6}$$

Although Eq. (2.6) describes the process of tube relaxation, it says nothing about conformational changes in the actual polymer chain or about the motion of its center of mass. Here we must distinguish between times that are short and times that are long compared to τ_{rep}. At short times, a chain segment is restricted to motion along the chain's tube, and is therefore executing a random walk along a Brownian path which remains fixed until the segment escapes from the initial tube. It is this persistence of the segment track that differentiates its motion from that of a monomer molecule dissolved in the bulk polymer. Whereas the mean square displacement of the monomer increases as the first power of the time, it is the mean square displacement along the tube that is proportional to t for the chain segment. However, the mean square displacement in space $g_1(t)$ goes as the first power of the tube displacement, and therefore as $t^{0.5}$:

$$g_1(t) = l^2(vt)^{0.5}, \qquad t \ll \tau_{rep} \qquad (2.7)$$

where l is the segment length. It should be noted here that $g_1(t)$ for a segment of a Rouse chain also increases as $t^{0.5}$ at short times, but in contrast to reptation, the coefficient of proportionality is independent of N.

Common sense suggests that, whatever the peculiarities of reptation at short times, at sufficiently long times the mean square spatial displacement of a polymer chain must become proportional to t. That this characteristic feature of Brownian motion does indeed emerge when $t \gg \tau_{rep}$ is a consequence of tube relaxation. Immediately after the first tube has been destroyed, the chain finds itself in a new tube, one end of which lies somewhere along the path of the previous tube. This tube is in turn destroyed, and so on (Fig. 2.2). Successive tube lifetimes can be considered to form a standard Markov sequence. The mean square displacement of a chain end during the tube lifetime is Nl^2, and after t/τ_{rep} lifetimes it is $Nl^2 t/\tau_{rep}$. The diffusion coefficient for a chain in strict reptation is therefore

$$D_{rep} = \frac{Nl^2}{6\tau_{rep}} = \frac{l^2 v_0}{3N^2} \qquad (2.8)$$

This inverse square dependence on the polymer molecular weight should be compared to the inverse first power behavior predicted by the Rouse model for diffusion of a polymer in a small-molecule solvent.

2. The Scattering Function

For strict reptation, the wave number and the time dependence of the dynamic structure factor $S(q, t)$ can be obtained by extension of these arguments. If we make the usual division of S into its coherent and incoherent portions S_C and S_I and define the correlation

$$\tilde{g}(q, t; m, n) = \langle \exp[i\mathbf{q} \cdot (\mathbf{r}_n(t) - \mathbf{r}_m(0))] \rangle \qquad (2.9)$$

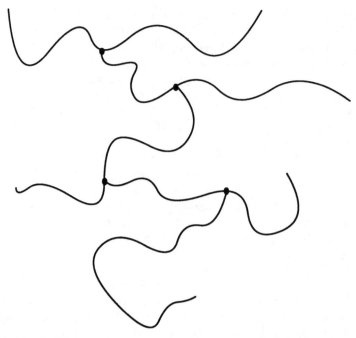

Figure 2.2. Series of successive primitive chains, showing how one chain end must always lie somewhere on the previous primitive chain.

then $S = S_C + S_I$, with

$$S_C(q,t) = \int_0^N \int_0^N \tilde{g}(q,t;m,n)\,dm\,dn \tag{2.10a}$$

$$S_I(q,t) = \int_0^N \tilde{g}(q,t;m,m)\,dm \tag{2.10b}$$

Here, $\mathbf{q}$ is the scattering vector and $\mathbf{r}_n$ is the position of the nth segment. Doi and Edwards[7] show that for wavelengths long compared to the length of a reptation segment ($lq \ll 1$), the correlation $\tilde{g}$, regarded as a function of n and t with parameters q, m, satisfies a diffusion-type equation such as Eq. (2.2), but with boundary and initial conditions

$$\frac{d\tilde{g}}{dn} = 0 \quad \text{at } n = 0 \text{ or } N$$

$$\tilde{g} = \exp\left(\frac{l^2 q^2\,|n-m|}{6}\right) \quad \text{at } t = 0 \tag{2.11}$$

(Actually they use radiative boundary conditions at the ends, but this makes little difference unless m or n is of order unity, in which case modeling the chain as a continuous rope becomes questionable.) For wavelengths long on the scale of a coil diameter, the resulting structure factor takes the expected exponential form

$$S = N^2 \exp(-D_{rep}q^2 t), \qquad Nl^2q^2 \ll 1 \qquad (2.12)$$

characteristic of a diffusing point particle. At shorter wavelengths ($Nl^2q^2 \gg 1$) the conformational changes produced by reptation are dominant:

$$S_C = \left(\frac{96N}{\pi^2 q^2 l^2}\right) \sum_{j,\text{odd}} \left[\frac{\exp(-\pi^2 j^2 v_0 t/2N^3)}{j^2}\right] \qquad (2.13a)$$

$$S_I = N \exp\left(\frac{q^4 l^4 v_0 t}{18N}\right) erfc\left[q^2 l^2 \left(\frac{v_0 t}{18N}\right)^{1/2}\right] \qquad (2.13b)$$

For the internal modes described by Eqs. (2.13), $S(q,t)/S(q,0)$ goes to zero only as $t^{-1/2}$ up to times of order τ_{rep}, much more slowly than the corresponding Rouse modes, which decay exponentially in $t^{1/2}$. Moreover the latter are uninfluenced by chain length, whereas the dependence of v on N ensures a nontrivial molecular-weight dependence for the reptational structure factor at all q.

3. Stress Relaxation

In the strict reptation model stress relaxation occurs simply by the escape of chain segments from the initial tube. If we think of each segment as a Hookean spring of zero rest length, then an instantaneously applied small strain **e** produces a change $\mathbf{e} \cdot \mathbf{h}_i$ in the end-to-end vector $\mathbf{h}_i$ of the ith segment. This affine deformation, and the contribution it makes to the stress response, persist until that segment has reptated out of the initial tube formed by its chain. The probability that a randomly selected segment has not yet escaped in this sense is simply the average of $P(n,t;0,N)$ over the interval $0 < n < N$. The relaxation modulus $G(t)$ at time t after the application of an initial deformation is therefore

$$G(t) = \left(\frac{8}{\pi^2}\right) G(0) \sum_{j,\text{odd}} j^{-2} \exp\left(-\frac{\pi^2 j^2 v_0 t}{2N^3}\right) \qquad (2.14)$$

Here $G(0)$ should be identified as the rubbery plateau modulus G_N^0; times preceding the onset of the plateau lie outside the scope of the strict reptation model.

We can of course simply take $G(0)$ from experiment, but a theoretical estimate is also possible. The stress $\sigma(t) \cdot \mathbf{n}$ across a plane normal to the n direction in a relaxing polymer is simply the sum of the contributions from all the chain segments crossing a unit area of that plane:

$$\sigma(t) = c\langle \mathbf{fh} \rangle + p\mathbf{I} = \left(\frac{c}{N}\right) \sum_{i=1}^{N} \langle \mathbf{f}_i \mathbf{h}_i \rangle + p\mathbf{I} \qquad (2.15)$$

where $\mathbf{h}_i$ is the end-to-end vector of the ith segment in a typical chain, $\mathbf{f}_i$ the tension in that segment, c the number of segments per unit volume, and p the pressure arising from van der Waals interactions and kinetic terms; the average $\langle \cdots \rangle$ is to be taken with respect to the nonequilibrium distribution prevailing at time t. The assumption of affine deformation behavior made here implies that the chain segments are Hookean entropic springs of zero rest length and root mean square end-to-end distance l, and it follows that $\mathbf{f}_i = (3kT/l^2)\mathbf{h}_i$. For small constant-volume shear deformations, the initial modulus G_N^0 then becomes just kTc, the value for an ideal rubber having a density c of network chains.

A sudden shear strain imposed on the polymer sample will deform all the chains and their associated tubes, and the question arises whether the contraction to equilibrium segment lengths occurs on the same time scale as the return to an equilibrium distribution of segment orientations. Doi and Edwards argue that there is a rapid (instantaneous in their theory) initial contraction of each chain within its tube, after which the mean tension in each segment has regained its equilibrium value of $3kT/l$; subsequent reptation produces only orientational relaxation. Any portions of the tube that are vacated as a result of this shrinkage are lost. (However, to first order in the strain, the loss has no effect on the stress.) At the end of the contraction the segments retain the nonequilibrium orientational distribution imparted by the applied strain, but their tensions have returned to those in the unstrained polymer. For small shear strains, Doi and Edwards show that this first relaxation reduces G_N^0 to $4/5kTc$.

If instead of applying an impulsive strain we subject the polymer to a constant unit strain rate, the stress response in the steady-state limit is the time integral of $G(t)$ from 0 to infinity. In the case of a steady rate of shear, we obtain in this way the viscosity of the polymer:

$$\eta = \frac{G_N^0 N^3}{6v_0} \qquad (2.16)$$

As with Eq. (2.14), Eq. (2.16) assumes that the material remains in its linear response range. It predicts N^3 behavior for the low-shear rate viscosity of polymer melts of high-molecular-weight polymers, provided that G_N^0 has no dependence on N.

4. Comments on Strict Reptation

One of the main virtues of the strict reptation model is its well-defined character and the readiness with which it lends itself to simple mathematical formulation. However, the simplicity and directness of the tube concept come at the price of much difficulty in relating the idealized tube to the physical polymer chain and its surroundings. Just what is meant by a tube segment? How is the segment density c in an actual polymer to be determined?

From the beginning there has been general awareness that, even in the absence of any small-molecule diluent, a real polymer chain is not nearly as confined by its neighbors as the strict reptation model assumes. The real tube is not so much a continuous sleeve for its chain as a sequence of widely separated entanglement points through which the chain must pass. Reptation then consists of the chain slipping back and forth through these entanglements, and occasionally out of them, the latter event corresponding to tube destruction. The segment length in this view should be closely identified with the entanglement length a, the root mean square end-to-end distance of a chain of molecular weight $M = M_e = \rho RT/G_N^0$, where G_N^0 is the rubbery plateau value of the relaxation modulus and ρ the mass density of the polymer. Typical values of M_e are of order 10^4, quite long enough to justify the Doi–Edwards representation of a reptation segment as an entropic spring. This loosening of the constraints of strict reptation is often summarized by assigning a diameter to the tube roughly equal to a; lateral displacements of order a are permitted even for the middle segments of a chain.

For $M < M_e$, reptational motion presumably becomes insignificant, and relaxation occurs entirely via the Rouse modes [i.e., as in Eq. (1.7)]. Rouse modes continue to govern stress relaxation at times less than the intrasegmental response time τ_e, even when $M \gg M_e$. It is only at times comparable to $\tau_{rep} \gg \tau_e$ that reptation becomes dominant; the range $\tau_e < t < \tau_{rep}$ is roughly the extent of the rubber plateau. As already stated in the preceding section, the strict reptation model covers only relaxation times beyond τ_e.

If we follow the motion of a single monomer which is not close to either end of the chain, then the additional freedom provided by the widened tube leads to a more complex evolution of $g(t)$ at intermediate times. At very short times, while its displacement is less than a subchain length, the monomer behaves as a free particle and $g(t) \sim t$. There then follows a period of Rouse-like relaxation over distances less than a, for which $g(t) \sim t^{0.5}$. If the chain were to slide back and forth in the tube as a rigid object, this Rouse regime would terminate in a plateau followed by a second interval of $t^{0.5}$ behavior, described by Eq. (2.7). However, if the lateral freedom in the tube is taken into account (which might correspond to chain contour length fluctuations or the defect mechanism originally proposed by de Gennes), the plateau becomes an intermediate interval where $g(t) \sim t^{0.25}$. Ultimately, of course, the monomer must follow the center of mass motion of the chain, and $g(t) \sim t$ again. These five time regimes are discussed again in Section III.B.4.

A further consequence of wide spacing between entanglements is that there is now a real distinction between the conformation of the tube (defined, say, as a broken line passing through the entanglement points, referred to as the primitive path by Doi and Edwards) and that of its chain. Theories seeking to include intrasegment relaxation in addition to reptation may therefore introduce forces or potentials that depend on some measure of the deviation between actual and primitive chain paths.

Finally there is a feature that reptation theory shares with many of its competitors: it really treats only the motion of a single chain. The rest of the liquid, which is after all made up of similar chains that are also in motion, is represented by the essentially static tube. The possibility that an entanglement or a section of tube might be set in motion or even disappear through movements in neighboring molecules is not contained in the strict reptation model, nor are any of the effects of probe chain motion on the surrounding fluid.

B. Modified Reptation Models

1. Additional Relaxation Mechanisms

It is immediately apparent that the most basic result of strict reptation, namely, $\eta \sim \tau_1 \sim M^3$, where τ_1 is the longest relaxation time of the chain, is not in accordance with a wealth of viscosity measurements. (Note that $\eta \sim \tau_1$ is also a prediction of the model). It is equally apparent that the most drastic assumption underlying the application of reptation to monodisperse melts is that the matrix is effectively frozen over time scales comparable to τ_1. Consequently there have been a variety of attempts to reconcile the experimentally observed 3.4 exponent with the reptation picture, and similarly a number of methods to account for matrix mobility have been proposed. In this section we discuss the two most successful of these modifications, the tube fluctuation argument of Doi[31] and the process of constraint release as developed by Graessley.[32] In addition, a few general comments about these issues are offered.

The concept of tube fluctuation may be viewed quite simply. The contour length of the primitive chain in the tube L will fluctuate on a time scale much shorter than τ_1, with a typical magnitude $\delta L/L \sim (M/M_e)^{0.5}$. Each fluctuation to a shorter value of L effectively destroys an extra portion of the tube, so that the relaxation of stress is enhanced. A detailed but still approximate analysis by Doi leads to

$$\eta = \eta_{rep}\left[1 - \kappa\left(\frac{M_e}{M}\right)^{0.5}\right]^3 \qquad (2.17)$$

where η_{rep} is the strict reptation result. The dependence of η on M in this case is not a power law, but is quite similar to $\eta \sim M^{3.4}$ over a range of M. A calculation estimates κ as about 1.5,[31] which means that the fluctuation contribution is significant even for $M/M_e = 100$. However, at sufficiently large M the reptation prediction is recovered. Because the fluctuation mechanism need not move the chain center of mass, the diffusion coefficient is unaffected, although this position has been questioned.[33] Rubinstein[34] has presented a discretized reptation model, where finite chain lengths lead naturally to a broad crossover region with a viscosity exponent of 3.2–3.5.

As mentioned, other ways to reconcile the 3.4 exponent have been proposed,[35] but we will not discuss them here. However, some additional comments are in order. In an important but largely ignored paper, Brereton and Rusli[36] consider the effect of entanglements on the dynamics of polymer chains in a very general way. In essence, they apply the second fluctuation-dissipation theorem to relate friction functions to random forces. With what appear to be relatively benign assumptions, it emerges that if the forces along the chain are uncorrelated over distance scales longer than the subchain size, Rouse-like behavior will be observed, but if the forces have a significant long-range correlation, for example, over an entanglement length, then $\eta \sim M^3$. This result is completely consistent with reptation, but does not require it. In other words, while strict reptation leads directly to the M^3 law, there may be many other physical processes which lead to the same result and, by extension, to $D \sim M^{-2}$. From this perspective, the experimentally observed 3.4 exponent suggests an additional correlation of forces along the chain. One possible origin for this is the fact that a given chain will be in close proximity to another chain in several places, not just one. This kind of argument has been pursued by Fixman,[19] for example, as discussed in Section II.C.2.

The basic ideas of constraint release were considered by Daoud and de Gennes,[37] for a chain of length N in a matrix of chains with length P. The main idea is simple; the obstacles which constrain a chain to execute strict reptation can themselves move if the matrix is not cross-linked. In the limit $P \gg N$, the matrix should act like a network, and the N chain should exhibit strict reptation. Conversely, when $P \ll N$, the limit of a chain in a low-molecular-weight solvent is obtained. The interesting regime, where $0.1 \lesssim P/N \lesssim 10$, is more complicated, although the initial conclusion was that constraint release would not be important.[37] (As an aside, we note that constraint release is also frequently referred to as tube renewal. We feel that this nomenclature is unfortunate and should be abandoned. First, many authors consider the process of forming a new tube by reptation to be tube renewal, which means that the same term refers to completely different processes. Second, tube renewal presupposes the existence of a tube, which is not a necessary concept either for strict reptation or for reptation-based theories.)

Graessley has developed a model for the constraint release contribution to the mobility and stress relaxation of a polymer.[32] Motion by constraint release occurs when a neighboring chain end moves past the test chain, thus permitting a lateral excursion of the test chain, as illustrated in Fig. 2.3. The rate constant for constraint release k_{cr} is therefore inversely proportional to the longest relaxation time of the neighboring chain,

$$k_{cr} = \frac{\mu}{\tau_1(P)} \sim P^{-3} \tag{2.18}$$

The P^{-3} dependence comes from the assumption of reptative motion of the P chain; μ is an adjustable parameter. If we further assume that the reptation and constraint release contributions are independent, we obtain a diffusion coefficient which is the sum of reptative and constraint release terms, proportional to $M^{-2}P^0$ and $M^{-1}P^{-3}$, respectively. Similarly, the relaxation modulus now incorporates the product of two functions,

$$G(t)/G_N^0 = F(t)R(t), \tag{2.19}$$

where $R(t)$ is taken to be the Rouse-like relaxation of the tube itself. The net result of incorporating a constraint release term is to decrease the viscosity, increase the breadth of the terminal relaxation zone, and enhance diffusion, all relative to the strict reptation case. As will emerge in Section III, this generally brings the model into closer agreement with experiment.

A variety of alternative approaches to treating constraint release has been developed, which differ in detail but not in the underlying picture, that is, chains reptate but lateral motion is not insignificant.[38-44] A thorough comparison of these models is beyond the scope of this chapter, but there are some subtleties inherent in the treatment of constraint release that should be noted, particularly revolving around the issue of self-consistency. First, the importance of the release of an individual constraint presumably depends strongly on its location along the test chain contour. The limiting process in stress relaxation by strict reptation is the reorientation of the chain middle, and thus the effect of lateral displacements may be more significant near the center of a test chain. Second, the rate of constraint release may vary along the chain contour; this is the basis for one recently proposed modification[41] and may also be crucial in the relaxation of branched polymers.[42] Third, the release

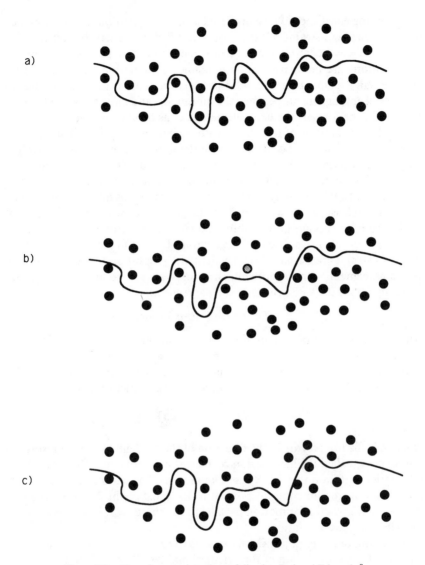

Figure 2.3. Idea of constraint release, following Doi and Edwards.[8]

of constraints may be correlated, in the sense that of the set of two-chain contacts along a test chain contour, many involve the same neighboring chain; this aspect of the problem is incorporated in the pictures of Klein[38] and Fixman,[19] for example. Fourth, the release of a constraint on a test chain is also potentially the release of a constraint on the neighboring chain;[43] whether this coupling effect should be considered is open to question. Fifth, for mono-disperse melts constraint release is only justified as a perturbation to strict reptation, not as a comparable or greater contribution. In situations where the contributions are comparable, the fundamental reptation hypothesis is invalid. Finally, as mentioned in the Introduction, there is an inconsistency in the view of entanglements between the tube model and most constraint release models. In the former, the entanglements act as a tube, or field, which constrains lateral motion of the chain, with a characteristic length scale (from experiment) on the order of 50 Å. In the latter, it is the motion of an individual neighboring chain which provides the release of a constraint. This implicitly assumes that there are particular two-chain contacts, or sites, which act as entanglements. While it may be possible to overcome this difficulty by viewing constraint release as a "widening" of the tube,[39] such approaches do not appear to be successful in describing the viscoelastic properties of blends of short and long chains, as discussed in Section III.A.4. On the whole, therefore, while constraint release models can help to bring reptation-based theories into better agreement with experiment, there may always be an underlying uncertainty about the significance of such agreement.

2.   The Curtiss–Bird Theory

An alternative way of formulating a reptation theory has been offered by Curtiss, Bird, and coworkers, who in fact do away with the tube concept altogether.[23,24] Instead, they postulate a chain of freely jointed rigid-rod segments, each with a locally anisotropic segmental friction tensor,

$$\zeta = \zeta[\mathbf{I} - (1 - \varepsilon)\mathbf{uu}] \tag{2.20}$$

where $\mathbf{u}$ is the unit vector in the direction of the segment, and the anisotropy parameter ε (called the link tension coefficient by Curtiss and Bird) runs from 0 for strict reptation to 1 for Rouse-like motion. The parameter ζ represents the friction factor for motions normal to the local chain direction; the segmental friction factor for motion parallel to the chain is $\zeta\varepsilon$. Thus, finite values of ε allow for lateral motion.

From this starting point, Curtiss and Bird develop a convective diffusion equation for the evolution of the orientational distribution function $f(\mathbf{u}, n, t)$ of the nth segment in a chain. As a consequence of simplifying approximations made in the course of the development, their result is, apart from a numerical factor, identical with that obtained by Doi and Edwards for strict reptation:

$$\frac{\partial f}{\partial t} = \left(\frac{v}{2}\right)\frac{\partial^2 f}{\partial n^2} - \frac{\partial}{\partial \mathbf{u}} \cdot ([\overset{\circ}{\mathbf{e}} \cdot \mathbf{u} - \overset{\circ}{\mathbf{e}}: \mathbf{uuu}]f) \tag{2.21}$$

where the term in square brackets is the velocity normal to $\mathbf{u}$ arising from the applied strain rate $\overset{\circ}{\mathbf{e}}(t)$, and $(\partial/\partial\mathbf{u})\cdot$ is the divergence operator on the surface of a unit sphere.

Although Eq. (2.21) predicts, in the absence of externally applied strains, the same conformational evolution as strict reptation theory, it does not follow that use of the anisotropic friction tensor has no effect on the predicted stress. For stress relaxation at small strains only $G(0)$ is affected, and the time dependence of $G(t)$ is still given by Eq. (2.14). In the case of a suddenly applied tensile strain, the Curtiss–Bird value of $G(0)$ is $(1 + \varepsilon/3)$ times that given by Doi and Edwards. At strains large enough to cause nonlinear viscoelastic behavior (the main focus of the Curtiss–Bird papers), the differences between Curtiss–Bird and Doi–Edwards theories are substantial, however. There are also numerous differences in both formalism and interpretation of the basic models. For a discussion of some of these differences, see References 45 and 46. In addition, Reference 46 presents extensive comparisons of rheological data with the predictions of both models.

3. The Models of Hess and Noolandi

A major weakness of the reptation picture, recognized early on, was that it remained a postulate, rather than the result of a more elaborate statistical mechanical calculation. Hess attempted to rectify this through a model that concentrated on chain diffusion in semidilute solution, and which predicted a transition from Rouse-like to reptation-like diffusion as either concentration or chain length is increased.[21] The starting point is a generalized Fokker–Planck equation for the polymer segments. It proceeds to an expression for the diffusion coefficient, which may be written

$$D = \frac{kT}{N[\zeta_0(c, N) + \Delta\zeta]} \tag{2.22}$$

where ζ_0 is a Rouse-like segmental friction and $\Delta\zeta$ is a friction function which is the main object of his paper. It is calculated in two steps. The first is to assume a potential for each continuous chain, which includes a delta-function excluded volume interaction between two neighboring chains. It is argued that this is the dominant term at higher concentrations, and because it is always orthogonal to the local tangent to the smooth curve representing a chain, it eventually leads to chains being constrained to move longitudinally. The second step is to decompose the Fokker–Planck operator into components parallel and perpendicular to the chain contour. Above a certain degree of entanglement, the perpendicular component of D vanishes, yielding

$$D = \frac{kT}{\zeta_0 N[1 + 2N/N_c(c)]} \tag{2.23}$$

where N_c is a critical molecular weight for entanglement. The result is a smooth crossover from Rouse-like to reptation-like diffusion, which (as will be discussed) has been observed in melts. The model itself applies to semidilute solutions.

In Hess's model reptation is introduced, not by a tube postulate or an anisotropic friction tensor, but rather through the imposition of a specific two-body interaction potential. This potential may be understood by considering an initial contact between two chains. If this situation is followed by a sliding motion of the two chains with respect to one another, by moving along their own contours, there is no change in their interaction because there remain two segments (of the same two chains) in contact. If on the other hand their motion is transverse to the chain contour, the contact is broken and a contribution to the two-body potential is obtained. A similar contribution is obtained when two chains come into contact.

The theory developed by Noolandi et al.[22] can be considered in two parts, the first providing the description of an entangled melt and the second the application of the generalized Rouse model (GRM) to the theory. In the former, the test chain is divided into spheres with diameters equal to the mean distance spanned by N_e bonds, where N_e is postulated to be the mean spacing between entanglements (see Fig. 2.4). The heavy solid line represents a portion of the test chain. A coordination number $\tilde{N}$ is defined as the number of chain segments (thin solid lines) that thread through and laterally constrain the probe chain within a test sphere. The thin broken lines are chain segments that are ineffective in the lateral constraint of the test chain. As N is lowered, the spheres have to be enlarged to include a sufficient number of constraints, so that the system is still entangled; for example, in going from the upper to the lower part of Fig. 2.4, the molecular weight of the surrounding chains has been halved. Eventually the spheres have to be enlarged to the point that they enclose the whole test chain and still have fewer than $\tilde{N}$ constraints, resulting in a transition to the unentangled regime. Via conservation of mass within a sphere in the test chain, the model interrelates $\tilde{N}$, N_e, and N. $\tilde{N}$ is expected to be universal, while N_e is predicted to be a function of N and to become constant as N becomes large. It follows from this description that an accumulation of topological constraints provides the transition from the unentangled to the entangled regime, which is a gradual function of N for the chains surrounding the test chain.

The generalized Rouse model consists of a three-dimensional Fourier analysis of the Brownian dynamics of a test chain. The effects of entanglements are represented by introducing longitudinal and transverse friction coefficients ζ_l and ζ_t. These directions are defined in terms of a tube axis, which in turn is defined with a set of vectors obtained with a truncated Fourier series which provides a tube with the same overall conformation as the test chain, but

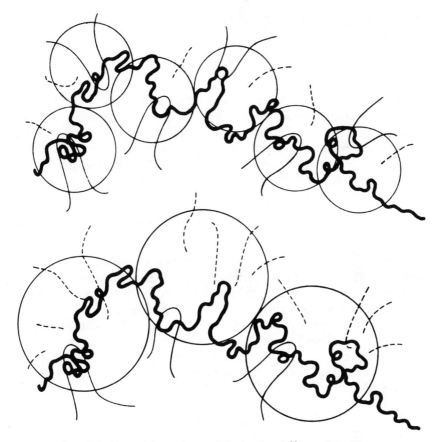

Figure 2.4. Entanglement picture of Noolandi et al.,[22] as explained in text.

where the short scale detail has been smoothed out. The statics and dynamics are now connected by smoothing over a length scale equal to the spacing between entanglements N_e. For $N \gg N_e$ the transverse friction diverges, enforcing a transition to reptation. Interestingly, the relaxation times of the longer modes scale with the mode index p to the fourth power, unlike the p^2 scaling in either the Rouse [Eq. (1.7)] or the reptation case [Eq. (2.9)].

C. Theories Based on Extended Langevin Equations

We may regard the Curtiss–Bird theory as the first example of a group of models that, while formally treating the motion of a single N-mer chain, introduce the dynamics of neighboring chains through a generalized N-particle Langevin equation. If inertia terms are neglected, this takes, in the absence of externally applied forces or strains, the form

$$-\hat{\mathbf{B}}\frac{d\mathbf{R}}{dt} + \mathbf{F}_{int} + \mathbf{F}_{rnd} = 0 \qquad (2.24)$$

where $\mathbf{R} = (\mathbf{r}_1, \ldots, \mathbf{r}_N)$ is the conformation of the probe chain, and $\hat{\mathbf{B}}(\mathbf{R}, t)$ a $3N \times 3N$ tensor operator that includes frictional interactions between probe beads as well as memory effects arising from interactions with the medium. In general, the operation of $\hat{\mathbf{B}}$ on a velocity vector $\mathbf{V}(\mathbf{R}, t) \equiv d\mathbf{R}/dt = (\mathbf{v}_1, \ldots, \mathbf{v}_N)$ will involve convolution in the time variable with a memory kernel $\mathbf{B}(\mathbf{R}, t)$:

$$\hat{\mathbf{B}}\mathbf{V} = \int_0^t \mathbf{B}(\mathbf{R}, t - t') \cdot \mathbf{V}(\mathbf{R}, t') \, dt' \qquad (2.25)$$

The normal history-independent frictional drag is recovered if the time dependence of $\mathbf{B}$ takes the form of a delta function. The 3N-dimensional vector $\mathbf{F}_{int}(\mathbf{R}) = (\mathbf{f}_1, \ldots, \mathbf{f}_N)$ represents the forces exerted on probe beads through their interaction potentials with other probe beads and with the beads of nearby chains in the matrix (and with solvent molecules, if present), and $\mathbf{F}_{rnd} = (\xi_1, \ldots, \xi_N)$ the random Langevin forces responsible for Brownian motion. The Langevin forces must at long times generate the equilibrium conformational and velocity distributions for the probe chain, which can be shown to imply the relation

$$B_{ij}(t) = \frac{\langle \xi_i(t)\xi_j(0) \rangle}{kT} \qquad (2.26)$$

1.   Ronca's Model

An early example of this approach was published by Ronca in 1983.[18] Ronca starts with the generalized Langevin equation for a Rouse chain in a viscoelastic medium,

$$\kappa(\mathbf{r}_{i+1} - 2\mathbf{r}_i + \mathbf{r}_{i+1}) + \zeta\mathbf{v}_i + \alpha\hat{K}\mathbf{v}_i = \xi_i \qquad (2.27)$$

where $\alpha\hat{K}(t - t')$ is a scalar memory kernel representing the delayed responses of the surrounding polymer matrix and ζ is a "local" friction constant. The stress relaxation function for a polymer of Gaussian chains containing N segments each is

$$G(t) = \kappa\left(\frac{c}{N}\right)\sum_{i=1}^{N-1}\left[\langle(\mathbf{r}_{i+1} - \mathbf{r}_i)(\mathbf{r}_{i+1} - \mathbf{r}_i)\rangle\right] \qquad (2.28)$$

where c is the concentration of segments. The bond autocorrelations on the right-hand side of Eq. (2.28) can be obtained by standard Fourier transform methods, and Ronca does this for a sinusoidally oscillating shear. The result is an integral relationship between G and K or, rather, between their Fourier transforms.

Instead of introducing now a physical mechanism by which the beads of the probe chain interact with each other and with neighboring chains, Ronca suggests that there should be a simple relationship between $K(t)$ and the macroscopic stress relaxation function $G(t)$, which can be used to impose a self-consistency condition. At short times, where both K and G are expected to be independent of molecular weight, Ronca takes $K = G$. At longer times he points out that K only involves more or less local relaxations, and therefore lacks the slowest modes of the macroscopic relaxation function, so that its long-time behavior is a purely elastic matrix response. The simplest means of introducing the missing slower modes is to postulate

$$\frac{dK}{dt} + \frac{K}{t_{max}} = \frac{dG}{dt} \tag{2.29}$$

which has the effect of adding a terminal relaxation time t_{max} to the spectrum of K. Either choice provides a second relation between G and K that is then combined with Eq. (2.28) to generate an integral equation for $G(t)$.

The results of Ronca's theory may be summarized as follows.

1. If G and K are set equal, and the plateau range of times is not exceeded, one obtains the relaxation behavior of a Rousean rubber network, that is, there is a transition region where $G \sim t^{-0.5}$, ending in a rubber plateau of indefinite extent. Attempting to pursue the calculation to longer times leads to unphysical consequences.

2. If Eq. (2.29) is used instead, the unknown dependence of t_{max} on molecular weight prevents evaluation of the exponent in the viscosity–molecular-weight relation, but Ronca is able to show that the terminal relaxation time must have the same molecular-weight dependence as the viscosity.

3. Deviations from Rouse behavior appear also in the single-chain scattering function $S(q, t)$, even assuming $G = K$. For long chains consisting of many entanglement lengths and for wave numbers comparable to the coil radius of an entanglement segment, the ratio $S(q, t)/S(q, 0)$ follows Rouse behavior at small times, but levels off at a limiting value $S(q, \infty)/S(q, 0)$, which decreases with increasing q. This is of course another manifestation of the elastic behavior assigned to $K(t)$ at times beyond the plateau range.

Ronca's generalized Langevin model thus appears to complement rather than to supplant reptation, at least in the $G = K$ implementation. Its unique feature lies in the use of a self-consistency requirement, which avoids to some extent (but unfortunately not completely) the need for more or less arbitrary assumptions about the form of the memory function.

2. Fixman's Model

Perhaps the most elaborate example of the Langevin approach is the theory recently published by Fixman.[19] In Fixman's model each bead of the probe chain is connected to a "center of force," representing entanglements with neighboring chains (Fig. 2.5). Essentially the path laid out by these centers is the primitive path of Doi and Edwards. During its lifetime it interacts with the actual path of the probe chain through a Hookean potential which is minimized when the two paths coincide. The whole assembly is considered to be immersed in a viscoelastic medium, which may itself be set in motion, through externally applied strains as well as through motion of the probe. This picture leads Fixman to introduce a hierarchy of interactions between the probe and its environment, which may, in the absence of imposed strain rates, be summarized as follows.

1. If a small-molecule solvent is present (and perhaps even in the dry polymer), a probe bead moving with velocity $\mathbf{v}_i$ experiences a simple viscous drag $b^f \mathbf{v}_i$, where b^f is a scalar friction factor. Hydrodynamic interaction effects

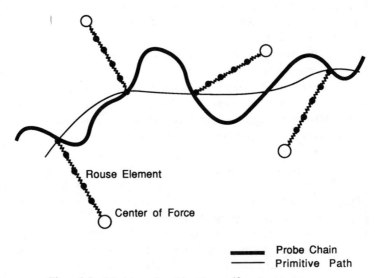

Figure 2.5.   Model employed by Fixman,[19] as discussed in text.

from this source, though present in principle, are ignored, since this is not a theory of dilute polymer solutions.

2. The interaction between probe bead i and its center of force is taken to be transmitted by a polymer chain (or a section thereof) and is assumed to have a full set of Rouse-type relaxation modes. If bead i is moved with a velocity v_i, the first result is a delayed force,

$$\mathbf{f}_i^s = \hat{b} \mathbf{v}_i \tag{2.30a}$$

$$b(t) = \frac{3kT\tau^f}{a^2 N} \sum_{j=1}^{N} \exp\left(-\frac{t}{\tau_j}\right) \tag{2.30b}$$

where $\tau_j = \tau^f \sin^2(pj/2N)$, the relaxation time of the jth Rouse mode, is proportional to j^2 if $j \ll N$, and the fastest relaxation time τ^f is independent of molecular weight.

3. The force $\mathbf{f}_j^s$ on bead j induces a velocity $\mathbf{v}_{ij}^u$ in the medium at the location of bead i, just as in the Kirkwood–Riseman theory of hydrodynamic interactions in dilute polymer solutions, except that we are now dealing with a viscoelastic rather than a purely viscous fluid. Fortunately the elastic and viscous contributions take the same form in an incompressible medium, and the usual preaveraging operation gives

$$\mathbf{v}_i^u = \sum_{j=1}^{N} \mathbf{v}_{ij}^u = \sum_{j=1}^{N} P_{ij}\left(y\frac{d\mathbf{f}_j^s}{dt} + w\mathbf{f}_j^s\right) \tag{2.31}$$

where $P_{ij} = 1/|i - j|^{0.5}$ are off-diagonal elements of the preaveraged viscoelastic interaction matrix ($P_{ii} = 0$), and y, w are constant coefficients, which determine the magnitudes of the elastic and viscous components, respectively. The total velocity $\mathbf{v}_i^u$ induced by the other beads of the probe chain makes a contribution $\hat{b}\mathbf{v}_i^u$ to the drag force $\mathbf{f}_i^s$ on bead i. In the matrix operator notation used at the start of this section, we may express the results up to this point in the form

$$\left[\mathbf{I} + \hat{b}\mathbf{P}\left(w + y\frac{d}{dt}\right)\right]\mathbf{F}^s = \hat{b}\mathbf{V} \tag{2.32}$$

so that the contribution of probe–matrix interactions to the friction operator $\hat{\mathbf{B}}$ of Eq. (2.24) is, formally,

$$\hat{\mathbf{B}}_M = b^f \mathbf{I} + \left[\mathbf{I} + \hat{b}\mathbf{P}\left(y\frac{d}{dt} + w\right)\right]^{-1}\hat{b} \tag{2.33}$$

4. In addition to $\hat{\mathbf{B}}_M$ there are further dissipative effects coming from interactions between beads of the probe chain. These appear in the theory as internal viscosity terms, involving only motions of probe beads relative to each other. Under this heading, Fixman includes rotations about backbone bonds and dissolution of entanglements by reptation. The former are important only at very short relaxation times and we omit them from the present discussion. Entanglement dissolution is represented approximately by the memory matrix

$$[\mathbf{B}_e]_{ij}(t) = (\delta_{i,j-1} - 2\delta_{ij} + \delta_{i,j+1})\gamma e^{-\lambda t} \qquad (2.34a)$$

or, in terms of the memory operator $\hat{B}_e$ acting on the probe velocity $\mathbf{V}$,

$$[\hat{\mathbf{B}}_e\mathbf{V}]_i = \gamma \int_0^t [\mathbf{v}_{i+1}(t') - 2\mathbf{v}_i(t') + \mathbf{v}_{i-1}(t')] \exp[-\lambda(t-t')]\,dt' \qquad (2.34b)$$

We may think of $\hat{\mathbf{B}}_e$ as resulting from multiple entanglements of the probe with the same set of vicinal chains. At short times, such entanglements act as supplementary springs in parallel with those arising from the primary valence structure (see item 5). The constant γ reflects the stiffness of this elastic interaction. However, these entanglement springs have only a finite lifetime $1/\lambda$, the mean time for an entanglement to dissolve by motion of either the probe chain or its partner. From the viewpoint of the reptation model, it appears that $1/\lambda$ must be comparable to the tube relaxation time τ_{rep}, but Fixman treats it as an unknown to be determined; in particular, he does not assume λ to be proportional to $1/N^3$.

5. The forces $\mathbf{F}_{int}$ in Eq. (2.24) consist of intramolecular interactions within the probe chain, the most important being of course the valence bonds holding the backbone together. These may be represented as Hookean springs of stiffness κ, just as in the Rouse model:

$$\mathbf{f}_i = \kappa(\mathbf{r}_{i-1} - 2\mathbf{r}_i + \mathbf{r}_{i+1}) \qquad (2.35)$$

We first consider the steady state of a probe chain all of whose segments are moving with the same constant velocity $\mathbf{v}_0$. If the forces required to maintain this motion are given by Eq. (2.30), the convolution is readily performed to give, in the large N limit, f_i^s proportional to N and a total force on the chain proportional to N^2. In effect, the probe chain appears to drag along with it a number of entangled neighbors proportional to N, and the size of this cluster increases as N^2. The Rouse-like character of the relaxation function $b(t)$ suggests that the total drag on the cluster varies as the number of segments in it, and the friction coefficient is therefore proportional to N^2 as well. This is just what the strict reptation model would predict, yet it has all been done with Rouse-like chains, with no mention of tubes.

However, Eq. (2.30) is only a first stage, and we now turn to the viscoelastic interactions [Eq. (2.32)]. The elastic term y on the left vanishes in the steady state, and the coefficient w is inversely proportional to some local viscosity that reflects the drag which the polymer matrix exerts on a probe bead. A typical element of the interaction matrix $\mathbf{P}$ will be of order $N^{-0.5}$, the sum of all its elements of order $N^{1.5}$. It follows that for large N the chain friction factor has the form

$$b^{\text{chn}} \sim \frac{N^2}{1 + hwN^{1.5}} \tag{2.36}$$

where h is a constant coefficient, independent of molecular weight. Fixman argues that for long chains w reflects primarily the ease with which entanglements can move relative to one another, and that this should go to zero rapidly enough with increasing molecular weight to make even the product $wN^{1.5}$ vanish as $N \to \infty$, so that the N^2 behavior of the friction coefficient is retained.

In contrast to simple translational motion of the probe, stress relaxation does involve intraprobe interactions of the types listed under items 4 and 5. We require therefore the entanglement lifetime $1/\lambda$, especially its molecular weight dependence. To this end, Fixman equates the reptational motion of probe bead i with the elastic component d_i of its displacement relative to its center of force. By arguments similar to those used in deriving Eq. (2.31), the retractive force by which the matrix responds to d_i is

$$\mathbf{f}_i^d \equiv \gamma_s d_i = \mathbf{f}_i^s + \gamma_s y \sum_{j=1}^{N} P_{ij} \mathbf{f}_j^s \tag{2.37}$$

where γ_s is the stiffness constant associated with a chain segment of length N_e ($\gamma_s y$ is of order unity), and f_i^s is related to v_i by Eq. (2.32). We have already seen that in steady translational motion of the probe chain through the matrix at speed v, $b^{\text{chn}} \equiv (1/v) \sum_{j=1}^{N} f_i \sim N^2$ at high molecular weights. Under the same conditions, the response of $b^{\text{rept}} \equiv (1/v) \sum_{j=1}^{N} f_i^d \sim N^{0.5} b^{\text{chn}} \sim N^{2.5}$ has an even stronger molecular-weight dependence.

To understand the physical meaning of Eq. (2.37), we recall that in translation the probe can move either by dragging along its entanglement partners or by moving relative to them as a bare, unencumbered chain. The two processes are parallel, competitive mechanisms, and both occur to some extent, but as N becomes large, the entanglement-cluster mode becomes dominant, and the friction factor increases as N^2. Conformational relaxation, however, cannot occur without cluster deformation and disentanglement. The reptational friction factor b^{rept} is essentially the friction factor that would be observed if the cluster mode of translation were somehow suppressed. In that case, the self-diffusion coefficient of the polymer would vary as $N^{-2.5}$ at large N, and the terminal relaxation time $1/\lambda$, which can be thought of as the time required for the center of mass of the probe to move by one coil diameter, would increase as $N^{3.5}$. This is what Fixman uses in Eq. (2.34) for the remainder of the development. It is perhaps worth noting that in terms of the reptation tube concept, $1/\lambda \sim N^{3.5}$ corresponds to the segmental jump frequency v_0 of Eq. (2.1), acquiring a molecular weight dependence, $v_0 \sim N^{0.5}$.

The stress relaxation modulus can now be obtained through Eq. (2.15). Fixman, using parameter values estimated for real polymers, obtains results which reproduce the main features of the experimental $G(t)$ curves. In particular, both Rouse and reptation behavior, separated by a rubbery plateau, are found if the molecular weight is sufficiently high. The zero shear viscosity can of course also be calculated from $G(t)$, and it shows the same molecular-weight dependence in the large N limit as the terminal relaxation time, $\eta \sim N^{3.5}$, which is certainly in better agreement with experiment than the strict reptation theory.

3. Schweizer's Model

We conclude this section with the Langevin model recently developed by Schweizer.[20] The starting equation used by Schweizer differs from Eq. (2.27) only in that the memory function is now a matrix $\mathbf{K}(t)$ rather than the scalar $K(t)$ in Ronca's theory:

$$k(r_{i+1} - 2r_i + r_{i+1}) + \zeta v_i + a \sum_{j=1}^{N} \hat{K}_{ij} v_j = \zeta_i \qquad (2.38)$$

Schweizer's selection of $\mathbf{K}$ is based on its relation to the random force autocorrelation [see Eq. (2.26)]. After a number of approximations he arrives at the result

$$K_{ij}(t) = b^* \int_0^{l^{-1}} q^4 g_{ij}(q, t) S_M(q, t)\, dq \qquad (2.39)$$

where the g_{ij} are the single-chain partial structure factors defined in Eq. (2.9), S_M is the structure factor for the matrix in the absence of the probe, and b^* is a constant of proportionality. The upper limit in Eq. (2.39) represents a cutoff at wave lengths below the segment length l.

As a first step, Schweizer next replaces $S_M(q, t)$ by the static structure factor $S_M(q) = S_M(q, 0)$ and uses the g_{ij} of the Rouse model in Eq. (2.39). This so-called renormalized Rouse theory leads to results that are intermediate between the classical Rouse model and reptation: for long chains, the terminal relaxation time increases as $N^{2.5}$ with increasing molecular weight, the self-diffusion coefficient decreases as $N^{-1.5}$, and the viscosity in steady shear rises as $N^{1.5}$. In contrast to reptation theory, the product of the latter two quantities is independent of molecular weight.

What appears to be missing at this stage are transient long-range interactions of the kind introduced by entanglement network models. Schweizer points out that the development to this point has not really taken into account the coupling between probe and matrix dynamics, particularly coupling to the slow, long-wavelength modes of the matrix. His arguments are too involved for repetition here, but it should perhaps be mentioned that he starts by using the segmental friction factor of the renormalized Rouse model, which increases as $N^{0.5}$ at large N, in place of the N-independent factors of the original. This step alone ensures the reptational N^{-2} behavior of the self-diffusion coefficient, although it is not enough to generate $\eta \sim N^3$. The memory matrix ultimately obtained by Schweizer is quite simple in form:

$$K_{ii} = K(t) - 2M(t), \qquad K_{i,i+1} = K_{i,i-1} = M(t), \qquad K_{ij} = 0 \quad \text{if } |i - j| > 1$$
$$(2.40)$$

where the characteristic times of the memory functions K and M increase as N and N^3, respectively, in the limit $N \rightarrow \infty$. The tridiagonal nature of $\hat{K}$ makes it possible to reduce the equation of motion [Eq. (2.38)], or rather its Laplace transform, to that of a simple Rouse chain, and so to work out its normal modes, stress relaxation behavior, and so on. The viscosity at high molecular weights now does indeed follow the N^3 behavior of the reptation model, thanks to the off-diagonal elements of $\hat{K}$, whereas in simple translation, with all probe beads moving at the same velocity, the $M(t)$ terms become inactive. Despite its complexity, Schweizer's model is particularly promising in terms of future development. For example, he attempts to compute the memory function explicitly from a microscopic theory, i.e., the static correlation function from the RISM integral equation, and the force-time correlation functions from mode coupling theory.

D. Models with Sparse Constraints

This class of models shares a common starting point that is essentially the opposite to reptation theories. Namely, entanglements are viewed as particular two-chain sites which enhance the friction of each site as it moves isotropically through the liquid. In other words the constraints of neighboring chains are felt as one chain drags another along over some finite distance or time. This reduces the chain mobility drastically, and enhances the chain-length dependence of the longest relaxation time. However, it is not obvious how to translate this physical picture into a computational scheme for the full dynamics of polymer chains; it has been attempted in several distinct ways. In this section we discuss briefly the approach of Bueche,[12] who first proposed this point of view, and then the more recent treatments of Shen et al.,[13] Fujita and Einaga,[15] Skolnick et al.,[16] and Ngai et al.[17] These models are clearly phenomenological, and the developments have uniformly been driven by the desire to describe known experimental results. Strict reptation coupled with the tube ansatz, on the other hand, is somewhat less phenomenological in origin, in the sense that once the initial postulate is made, the calculational scheme seems less arbitrary. However, when combined with the additional processes described in Section II.B.1, the reptation picture becomes equally phenomenological in practice. It is certainly conceivable that a significant contribution to the popularity of reptation (as opposed to the approaches described thereafter) as a framework to interpret experimental results is the better defined nature of the basic postulate; however, that is a question distinct from the one posed in the title of this chapter.

1. Bueche's Model

In this picture the central quantity is the chain friction coefficient, which for unentangled polymers can be written as $N\zeta$, where N is the number of subchains and ζ the subchain friction; this is identical to the Rouse result. Furthermore, in Bueche's calculation $\eta \sim N$ and $D \sim N^{-1}$. When the chains become entangled, the total chain friction is enhanced and may be written as $N^*\zeta$. The calculation of N^* is not transparent, but the physical idea is that a number of neighboring chains (determined by R_g for the test chain) are dragged along with a velocity that is a fraction of the test chain velocity; this fraction is called the slippage factor. In the initial treatment, $N^* \sim N^{2.5}$ as $N \to \infty$, leading to $\eta \sim N^{2.5}$ and $D \sim N^{-2.5}$, but subsequently, prompted by increasing experimental evidence, the calculation was modified in such a way that $N^* \sim N^{3.5}$. The resulting prediction $D \sim N^{-3.5}$ is now known to be well outside the experimental uncertainty. In a further extension to branched polymers, the main result was that for a given total molecular weight, the viscosity would decrease with the number of arms, as a power of R_g. Although this qualitative prediction is correct for low molecular weights, at sufficiently high molecular weights star-branched polymers exhibit a higher viscosity, as discussed in Section III. Thus, in terms of describing experimental results, the Bueche treatment fails. However, this does not necessarily refute the underlying physical concept. On the contrary, this prototype model indicates clearly that isotropic motion and strong molecular-weight dependence are not incompatible. In addition, it is one model that accounts for a rather sharp transition between entangled and unentangled regimes, in terms of a percolationlike argument or temporary network. Finally, this model stood for a long time; it was not until late in the 1970s that extensive experimental studies of diffusion began to appear, in response to the reptation postulate.

2.   A Rouse Chain with "Heavy Beads"

The concept of an entanglement point as representing the dragging of one chain by another, coupled with the success of the Rouse–Zimm bead–spring formalism for describing dilute solution dynamics,[29,30] leads rather naturally to the idea of exploring the effect of heavy beads on the dynamics of a Rouse chain. This approach has been taken by several workers, with the treatment of Shen et al.[13] offering a particularly transparent and computationally direct approach. In their model, some fraction of the beads along the chain are assigned an enhanced friction coefficient; otherwise the model follows the Rouse scheme exactly. This allows the spectrum of relaxation times to be calculated numerically, and therefore the viscoelastic functions. The presence of heavy beads leads directly to a plateau in $G(t)$, by virtue of the separation of bead motion time scales, but in many other respects this calculation is not in even qualitative accordance with experiment. The next level of complexity is to introduce an elastic coupling between the heavy beads on the test chain, that is, by connecting a weak, extra spring between each pair of heavy beads. This step is motivated by the temporary network models, in that these springs represent the effect of elastic response of the medium on the chain. However, the model is clearly still a single-chain treatment. Mathematically this addition introduces paths for the transmission of forces between beads remote along the chain, but it is still possible to compute the exact relaxation time spectra of the enhanced model. The result is that with this elastic coupling, and with a variable degree of enhanced friction for the heavy beads, the model is able to imitate the experimentally observed relaxation spectrum for melts quite well, particularly if the heaviest beads are in the chain middle. While this kind of treatment clearly has limitations, it does provide a straightforward way to simulate the effect of modifications to the test chain dynamics, in a computationally direct manner.

On the basis of extensive computer simulations of concentrated polymer systems, Skolnick and coworkers[16] have recently proposed a phenomenological model which is really very similar to that of Shen et al., although more emphasis is placed on chain diffusion. The simulations are discussed in more detail in Section III, but the physical picture which the authors advance on the basis of these simulations is one of isotropic motion with entanglements being two-chain contacts which persist over the time scale of the longest relaxation time. Such long-lived "dynamic entanglements" are infrequent, and so the model treats the entangled polymer as a dilute solution of heavy beads. Borrowing the formalism of Hess, and taking the dynamic evolution of pair contacts as a weighted sum of contributions from heavy-bead diffusion and chain diffusion, the result for the self-diffusion coefficient becomes

$$D = D_{\text{Rouse}} \left(1 + \frac{N}{N_e} \right)^{-1} \tag{2.41}$$

where N_e is the spacing between heavy beads. This procedure guarantees a diffusion coefficient in reasonable agreement with the experiment, that is, one that crosses over from $D \sim M^{-1}$ to $D \sim M^{-2}$ as the number of entanglements per chain increases. Note that this result has the same form as the Bueche expression, albeit with a quite different N dependence.

The viscoelastic response is generated with two additional assumptions. First, the initial heavy beads appear in the chain middle (i.e., for $N/N_e \approx 1$); this is similar to the model of Shen et al. The distinction is that while Shen et al. found that varying the friction of the heavy beads as a function of placement along the chain gave the best result, Skolnick and coworkers make the friction of each heavy bead increase with N. Second, the stress at short times is taken directly from the theory of rubber elasticity. This is identical to the approach of Doi and Edwards, and again amounts essentially to grafting a temporary network model onto an explicit picture for single-chain relaxation at long times. The results are shown to be at least in qualitative accordance with some experiments. However, this is not surprising in the sense that the spirit of the model is to ask "assuming isotropic motion, what is required of dynamic entanglements on a test chain in order to produce results for diffusion and viscoelasticity which resemble experiment?"

Fujita and Einaga[15] have also presented a treatment of entangled chain diffusion which echoes Bueche's approach. Again, chains are presumed to move isotropically, with an enhanced friction relative to the Rouse case. In this instance, the N dependence of the enhanced friction results in $D \sim N^{-2.5}$, but with a concentration dependence of $c^{-1.5}$, which is certainly much less than observed in entangled solutions. The model has also been extended to the case of binary blends, and to an estimate of the N dependence of the viscosity.

In summary, these results underscore the two points made above, namely, that the physical picture underlying Bueche's treatment and adopted (in various forms) can generate molecular-weight scaling exponents much higher than the Rouse results, and in reasonable agreement with experiment, but that the calculational schemes are not always transparent, and thus it is not always clear how to evaluate the legitimacy of any particular development. One difficulty which all these models encounter is in attempting to explain the dynamics of high-molecular-weight star-branched polymers, as discussed in Section III. The basic problem is that stars become considerably less mobile than linear polymers of either equivalent total molecular weight or equivalent

R_g, when the molecular weight is large. This may be explained as due to the presence of "very heavy beads," but it is not clear how to justify this on physical grounds. Another difficulty, as noted by Graessley,[47] is that the experimental behavior of the steady state compliance cannot be recovered merely by introducing a distribution of friction coefficients along the chain.

3. Coupling Model

The coupling model of Ngai et al.,[17] although classified with the cooperative motion models in Section I, is in fact unique among the various ways of interpreting entangled polymer behavior in that it does not attempt to describe what polymer molecules do in any detailed way. Rather, it addresses the effect of coupling a specific dynamic mode to an environment in which many-body cooperative motion is an essential feature. Thus its potential applicability extends far beyond polymer liquids; the concomitant drawback is that it cannot be used to provide a first principles theory for polymer liquids, nor has a constitutive equation based on the coupling model been proposed. Nevertheless, there are several experimental observations which are apparently accounted for by this approach, and not as yet by other models, so some discussion is appropriate.

The starting point is to select a "primitive species," which embodies the dominant response in the uncoupled mode. For polymers, the natural choice is the Rouse chain, with its M scaling for η, D, and τ_1. The relaxation of the primitive species follows the simple relation

$$\frac{d\phi}{dt} = -W_0\phi \tag{2.42}$$

where ϕ is a relaxation function and W_0 the constant, uncoupled relaxation rate. When polymer chains are sufficiently long for the effects of entanglement to be felt, the relaxation of the chain becomes coupled to its environment, but only for times longer than a crossover time t_c. Thus at short times the constraints of the neighboring chains are not felt. The result is that the relaxation rate becomes a function of time:

$$\frac{d\phi}{dt} = -W_0 f(t)\phi, \qquad t > t_c \tag{2.43}$$

where $f(t) < 1$ for $\omega_c t \gg 1$ ($\omega_c = 2\pi/t_c$). The form of $f(t)$ is important to the ultimate results, and can be obtained in several distinct ways. One route which is more natural to polymers is to require that the relaxation behavior obey time–temperature superposition (thermorheological simplicity). This leads directly to

$$f(t) = (\hat{\omega}_c t)^{-n}, \qquad \omega_c t \gg 1 \tag{2.44}$$

where $\hat{\omega}_c$ is proportional to ω_c (with a proportionality constant of order unity) and n is the coupling parameter. The basic predictions of the coupling model may then be summarized in three relations:

$$\phi(t) = \exp\left(-\frac{t}{\tau_0}\right), \qquad \omega_c t < 1 \tag{2.45a}$$

$$\phi(t) = \exp\left[\left(-\frac{t}{\tau^*}\right)^{1-n}\right], \qquad \omega_c t > 1 \tag{2.45b}$$

$$\tau^* = [(1-n)\omega_c^n \tau_0]^{1/(1-n)} \tag{2.45c}$$

where $\tau_0 = W_0^{-1}$. The main result is Eq. (2.45b), which is the empirical Kohlrausch–Williams–Watts (KWW) stretched exponential relaxation function. However, there are some key additional points. First, the KWW form is a result, not an assumption. Second, the mean relaxation time τ^* is directly related to the uncoupled τ_0, providing a direct test and a direct link between uncoupled and coupled behavior. Third, there is a temporal crossover from single exponential to KWW relaxation, with the crossover time also incorporated in τ^*.

The value of the parameter n can only be determined by comparison with experiment, such as by fitting Eq. (2.45b) to $G''(\omega)$ in the terminal regime. For a variety of different polymers, the fit is very good and gives $n \approx 0.4$.[48-50] The Rouse τ_0 is proportional to $M^2\zeta(T)$, so Eq. (2.45c) predicts that in the entangled regime $\tau_1 \sim M^{3.3}$, in reasonable agreement with experiment. For diffusion there is no equivalent to G'' from which n can be determined. Using the observation that $D \sim M^{-2}$ (see Section III.A.1), an n for diffusion of 0.33 is inferred. Thus the strength of the coupling is different for diffusion and viscosity, which is plausible. However, this proposition can be tested via the temperature dependence; the local friction is also incorporated in τ^*. A different n for viscoelastic properties and for diffusion requires a different temperature dependence, or apparent activation energy, with the two values related by $(1 - n)_{viscosity}/(1 - n)_{diffusion}$. This prediction is in agreement with some data[50] but perhaps not with others;[51,52] however, the prediction of a different temperature dependence is unique to this model. The combination of Eqs. (2.45) has been successful in reconciling other features of entangled polymer behavior, as discussed elsewhere.[17,53] The essential philosophy underlying its application is that since the many-chain problem is too complicated to solve rigorously, one should examine general features of the dynamics of entangled polymers rather than make drastic simplifying assumptions such as reptation, which will inevitably require additional modification to obtain quantitative agreement with experiment.

III. EXPERIMENTAL TESTS

A. Experiments for Times Longer Than τ_1

1. *Translational Diffusion of Linear Chains*

For several reasons polymer chain diffusion is the most appropriate place to begin an assessment of the success of reptation in describing the dynamics of polymer melts. First, the scaling laws derived in Section II.A.1 are the most direct predictions of the reptation model, in the sense that no additional assumptions are required. This situation may be contrasted, for example, with predictions for some of the rheological functions, where additional nontrivial assumptions are required to arrive at a constitutive equation. In other words, failure of such a constitutive equation to describe a given experiment may not indicate that the reptation postulate is invalid. Second, the introduction of the reptation hypothesis has stimulated an enormous amount of experimental effort to measure diffusivities over the past 15 years, including the development of a variety of elegant new techniques. Prior to this there were almost no entangled polymer diffusion studies reported in the literature. Thus the M^{-2} prediction emerged without the guidance of experiment, unlike, for example, the already experimentally well-established $M^{3.4}$ law for the viscosity. Third, the reptation hypothesis and subsequent developments emphasized the natural relationship between chain diffusion and relaxation, a relationship that had been largely ignored since Bueche's work.[12] Computation of reduced quantities, such as the product $D\eta$, can serve to eliminate parameters and lead to more robust examination of model predictions, as in the case of the Rouse model applied to low M melts. Fourth, in addition to the M^{-2} scaling law, the reptation model leads to strong, testable predictions about diffusion in networks, branched polymer diffusion, and the dependence on matrix molecular weight.

Before proceeding, some discussion of experimental issues is in order. The diffusion coefficient of interest is that defined by

$$\lim_{t \to \infty} \langle [\mathbf{R}_j(t_0 + t) - \mathbf{R}_j(t_0)]^2 \rangle = 6Dt \qquad (3.1)$$

where $\mathbf{R}_j$ is the vector from a laboratory-fixed origin to the center of mass of polymer j. In the case where the sample contains only one chemical species, D is commonly referred to as the self-diffusion coefficient D_s, whereas in the case of a mixture, D becomes the tracer diffusion coefficient D_t. In either case the quantity of interest concerns the Brownian motion of a test chain, in the absence of macroscopic gradients of chemical potential. Equation (3.1) suggests some experimental obstacles to be overcome. First, it is usually necessary to label some fraction of the molecules of interest, in order to monitor their spatial excursions; at the same time, the labeling scheme should have no effect on the property to be measured. Second, the choice of distance (or time) scale to be examined is crucial. If the characteristic distance of the experiment d is less than $5-10R_g$, the measured D may contain contributions from segmental motions relative to the center of mass. Conversely, if d is too large, for example on the millimeter scale, the diffusion times can extend to years. Both of these difficulties have been successfully overcome by a variety of techniques, including field-gradient NMR, forced Rayleigh scattering (FRS), forward recoil spectrometry (FRES), small-angle neutron scattering (SANS), and infrared microdensitometry (IRMD). The basic principles and comparative features of these experiments have been reviewed elsewhere.[54] Together they offer an extremely wide dynamic range ($10^{-5} \geq D \geq 10^{-18}$ cm^2/s) and typical relative precisions of 10–50%.

There are a number of additional issues which can have a direct bearing on the discussion to follow. Although all are reasonably familiar to polymer scientists, they are not always remembered.

1. Much of the ensuing analysis involves power-law exponents, particularly in the dependence of a dynamic property on M. It is important to bear in mind that such exponents can be very sensitive to experimental uncertainty, and that the commonly employed double logarithmic format can be quite successful at masking subtleties in the data. We feel that an exponent obtained from data that do not span at least one order of magnitude in the independent variable should not carry much weight.

2. It is unlikely that the absolute value of M is ever known with a relative accuracy better than 10%. Given a similar uncertainty in the measured quantity (which is usually optimistic), four data points spread over one decade on the abscissa give at least a $\pm 10\%$ uncertainty in the slope.

3. Polymer samples are always polydisperse, although more recent synthetic methodologies give ratios of $M_w/M_n < 1.1$. The quantity ($M_w/M_n - 1$) is the variance of the molecular-weight distribution (MWD). The stronger the M dependence, the more sensitive the exponent will be to polydispersity. Furthermore, if the shape of the molecular-weight distribution varies with M_w, the apparent exponents can be affected substantially.[55]

4. The glass transition temperature of amorphous polymers, T_g, is a function of M, generally increasing with M until the degree of polymerization is between 10^2 and 10^3. One manifestation of this is that at a given

temperature shorter chains are inherently more mobile on a local scale than longer chains, leading to an extra chain length dependence for measurable quantities such as D and η. This effect is commonly, but not always, corrected for by reducing to the "isofree volume"or "isolocal friction" state[2]; the omission can lead to misleading results.

To reiterate, the main point of the preceding discussion is that power-law exponents obtained from experimental measurements should always be viewed with some skepticism.

In this section we consider representative diffusion data for linear chains in melts, in light of three reptation-based predictions. These are the canonical

$$D = k_D M^{-2} \tag{3.2}$$

the expected independence of matrix molecular weight P

$$\frac{\partial D}{\partial P} = 0 \tag{3.3}$$

and, a consequence of the Doi–Edwards development[56]

$$k_D = \left(\frac{G_N^0}{135}\right)\left(\frac{\rho RT}{G_N^0}\right)^2\left(\frac{\langle h^2 \rangle}{M}\right)\left(\frac{M_c}{\eta(M_c)}\right) \tag{3.4}$$

Equation (3.3) embodies the basic reptation postulate, that the constraints around a test polymer chain are effectively immobile on the reptation time scale, and thus D for a reptating chain cannot depend on the molecular weight of the surrounding chains. Equation (3.4) is an expression for the prefactor k_D in Eq. (3.2), involving only measurable quantities such as the plateau modulus G_N^0 and the viscosity for $M = M_c$, where M_c is the critical M for the onset of the 3.4 power in η.[56] The predictions and results for branched polymers and for linear chains trapped in networks are considered in Sections III.A.5 and 6, respectively.

The large majority of melt diffusion data have been obtained for polystyrene (PS)[57-71] and polyethylene (PE).[52,59,60,70-79] Both are chemically robust and available in a variety of molecular weights, although PS is much more readily obtained with a narrow MWD. Values of D_s for PS melts are presented in Figs. 3.1 and 3.2, and for PE in Figs. 3.3 and 3.4. Several explanatory comments about the plots are appropriate. First, the abscissas are $\log(M/M_c)$, where M_c has been taken as 35,000 and 5000 for PS and PE, respectively. This selection of normalization is somewhat arbitrary, but certainly one should not expect to observe evidence for reptation much below this cutoff. (Although -2

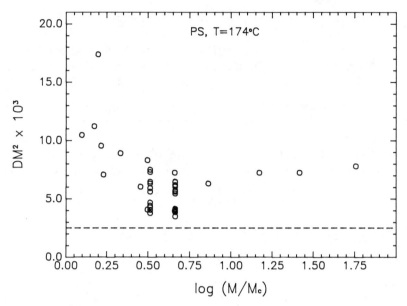

Figure 3.1. Polystyrene self-diffusion at 174 °C, for $M > M_c = 35,000$. Horizontal line indicates prediction of Eq. (3.4). From Refs. 57, 61–64, 67–69.

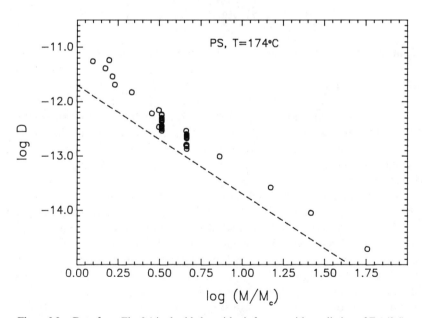

Figure 3.2. Data from Fig. 3.1 in double logarithmic format, with prediction of Eq. (3.4).

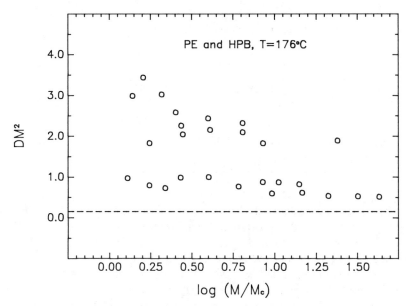

Figure 3.3. Polyethylene and hydrogenated polybutadiene self-diffusion at 176 °C, $M > M_c = 5000$. Horizontal line indicates prediction of Eq. (3.4). From Refs. 52, 59, 60, 70, 71, 76–79.

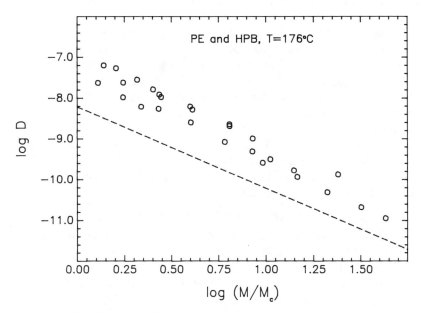

Figure 3.4. Data from Fig. 3.3 in double logarithmic format, with prediction of Eq. (3.4).

TABLE 3.1
Parameters Used to Calculate k_D via Eq. (3.4)

	PS (174 °C)	PE (176 °C)	PI (25 °C)
G_N^0, dyn/cm^2	2×10^6	2×10^7	3.5×10^6
ρ, g/cm^3	0.96	0.767	0.90
M_c	35,000	5200	10,000
$\eta_0(M_c)$, P	2930	1.15	54
$\langle h^2 \rangle / M$, cm^2/Da	0.46×10^{-16}	1.06×10^{-16}	0.64×10^{-16}
k_D, (Da)$^2 \cdot$ cm^2/s	0.0025	0.15	0.0125

exponents have been reported at lower M, this is probably due to an approximately M^{-1} dependence of chain-end free volume combining with the M^{-1} Rouse behavior, as mentioned before.) Thus for both polymers the data extend over a factor of 30 in reduced chain length. Second, in Figs. 3.1 and 3.3 the diffusivities have been scaled by M^2, so that the data should be independent of M if reptation is dominant. Note also that these ordinates are linear, not logarithmic, which provides a more rigorous test of Eq. (3.2). Figures 3.2 and 3.4, on the other hand, follow the more commonly utilized double logarithmic format. Third, the data presented are all self-diffusion, in the sense that $M = P$, although in all but the NMR experiments some chemical labeling has been employed. Fourth, in all four figures the prediction of Eq. (3.4) has been included as a dashed line. The parameter values used to generate these predictions are listed in Table 3.1. Fifth, in several cases the data have been reduced to the indicated temperatures (174 °C for PS and 176 °C for PE) by utilizing a Vogel temperature dependence for PS[2,66] and an Arrhenius correction for PE.[52] This procedure inherently, but unavoidably, introduces additional uncertainty, but in several cases the detailed T dependence of D was available. Furthermore, the selected T were those for which the largest amount of data was available, and which fell reasonably within the range reported. Sixth, in Figs. 3.5 and 3.6 the data in Figs. 3.1 and 3.2 are replotted with tracer diffusion data for $M \leq P$ included. Before examining the data in more detail, four general conclusions are apparent.

1. The M^{-2} dependence is a very reasonable representation of these data.
2. The magnitude of k_D predicted by Eq. (3.4) is less than that observed experimentally, by a factor of 2–3 for PS, but by up to an order of magnitude for PE.
3. The scatter in the magnitude of DM^2 is substantial. For PS the data range over a factor of 3, while for PE the range is at least twice as large. Given a typical relative precision of 10–20% in both M and D, this scatter raises the possibility of systematic error.

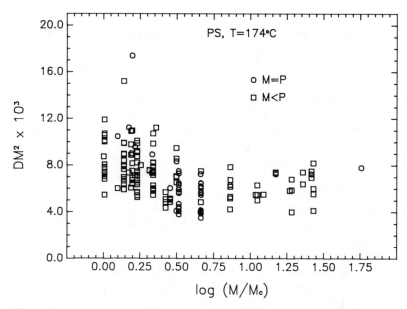

Figure 3.5. Polystyrene self-diffusion and tracer diffusion for $M < P$, at 174 °C. From Fig. 3.1 and Green and Kramer.[65]

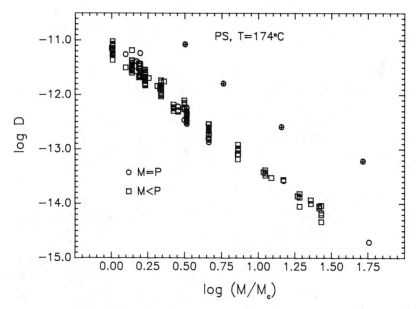

Figure 3.6. Data from Fig. 3.5 in double logarithmic format, plus data from Kumagai et al.[58] not shown in previous figures.

4. Broadly speaking, the tracer diffusion data are more or less indistinguishable from the self-diffusion case, and thus Eq. (3.3) is in good agreement with the results.

The data for PE in Figs. 3.3 and 3.4[52,59,60,70,71,76-79] display more scatter than one might expect, although for individual studies the M^{-2} dependence is followed extremely well. However, the origin of this scatter is not yet apparent. From Fig. 3.3 the data appear to fall into two groups, one with $k_D \approx 2.5 \pm 0.5$ cm$^2 \cdot g^2 \cdot s^{-1} \cdot$ mol^{-2}, the other with $k_D \approx 0.7 \pm 0.3$ cm$^2 \cdot g^2 \cdot s^{-1} \cdot$ mol^{-2}. The former group consists entirely of NMR measurements,[59,60,77-79] so one might be tempted to suspect that the difference reflects the experimental method. However, the lower group also includes extensive NMR measurements.[70,71] Furthermore, it is well known that both polydispersity and time scale can influence NMR measurements of D_s in rather subtle ways, and in several of these studies the authors have evidently taken great care to account for these. Sample polydispersity is clearly an important consideration. However, it also does not correlate clearly with either of the two data sets. For example, Pearson et al. in the upper group used quite narrow samples ($M_w/M_n \leq 1.2$),[79] while Bartels et al. and Klein et al. in the lower group had $M_w/M_n \leq 1.05$.[52,76] Fleischer[70,71] has corrected the data of Bachus and Kimmich[59,60] for the effects of polydispersity, which brings the highest 5 points in Fig. 3.3 down to the lower group. Yet, essentially the same procedure was employed by Pearson et al.,[79] and these data remain well above the prediction of Eq. (3.4). Another potentially important point is that several studies did not actually use PE per se, but essentially fully hydrogenated (or deuterated) 1,4-polybutadiene (HPB).[52,76,79] Thus due to the inevitable presence of some 1,2 microstructure (typically $< 10\%$), HPB has approximately 18 ethyl branches per 1000 backbone bonds. (The use of HPB is highly desirable because PB is readily polymerized by anionic methods, giving a very narrow MWD, and because the hydrogenation/deuteration step is straightforward and produces labeled and unlabeled molecules of identical size.) It is possible that these short branches contribute to the lower values of k_D, but this does not explain why Fleischer's NMR data for PE are in such close agreement.[70,71] In an earlier study of Klein and Briscoe,[75] the diffusant was also PE, not HPB, but the polydispersities were so large ($M_w/M_n \approx 2-3$) that these data should probably not be given as much weight and are not included in the plots. Also, recent measurements of D for HPB are in close agreement with those obtained for PE by the same technique.[51] In sum, therefore, there persists a substantial and unexplained variation in the magnitude of D for PE. If we assume that this is due to an (unidentified) systematic error which is independent of M, the PE data conform very closely to the reptation scaling laws $D \sim M^{-2}P^0$, but the magnitude of k_D differs significantly from the Doi-

Edwards-based prediction and always indicates more chain mobility than the model predicts.

The data on PS diffusion present a more consistent picture than do those for PE, although a series of measurements by Kumagai et al.[58] have been omitted from Figs. 3.1 and 3.2 because they fall almost a factor of 10 above the data presented, when adjusted to 174 °C with the same parameters used for the other data. However, they are included in Fig. 3.6. The dependence of D_t on P has also been studied in more detail for PS, with the most extensive set of measurements being those of Green and Kramer via FRES.[65] These data are included in Figs. 3.5 and 3.6, but are replotted in Fig. 3.7 to emphasize an important point. The data are plotted in the double logarithmic format as DM^2 versus P/M, and for $P/M > 1$, D scales clearly as Eqs. (3.2) and (3.3) predict; D is independent of P over at least two orders of magnitude of P/M.

For PS self-diffusion, the crossover from $M < M_c$ to $M > M_c$ has been examined very closely. From $M \approx 0.3M_c$ to $M \approx 3M_c$, the apparent M exponent varies from the -1 value associated with Rouse chains to about -2, but over a narrow region it appears to have a value close to -2.5. In other words the instantaneous M dependence is actually stronger than the reptation prediction. This has been attributed to a crossover effect; essentially because the transition from Rouse-like to reptation-like is not instantaneous, the extra

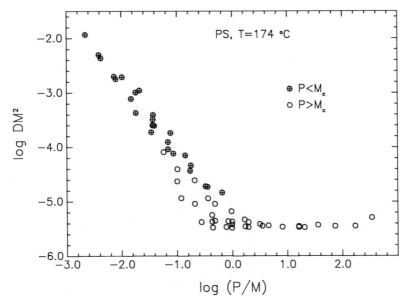

Figure 3.7. Polystyrene tracer diffusion as a function of matrix molecular weight. From Green and Kramer.[65]

relaxation modes enhance D over a narrow range in M.[69,80] However, in consideration of the remarks made about power laws, it may be inappropriate to interpret an apparent exponent in this much detail.

Melt diffusion measurements have been reported for other chemical systems, although none so extensive as the collections in Figs. 3.1 to 3.7. In the case of polyisoprene (PI) there are two features to the data worthy of consideration, as shown in Fig. 3.8.[81] First, an apparent M exponent approaching -3 is observed at values of M/M_c over the range of 5–20. As yet, no explanation for this behavior has emerged. Second, the necessary parameters to examine Eq. (3.4) are available and listed in Table 3.1. The result is indicated by the dashed lines in Fig. 3.8. As with PE and PS, the predicted value of k_D is generally below the data, but is always within a factor of 2.

In conclusion, there is ample evidence from melt diffusion studies that $D \sim M^{-2}$, in accordance with the reptation prediction; however, slightly larger exponents cannot be excluded. For $M \leq P$, D is substantially independent of P. At the same time, individual chains are typically a factor of approximately 2–5 times more mobile than the strict reptation prediction. Since strict reptation could be viewed as a lower bound on chain mobility, the inclusion of other relaxation processes might bring the theory into line with the experiment. But any such modifications must not alter the M depen-

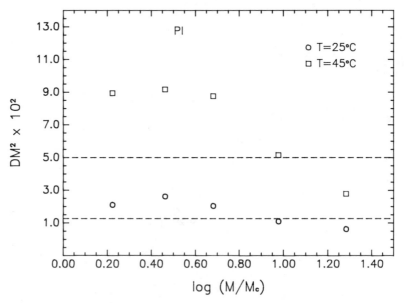

Figure 3.8. Polyisoprene self-diffusion at 25 and 45 °C, for $M > M_c = 10,000$. Horizontal lines indicate prediction of Eq. (3.4). From Landry.[81]

dence very much, and must not depend strongly on P. Since the -2 exponent is by no means unique to reptation, these diffusion results taken as a whole are consistent with reptation as a substantial contributor to chain mobility, but do not establish it as a necessary concept.

2. Melt Viscosity

As indicated in Fig. 1.4, the strong dependence of η on M for $M > M_c$ has been established for a wide variety of polymers. Parenthetically, the observed power-law exponent of 3.4 ± 0.2 is also observed for rather polydisperse samples, when $\log M_w$ is used as the abscissa, and for concentrated solutions.[2,6] In Fig. 3.9 we have selected a variety of more recently acquired data for seven different polymers [PS, PE, PI, PB, HPI, HPB, and poly(α-methylstyrene)].[82-91] These data have been chosen as a representative subset, emphasizing measurements with very narrow MWD samples. The format in Fig. 3.9 serves to illustrate the extent of the $M^{3.4}$ regime. As with the diffusion data, the abscissa is taken as $\log(M/M_c)$. The ordinate is also logarithmic and is the measured zero-shear rate viscosity scaled by $M_*^{3.4}$, normalized to the same quantity at $M_* = 10M_c$. Thus the data for the different polymers are forced to coincide at $\log(M/M_c) = 1.0$; no particular temperature scaling has been invoked. (The data do not converge exactly to a point at

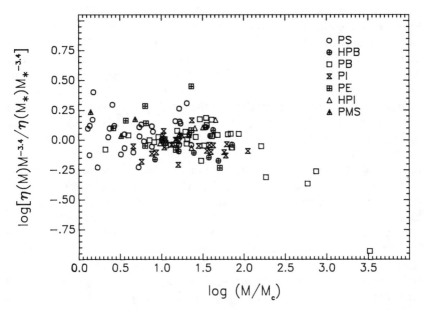

Figure 3.9. Reduced melt viscosity for seven different polymers, for $M > M_c$. From Ref. 5, 79, 82–91.

$\log(M/M_c) = 1.0$ because the values of $\eta(M_*)$ were interpolated from the range of measurements in each data set.] This enforced coincidence of the various data sets obscures any systematic errors that might be present in the data; the remaining scatter is entirely consistent with inherent relative uncertainties of 20% in M and η. Up to at least $\log(M/M_c) = 2.0$, there is no suggestion of a deviation from the 3.4 exponent.

The reptation predictions for the melt viscosity are the basic scaling law

$$\eta = k_\eta M^3 \sim \tau_1 G_N^0 \qquad (3.5)$$

and the Doi–Edwards-based[56]

$$\eta = \frac{15}{4} \left(\frac{G_N^0}{\rho R T} \right)^2 \left[\frac{M^3 \eta(M_c)}{M_c} \right] \qquad (3.6)$$

Clearly, there is a distinct difference between experimental and predicted exponents over at least two decades in M/M_c. The prediction of Eq. (3.6) cannot be compared with the data in Fig. 3.9, however, due to the normalization factor. The remaining discussion will consider the evidence for a crossover to an M^3 power law at $M/M_c > 100$, and the magnitude of η in comparison with Eq. (3.6).

The reptation model clearly requires that $\eta \sim M^3$ in the limit of infinite M. However, there are other considerations which suggest that such a crossover should be observed. First, as has been discussed in Section II, several other models lead to the M^3 dependence in the high M limit, while none provide a rigorous prediction of $M^{3.4}$ at arbitrarily large M. Second, following an argument presented by Colby et al.,[91] it is reasonable to expect that the high M limiting behavior of at least one of the set D, η, R_g, and J_e^0 must change. (J_e^0 is the recoverable compliance, discussed in Section III.A.3.) Briefly, the difficulty is this. The reorientation time for the whole chain τ_1 may be estimated as $J_e^0 \eta$, where experimentally J_e^0 is independent of M at high M. A diffusion time τ_D may also be estimated as $R_g^2/6D$. For flexible chains there is every reason to expect that these two times should be comparable, and particularly that τ_1/τ_D should not grow indefinitely with M. However, as long as the relevant four experimental scaling laws ($J_e^0 \sim M^0$, $\eta \sim M^{3.4}$, $R_g \sim M^{0.5}$, and $D \sim M^{-2}$) persist, the ratio of characteristic times will increase with $M^{0.4}$. This would imply that for progressively longer chains, a single molecule could diffuse many times its own average size without complete randomization of its end-to-end orientation. While such a result may be eminently plausible for a semidilute solution of rigid rods, it is much less so for flexible chains. Therefore at least one of the four exponents must change. For R_g and J_e^0 at least, this seems unlikely. For D no evidence of a change in exponent at high M is yet apparent, but the possibility cannot be ruled out. However, recently the possible onset of a departure from $\eta \sim M^{3.4}$ at ultrahigh M has been reported in an experimental tour de force.[91]

The polymer of choice was PB, which is polymerizable with very narrow MWD and to very high M. Furthermore, the small M_e (1850) and M_c(6380) meant that with the highest M synthesized (1.6×10^7), chains with over 10^3 entanglement lengths were available. The data are shown in Fig. 3.10, reproduced from Colby et al. The plot shows η/M^3 versus M on a double logarithmic plot, with the indicated error bars. The smooth curve represents the empirical correlation $\eta = 4.0 \times 10^{-12}M^{3.41}$. The point for the highest M may be subject to additional uncertainty, as it was obtained from a creep measurement which

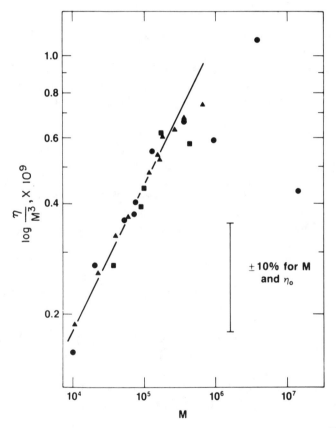

Figure 3.10. Reduced melt viscosity for polybutadienes. From Colby et al.[91]

may not have reached its long-time asymptotic dependence. Nevertheless, the data for the seven highest M samples shown fall below the line established for $10^4 < M < 3.6 \times 10^5$. At least two important conclusions may be drawn from these results. First, these data represent the only evidence, of which we are aware, of a change in the high M scaling of the four quantities listed; that it should be in the viscosity is consistent with various models, including reptation. Second, on the more practical side, as no commercial polymer is likely to approach this many entanglement lengths, the $M^{3.4}$ scaling law essentially holds for all entangled polymer melts and must therefore be incorporated in any useful constitutive equation.

In absolute magnitude, the diffusion data lay consistently above the strict reptation prediction of Eq. (3.4), implying in the reptation picture that other processes contribute to the mobility. However, such processes apparently do not exert a noticeable influence on the M exponent. Comparison of Eq. (3.6) with viscosity data also indicates that chains are able to relax more readily than by reptation alone; for example, the prediction of Eq. (3.6) for the infinite M plateau in Fig. 3.10 would be $6.4 \times 10^{-9} P$ mol$^3 \cdot$ g^{-3}, a factor of approximately 5 greater than the asymptote implied by the data. At the other extreme, $M = M_c$, and recognizing that $\rho RT / G_N^0 = M_e \approx 0.5 M_c$, the prediction of Eq. (3.6) is about an order of magnitude too large. Any additional process(es) contributing to the viscosity must also change the apparent exponent significantly, in contrast to the diffusion case. As discussed in Section II.B.1, one such process is the tube fluctuation argument of Doi,[31] in which fluctuations in tube length (due to enhanced mobility of chain ends) reduce τ_1 in a manner that becomes insignificant for sufficiently large M. This process is presumed not to contribute to D, although this interpretation has been questioned.[33] A second process is constraint release, also discussed in Section II.B.1, in which the ability of neighboring chains (which constrain the test chain) to move away enhances the test chain mobility.[32,37-44]

An alternative explanation for the results in Fig. 3.10 has recently been proposed by Kavassalis and Noolandi,[22] as discussed in Section II.B.3. For the purposes of this discussion, the crucial result is that the parameter M_e is taken to be a function of molecular weight, particularly at lower molecular weights. In other words, while a chain with $M = 20 M_e$ may be assumed to have 19 effective entanglements, a chain with $M = 3 M_e$ should not be assumed to have two effective entanglements due to the enhanced mobility of the chain ends. The result is a crossover to $\eta \sim M^3$ with increasing M, as illustrated in Fig. 3.11 for the data of Fig. 3.10. It can be seen that the broad features are well captured, but the quantitative agreement is not compelling. Nevertheless, the concept of a molecular-weight-dependent entanglement length has some physical appeal, and it leads to an excellent correlation with G_N^0 for different polymer structures, as discussed in the next section.

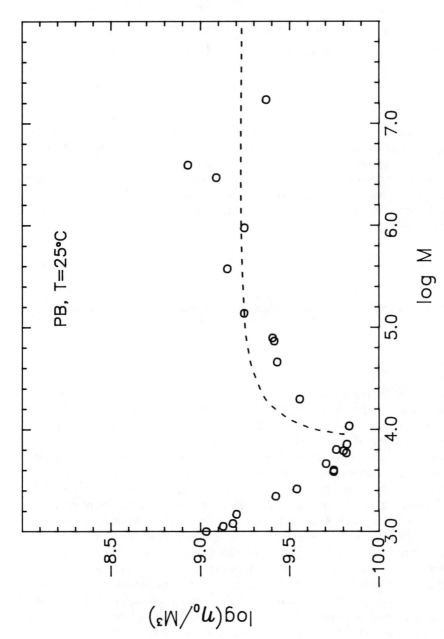

Figure 3.11. Data from Fig. 3.10 compared to model of Kavassalis and Noolandi.[22]

3.   Viscoelasticity of Monodisperse Melts

In this section we consider the linear viscoelastic properties of homopolymer melts, specifically the shear moduli $G'(\omega)$, $G''(\omega)$, $G(t)$, and the hallmarks of entangled systems, the plateau modulus G_N^0 and the steady-state compliance J_e^0.

a.   Plateau Modulus and Steady-State Compliance. As mentioned in Section I.B, the appearance of a plateau in $G(t)$ and $G'(\omega)$ leads naturally to the idea of an entanglement network which exists over a significant range of time scale. Also, the evidence is overwhelming that this is a universal feature of flexible polymer systems.[2] The experimentally observed plateau value G_N^0 can be converted to a parameter characteristic of a given polymer structure, the molecular weight between entanglements M_e

$$M_e = \frac{\rho R T}{G_N^0} \tag{3.7}$$

by direct analogy with the expression from the theory of rubber elasticity

$$G = \frac{\rho R T}{M_x} \tag{3.8}$$

where M_x is the molecular weight between cross-links.[3] Thus G_N^0 may be thought of as representing an entanglement density. The quantity G_N^0/kT is essentially independent of T, lending considerable additional support to the interpretation of entanglements as topological in nature, and not dependent on any specific interaction. The absolute value of G_N^0 varies considerably among the various polymers examined, for example, from 3.3×10^5 dyn/cm^2 for poly(n-octyl methacrylate) to 2.1×10^7 dyn/cm^2 for polyethylene.[2,92] However, it has recently been demonstrated that G_N^0 for different polymers correlates very well with the length of chain contour per unit volume,[92] as illustrated in Fig. 3.12. Slightly different approaches to this correlation have been proposed and the origin of the differences identified; the expression of Kavassalis and Noolandi, for example, expresses M_e as a function of ρ, the Kuhn step length, the characteristic ratio, and a coordination number $\tilde{N}$.[22] The value of $\tilde{N}$ has been shown to be very narrowly distributed (standard deviation $\approx 8\%$) about 8.1 for different polymers. Nevertheless it remains an unsolved problem to predict the value of $\tilde{N}$ (or, equivalently, M_e) from first principles.

The steady-state compliance is best defined from a creep experiment, that is, one in which a polymer is subjected to a constant stress σ, and the time

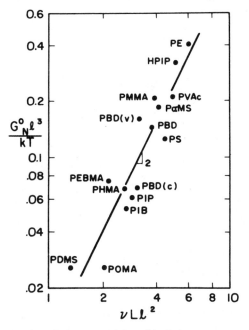

Figure 3.12. Correlation of plateau modulus with chain contour per unit volume. From Graessley and Edwards.[92]

evolution of the strain $\gamma(t)$ is measured. A schematic illustration of the response of a typical polymer melt is shown in Fig. 3.13. The steady-state compliance reflects the ability of the sample to recover its original shape upon removal of the stress and thus, like G_N^0, is elastic in origin. The interrelationships among the various quantities (in addition to those given in Section I.B) may be written

$$\eta = \int_0^\infty G(t)\,dt = \lim_{\omega \to 0} \left(\frac{G''}{\omega}\right) \tag{3.9}$$

$$J_e^0 = \lim_{\omega \to 0} \left(\frac{G'}{G''^2}\right) = \left(\frac{1}{\eta}\right)^2 \int_0^\infty t G(t)\,dt \tag{3.10}$$

The steady-state compliance is thus a normalized first moment of the modulus, and therefore of the relaxation spectrum. Following Graessley,[56] the Doi–Edwards expression for J_e^0 may be written

$$J_e^0 = \frac{1.2}{G_N^0} \tag{3.11}$$

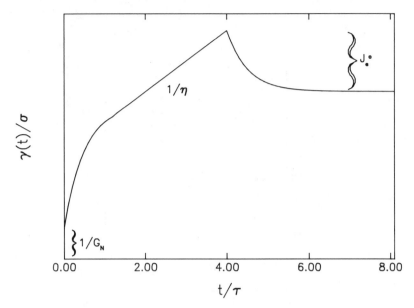

Figure 3.13. Schematic illustration of creep compliance for molten polymers. (Note that the $t = 0$ intercept is zero.)

J_e^0 is thus predicted to be independent of M, which is experimentally well established. However, the M independence is not established until $M > M_c'$, where M_c' is typically a factor of approximately 3 larger than M_c from η versus M.[2] Furthermore, J_e^0 is quite sensitive to polydispersity, as can be seen from Eq. (3.10), while G_N^0 is not. Experimentally, the product $G_N^0 J_e^0$, which may be thought of as the breadth of the distribution of relaxation times in the terminal zone [i.e., when $G(t)$ decreases from G_N^0 at long times] is typically 2–3, as opposed to 1 for a single relaxation time model. Thus the Doi–Edwards theory substantially underestimates this quantity. However, this should not be viewed as too surprising in view of the results for both D and η, from which it is apparent that the strict reptation contribution does not describe all the relaxation processes exhibited by the polymer molecules.

b. $G^*(\omega)$ and $G(t)$. Because these two quantities are directly related, we only discuss the reptation predictions for $G(t)$. In the simplest possible prediction, the reptation model gives $G(t) \approx G_N^0 \exp(-t/\tau_{\text{rep}})$, where τ_{rep} is the longest relaxation time of the chain. Although obviously only an estimate, this relation emphasizes an important point. The reptation process by itself only applies to the very long time (terminal) regime in stress relaxation. In particular, the plateau modulus is assumed (albeit subsumed in the tube diameter a) and a single exponential decay used to describe the subsequent relaxation. In the Doi–Edwards development based on this picture, the plateau modulus is still assumed (actually replaced by a confining tube of diameter a), but a slightly broader distribution of relaxation times is found [recall Eq. (2.14)]:

$$G(t) = G_N^0 \left(\frac{8}{\pi^2}\right) \sum_{j,\text{odd}} j^{-2} \exp\left(-\frac{j^2 t}{\tau_{\text{rep}}}\right) = G_N^0 F(t) \qquad (3.12)$$

Nevertheless, this function is still not in good quantitative agreement with experiment, which is to be expected in light of the viscosity results described previously. However, the tube model as laid out by Doi and Edwards, and as subsequently analyzed in more detail by Osaki[93] and Lin,[94] among others, can be modified to account for the experimentally observed behavior of $G(t)$ (restricted to small strains as assumed throughout the development so far). Lin, for example, proposes an expression for $G(t)$ comprising contributions from four distinct processes[94]:

$$G(t) = G_N^0 \left[1 + u_a\left(\frac{t}{\tau_a}\right)\right]\left[1 + 0.25 \exp\left(-\frac{t}{\tau_x}\right)\right]\left[B u_b\left(\frac{t}{\tau_b}\right) + C u_c\left(\frac{t}{\tau_c}\right)\right]$$

$$(3.13)$$

where u_b is a Rouse-like process with $\tau_b \sim M^2$, u_c is the reptation component corrected for the tube fluctuation process as considered by Doi, and u_a is the Rouse-like motion on distance scales less than the entanglement length. Osaki et al. have reported evidence of an intermediate relaxation time scaling with M^2.[93] The fourth process, u_x, refers to an intermediate time relaxation which allows the modulus to decrease from $\rho RT/M_e$ to 0.80 of that value in the plateau region. It corresponds to equilibration of the chain contour length within the deformed tube, as discussed in Section II.A.3. The functional forms of u_b and u_c are identical, differing only in the expression for the fundamental relaxation time, u_a is the sum of exponentials associated with a Rouse chain, and u_x is assumed to be a single exponential. The result of Eq. (3.13) is that $G(t)$ for a monodisperse melt can be described very well, even at much shorter times than the reptation time. An example is shown in Fig. 3.14, in which the contributions of the four processes are identified.

As with the results for diffusion and viscosity, the strict reptation model captures the main features of the viscoelastic response of polymer melts in the long-time region, but the agreement between theory and experiment is not quantitative. Better agreement has been achieved by considering other relaxation processes, but further experimental and theoretical work will be required to establish the self-consistency of such modifications, as well as their success or failure in describing results for different polymer systems.

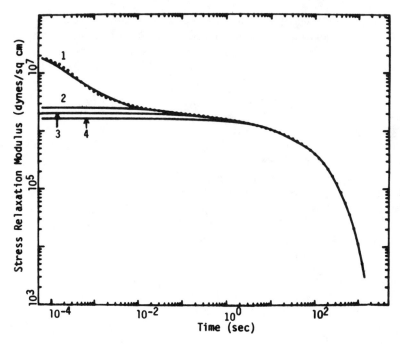

Figure 3.14. Comparison of model of Lin with shear modulus for molten polystyrene, $M = 4.2 \times 10^5$, at 127.5 °C. Smooth curves represent contributions of four processes described in text. From Lin.[94]

4. Viscoelastic Properties of Molecular-Weight Blends

The extension of models for monodisperse polymer melts to the polydisperse case is obviously important because almost all commercial polymer samples are polydisperse. It is also of interest in probing the limits of applicability of the reptation concept. In this sense, the most attention has been directed to the simplest polydisperse system, a binary blend of two monodisperse samples differing only in molecular weight. Furthermore, we will restrict the discussion to the case where M_L and M_S are both significantly greater than M_e, where M_L and M_S are the molecular weights of the long and short components, and focus primarily on the relaxation modulus. What is sought is a "blending law," which will allow the prediction of $G(t)$ (and by extension, the other linear viscoelastic functions) from the properties of the pure components and the blend composition.

Because the reptation hypothesis amounts to a direct consideration of the behavior of a single polymer chain, it is certainly natural to try and extend it to mixtures. The major difficulty in doing so, however, is the inherent single-chain character of the basic concept. Once one begins by assuming reptation, as is the case for the Doi–Edwards treatment, and introducing the entanglement correlation length as a parameter, it is difficult to proceed directly to a multichain problem. The main route that has been adopted is the assumption of the process called constraint release, in which a test chain acquires additional mobility due to the fact that the constraints (entanglements) provided by neighboring chains are not infinitely long lived, as discussed in Section II.B.1.

The simplest reptation-based law for a two-component blend is a direct result of the Doi–Edwards model and reads

$$\frac{G(t)}{G_N^0} = \phi_L F_L(t) + \phi_S F_S(t) \tag{3.14}$$

where ϕ_i is the volume fraction of component i [$i = L$ (long) or S (short)] and $F_i(t)$ is the relaxation function defined in Eq. (3.12). In this case, the complete neglect of chain–matrix interactions leads to an expression which is not in agreement with experiment. For example, Eq. (3.14) implies that the longest relaxation time for each component is unaffected by the presence of the other. While in some cases (but not all) $\tau_{1,L}$ is essentially independent of ϕ_S, $\tau_{1,s}$ increases with increasing ϕ_L (see Ref. 89 and references therein). The change in $\tau_{1,s}$ from the melt ($\phi_L = 0$) to the tracer case ($\phi_L = 1$) is typically a factor of 3–5, which is similar to the difference between τ_1 for a chain in a monodisperse melt and the same chain trapped in a network (see Section III.A.6). Interestingly, such a large change in $\tau_{1,s}$ is not observed in the tracer diffusion, where the decrease in D_t from the self-diffusion case ($M = P$) to the $M \ll P$ limit is at most a factor of 2. A detailed experimental study of binary blends of PB was recently carried out by Struglinski and Graessley[89]; an example of the loss modulus G'' is shown in Fig. 3.15. The prediction according to Eq. (3.14) for one particular composition is shown in Fig. 3.16, where it clearly does not describe the results.

A wide variety of blending laws have been proposed, both prior to and since the introduction of the Doi–Edwards model, although the earlier ones were largely empirical. Modifications to Eq. (3.14) have been proposed, for example, by Marrucci,[39] Viovy,[40] and Rubinstein et al.[95] A detailed comparison of these various approaches is well beyond the scope of this chapter, but the complexity of the problem is revealed in Fig. 3.16. In addition to the prediction of Eq. (3.14), the three other approaches are compared. The one that fits the data extremely well involves inclusion of constraint release via the Viovy–Marrucci form plus the tube length fluctuations as considered by Doi,[31] in addition to reptation. Although the agreement is good, the curve is generated with three additional, free parameters relative to Eq. (3.14), and the

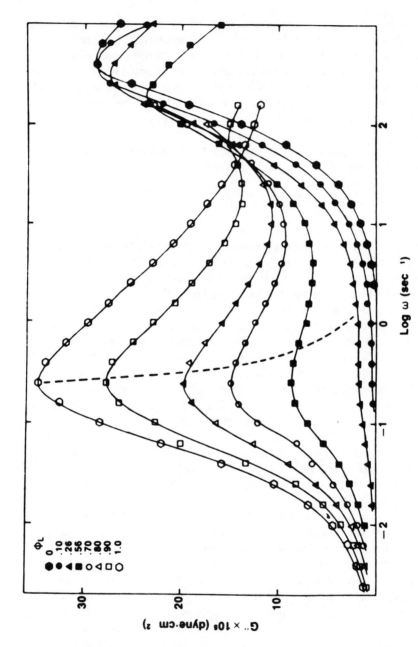

Figure 3.15. Loss modulus for binary blends of polybutadiene, $M_S = 4.1 \times 10^4$, $M_L = 4.35 \times 10^5$, at 25 °C. From Struglinski and Graessley.[89]

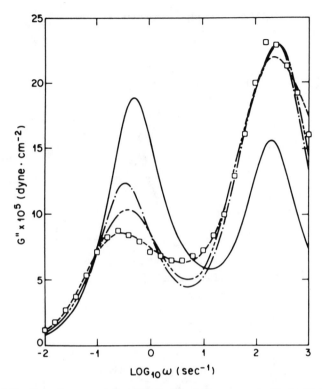

Figure 3.16. Data from one composition in Fig. 3.15, compared with the Doi–Edwards[7] (solid curve), Marrucci[39] and Viovy[40] models (short dashed curve), and model described in Rubinstein et al.[95] (dot-dash curve and short dash-long dash curve). From Rubinstein et al.[95]

plausibility of the parameter values has been questioned.[95] It is therefore difficult to say at this time whether this kind of treatment contains physics that is essentially correct, or if it is more an exercise in curve fitting.

More recently Rubinstein and Colby[96] have reviewed the various constraint release models and proposed a modification that treats the process in a more self-consistent way. Furthermore, they have acquired additional data for PB binary blends, over a wider frequency range than heretofore available. The newer data and the comparison with the model are shown in Fig. 3.17. Alternatively, des Cloizeaux[43] has proposed a picture dubbed "double reptation," in which it is sufficient to consider only the fact that the release of a constraint on one chain is also the release of a constraint on the other. As des Cloizeaux points out, while this self-consistency also underlies the Rubinstein–Colby picture, it is more direct in the double reptation treatment; in particular, no new parameters are introduced. The result of this calculation for the data

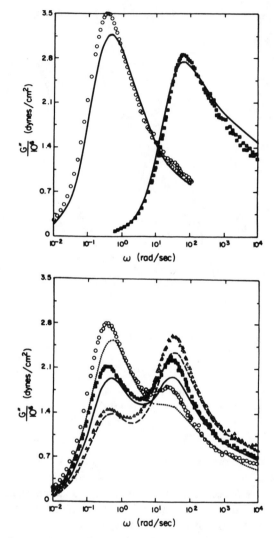

Figure 3.17. Loss modulus for (a) two monodisperse polybutadiene melts ($M_S = 7.1 \times 10^4$, $M_L = 3.6 \times 10^5$) and (b) blends, compared with the predictions of the Rubinstein–Colby model. From Rubinstein and Colby.[96]

Fig. 3.17 is shown in Fig. 3.18. The agreement is quite good, particularly in view of the fact that no tube fluctuations are incorporated, in contrast to Fig. 3.17. Despite the apparent success of these approaches, as illustrated in Figs. 3.16 to 3.18, it is clear that the viscoelastic properties of blends are not trivial to model, and a good deal of further work will be required. It is already apparent that strict reptation does not describe the breadth of the relaxation spectrum in the terminal zone even for monodisperse melts; neither does it describe relaxations at higher frequencies adequately. The binary blend case is clearly much more complicated, as it must involve cross-terms which the reptation hypothesis alone cannot address. On the experimental side, with M_L, M_S, and ϕ_L as variables, a huge amount of work would be required to provide a complete picture. Furthermore, rheological experiments on multi-component systems are inherently difficult to interpret in terms of the contributions from the individual components, because one can only measure the response of the material integrated over the sample. In this instance, rheo-optical approaches offer a great deal of promise.

When a polymer liquid is deformed, it will in general exhibit birefringence and/or dichroism. In essence, the strength of either property will be given by the product of the optical anisotropy of the monomer unit and the degree of orientation of the monomers induced by the flow. For birefringence, it is anisotropy in polarizability which is important, and this is best examined at a nonabsorbing wavelength. Dichroism, on the other hand, reflects anisotropy in the extinction coefficient, and must be examined within an absorption band. In many models for polymer liquids, including the Doi–Edwards theory, a stress-optical law holds. This relation requires that the principal axes of the stress and refractive index tensors coincide, and furthermore that the differences in principal values be proportional; the constant of proportionality is called the stress-optical coefficient and is characteristic of a given polymer structure.[97] It is largely independent of M and T. There is ample evidence that the stress-optical law is obeyed for homopolymer melts at low and moderate shear rates, including the region where the viscosity is shear-rate dependent.[97] Intuitively, this result is very reasonable in that the stress is reflected in the orientation of polymer subunits. Rheooptics, the measurement of flow-induced birefringence or dichroism, offers another route to rheological characterization with particular advantages in certain situations. Recently this approach has been exploited in the case of binary blend stress relaxation.[98] Two PI samples were blended, with one component deuterated to some extent. The flow birefringence is insensitive to deuterium substitution, so this quantity gives the optical equivalent of $G(t)$ for the blend. The dichroism at the wavelength of the $C-D$ stretching mode, on the other hand, can be interpreted in terms of the relaxation of the labeled component.

With increasing ϕ_L it was found that the relaxation of the long chains

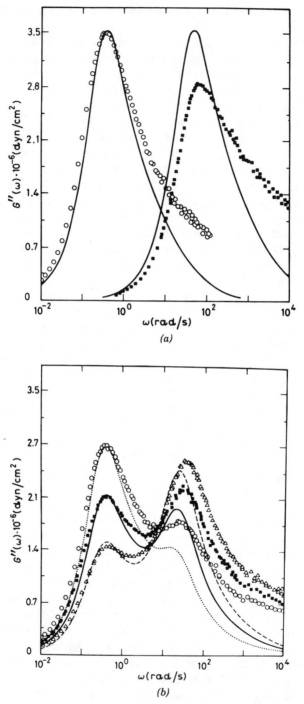

Figure 3.18. Data from Fig. 3.17 compared with des Cloizeaux model. From des Cloizeaux.[43]

slowed considerably, as did the relaxation of the short chains. These observations are also clearly beyond the capacity of Eq. (3.14), but furthermore the additional processes considered are not sufficient to account for these results. In particular, the direct measurement of chain orientation revealed that the short chains exhibited orientation at times much longer than $\tau_{1,s}$ obtained either for $\phi_L = 0$ or for the blends examined by viscoelastic measurements. One possible contributing factor is orientational coupling, in which a considerable degree of local order is imparted to a "solvent" in the neighborhood of an oriented long chain.[98-100] In the strict reptation picture, this can be thought of in the following way. When a short chain end moves out of its tube, it still experiences an oriented environment of long chains, and thus its random walk may be biased. Therefore it takes longer to completely randomize the orientation of the whole short chain. Attempts to quantify this process have recently been made,[99,100] but it is too soon to assess its appropriateness. One appealing feature of this picture, however, is the fact that it does not enter into the calculation of the diffusion coefficient of the short chain at equilibrium; thus there is no inconsistency with the observation that $D(M_S)$ does not depend on M_L(i.e., P). At the same time, along with tube fluctuation and constraint release, this approach represents an attempt to make the effective mean field experienced by the test chain more representative of reality. The inherent limitations of the single-chain models are thus underscored. In addition, there is a certain inconsistency in the view of entanglements taken in this kind of model, as mentioned in the Introduction and in Section II.B.1. Namely, the tube model represents the entanglement effect as a field with a characteristic length scale, while constraint release and orientational coupling are strictly chain–chain processes. Further work will be required if this apparent discrepancy is to be removed; clearly it is intimately connected to the problem of identifying what constitutes an entanglement.

5. Diffusion and Rheology of Stars and Rings

The dynamics of polymers with nonlinear architecture is a fascinating problem and could merit a lengthy discussion of its own. However, that would be beyond the scope of this chapter. Nevertheless, the behavior of these molecules, particularly regular stars (f equal-length arms emanating from a central linking agent) and cycles, is crucial to the main focus of this work. The reason for this is clear: of all the proposed methods for describing the dynamics of entangled polymer liquids, reptation is unique in ascribing a special significance to molecular architecture. In strict reptation, molecular reorientation requires random motion of chain ends, coupled with longitudinal motion of the chain middle. For cycles there are no ends, whereas for stars the large-scale longitudinal motion must surely be suppressed; in short, neither stars nor rings can reptate. Therefore if reptation is the dominant mode for linear polymers, it is reasonable to expect some difference between the dynamics of linear chains on the one hand, and rings and stars on the other. This point has been recognized for well over a decade, and has stimulated a great deal of elegant synthetic work to provide the necessary samples. The following discussion will consider, respectively, the tracer diffusion, melt viscosity, and viscoelastic properties of regular stars, followed by a briefer summary of the rheological properties of rings.

a. *Star Diffusion.* The first consideration of star diffusion in the context of the reptation model was by de Gennes,[101] which was followed by several reformulations of the problem.[32,42,102–105] For a three-arm star in a "tightly" cross-linked matrix, the central monomers, or node, are almost immobile. To enable a diffusive step on the order of the mesh size, one of the three arm ends must retract along the contour of its own arm to generate a quasi-linear molecule. This process is clearly entropically very unfavorable, and the probability decreases exponentially with chain length. This can be seen as follows. The distribution function for the end-to-end distance of an arm is approximately Gaussian:

$$\Psi_{eq}(h) \sim \exp\left[-\frac{3(h - \langle h \rangle)^2}{2N_a b^2}\right] \qquad (3.15)$$

where N_a is the number of subchains in an arm, and thus the motion of the free end can be viewed as that of a Brownian particle in a harmonic potential. The calculation of the first passage problem, that is, the average time for the free arm end to reach the node for the first time, gives a disengagement time proportional to $\exp(-\alpha M_a)$. The result for the star diffusivity is

$$D_{star} \sim M_a^{-x} \exp\left(-\frac{\alpha M_a}{M_e}\right) \qquad (3.16)$$

where x takes the value 1,[103] 2,[32,104] or 3,[101] depending on the detailed treatment. The exponential quickly dominates the power law as M_a increases, however. This prediction is an extremely powerful one, for the following reason. For linear and star polymers of comparable size diffusing in a high M matrix, the ratio D_{linear}/D_{star} will increase strongly and monotonically with M. No other molecular model of which we are aware makes this prediction (although it can be reconciled by the coupling model view[53]). For example, in the case of the cooperative motion models, chain architecture does not enter the picture, except possibly weakly through the radius of gyration. It is certainly possible to consider assigning extra friction to the star node, but as M_a increases this extra contribution must be diluted, and the ratio D_{linear}/D_{star}

returns to near unity. To state this point another way, if a model depends only on the number of entanglement (or high friction points) per molecule, and not on the mechanism by which these entanglements can be released, the ratio of $D_{\text{linear}}/D_{\text{star}}$ cannot vary far from unity. This point was clearly recognized by Berry et al. in considering the viscosity of branched poly (vinyl acetate).[106]

Measurements of three-arm star diffusion have been conducted for at least four systems. Klein et al. examined three-arm star DPB in a high M PE matrix, and found a regime where the dependence on M_a was consistent with Eq. (3.16); certainly, D_{star} was much lower than for linear polymers in the same matrix.[76] These results are illustrated in Fig. 3.19. As M_a increased further, however, D_{star} began to level off. This was attributed to motion of the linear matrix chains (i.e., constraint release) becoming the rate-limiting step. In contrast, Bartels et al. examined D_{star} in a matrix of stars for the same chemical system.[104] Again, an exponential dependence on M_a was observed, but the absolute magnitude of D was increased by a factor of approximately

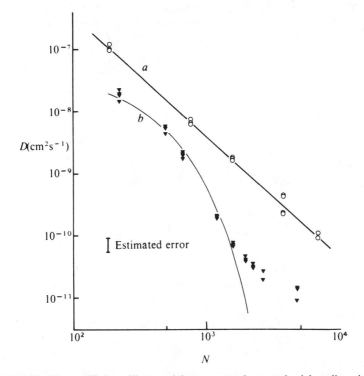

Figure 3.19. Tracer diffusion of linear and three-arm star deuterated polybutadienes in linear polyethylene, $M = 1.6 \times 10^5$, at 176 °C. Straight line corresponds to a slope of -1.95 ± 0.1, and curve to $D = D_0 \exp\{-\alpha N\}$. From Klein et al.[76]

20, underscoring the importance of matrix relaxation in determining D_{star} in this case. These results are also in agreement with the observation of Ferry and coworkers that a star trapped in a network relaxes much more slowly than one in a monodisperse melt.[107] Furthermore, when the product ηD was computed, it was found to increase with roughly $M^{1.4}$ for linear chains and to decrease roughly linearly with M for star melts. This exemplifies the difference in mechanism for linear and branched polymers. Antonietti and Sillescu[108] and Shull et al.[109] measured D_{star} for PS in cross-linked matrices, thus suppressing the matrix effect, and also found direct support for the exponential dependence; some of these results are shown in Fig. 3.20. Although several subtleties of these results and their interpretation are being glossed over, the main conclusion is unambiguous: star polymers are considerably less mobile than their linear counterparts, D displaying an approximately expo-

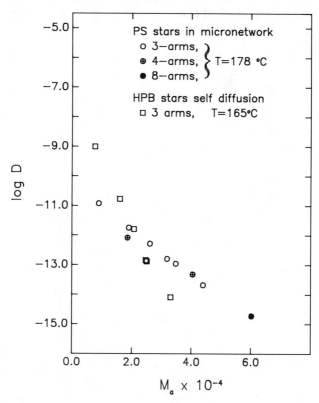

Figure 3.20. Tracer diffusion of polystyrene stars in micronetworks[109] and self-diffusion of hydrogenated polybutadiene stars,[104] plotted semilogarithmically against arm molecular weight.

nential dependence on arm length, and reptation is so far the only postulated mechanism of which we are aware that can account for this.

b. *Star Rheology.* The rheological properties of branched polymers in general, and the viscosity in particular, have long been a subject of great interest. The main experimental observations for the melt viscosity can be summarized as follows.[110]

1. At low but equal total M, η_{star} is less than η_{linear}, but at some $M > M_c$ the curves cross, and η_{star} ($> \eta_{\text{linear}}$) rises much more steeply than the 3.4 power law.
2. If the star and linear polymer viscosities are compared at equal M_a (i.e., viewing a linear polymer as a two-arm star), then $\eta_{\text{star}} \geq \eta_{\text{linear}}$ for all M.
3. If stars with different f are compared at equal M_a, η is almost independent of f.

These results are also at least qualitatively consistent with the suppression of reptation for stars. The same exponential suppression factor enters the expression for η as for D, because stress relaxation requires "escape from the tube," so that

$$\eta \sim \exp\left(\frac{\alpha M_a}{M_e}\right) \tag{3.17}$$

Note that in this picture the f arms relax independently, so that a factor of f does not appear in the exponential. A more detailed treatment has been presented by Pearson and Helfand.[105] This expression is a reasonable description of the observed M dependence, as illustrated in Fig. 3.21 for a recent series of measurements on PI stars ranging from $f = 3$ to $f = 33$.[111] With the exception of the stars having $f = 3$, which are slightly less viscous than the others, the independence of f predicted by Eq. (3.17) is amply demonstrated. However, Doi–Edwards theory calculations for α predicted a value of about 1.9,[8] whereas experimentally both η and D are more consistent with $\alpha \approx 0.6$, as is a computer simulation of D_{star} in a fixed network.[112] A possible explanation for this has recently emerged, invoking the so-called dynamic dilution of the tube as arm relaxation proceeds.[113] Thus at this stage the main features of the η behavior of regular stars are well accounted for by the reptation model, and quantitative agreement is possible.

The plateau modulus for stars, G_N^0, is essentially identical to that of their linear counterparts, although the shape of G'' in the plateau region is different. The former observation is completely in accordance with expectation, in that G_N^0 is felt to reflect the density of entanglements, and should thus be roughly

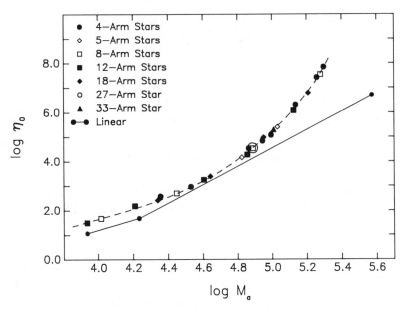

Figure 3.21. Melt viscosity of branched polyisoprenes as a function of span molecular weight, compared to results for linear polyisoprenes.[111]

independent of architecture and molecular weight. The recoverable compliance, however, is different; for entangled linear polymers J_e^0 is independent of M, while for stars it increases apparently linearly with M_a.[32,105] This, too, is in accordance with the derivation of Eq. (3.17). Examples are shown in Fig. 3.22 for a variety of polymer melts and solutions; as with D and η, the best agreement is obtained for $\alpha \approx 0.6$. It should be noted, however, that the prediction that $J_e^0 \sim M$ for stars is not as unique as the exponential dependence for D and η, because it is also a prediction of the Rouse model.

c. Ring Polymers. As pointed out above, polymer rings share with stars the inability to execute strict reptative motion. However, before discussing the experimental results on rings, two important differences from the star situation should be emphasized. First, unlike stars, there have not been extensive theoretical treatments of how rings actually should move. Second, despite herculean synthetic efforts there remain questions about the integrity of the samples. While the synthesis of regular stars with narrow MWD and very little polydispersity in arm number requires a great deal of skill and care, it does appear to be a solved problem. For rings, on the other hand, less experience has been gained, and even though the ring-closing step is performed in dilute solution, it is not yet possible to state unequivocally that a ring sample

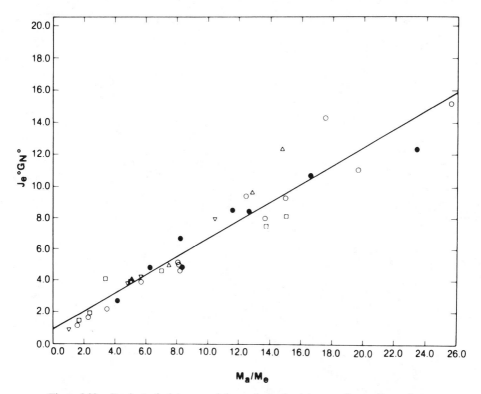

Figure 3.22. Product of plateau modulus and steady-state compliance. For polystyrene, polyisoprene, and polybutadiene stars as a function of arm molecular weight. From Pearson and Helfand.[105]

contains no self-knotted structures, concatenated rings, or traces of the linear precursor. Both of these distinctions between stars and rings are relevant to the issue at hand.

PS rings have been the most extensively studied; they have been synthesized by several different groups.[114-116] The results among the various samples, however, are not completely consistent. Nevertheless, several general conclusions emerge. For low M rings, η corrected for free volume effects follows the Rouse M^1 scaling, with a magnitude approximately one-half that for a linear polymer of the same M. For higher M rings, η increases steeply with M, probably more steeply than for linear polymers. However, over the range of M examined, η for rings is comparable to or below that for a linear polymer of identical M. Some of the data are shown in Fig. 3.23, reproduced from McKenna et al.,[116] along with the results for the corresponding linear PS. These results might be interpreted as evidence against reptation, because the

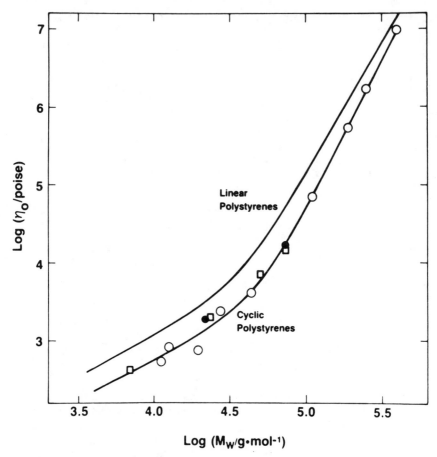

Figure 3.23. Melt viscosity as a function of molecular weight for linear and cyclic polystyrenes. From McKenna et al.[116]

nonreptating rings should move less rapidly than linear polymers, if the latter have a substantial reptation component to their motion. However, this interpretation would be premature. First, the data lie in the unentangled/entangled crossover region, where it is clear from the linear polymer results that reptation cannot be the only available mode. Second, the critical M for entanglement for rings, M_c, is roughly a factor of 2 larger than for linear polymers. If the horizontal axis in Fig. 3.23 were scaled by M_c, η for rings would lie on or above the η curve for linear polymers for all M. Thus when compared at an equal degree of entanglement, rings are less mobile than linear polymers. This is similar to the case for regular stars, where in terms of total M, the crossover to entangled behavior occurs at higher M than for linear polymers. It is also interesting to note that PS rings exhibit a plateau modulus approximately one-half that for linear polymers, again suggesting that rings are less effective at forming entanglements.[115] We are not aware of any theoretical explanation for this factor of 2. However, it could prove to be an excellent test for any attempts to gain deeper insight into the entanglement phenomenon. For example, the correlation of G_N^0 with length of chain contour per unit volume suggested by Graessley and Edwards[92] may not account for this observation. On the other hand, the factor of 2 is suggestive that the projected length or the coil radius of the ring may be the important quantity.

The behavior of the steady-state recoverable compliance J_e^0 remains uncertain for rings. In one study it increases steadily with increasing M,[115] while in another it appears to level off at high M.[116] Both η and J_e^0 are very sensitive to contamination by small amounts of linear polymers, as has been unequivocally demonstrated by the blending studies of McKenna et al.[116] It is also worth noting that in a recent study with PB rings, the results were quantitatively quite different than for PS. In particular, the plateau modulus was reduced by a factor of 5 from the linear case, and η was almost an order of magnitude lower than a linear PB for $M_{\text{ring}}/M_{c,\text{linear}} \approx 10$.[117] However, if it is assumed that $M_{c,\text{ring}} \approx 5M_{c,\text{linear}}$, at equal values of M/M_c the ring melt would also have a higher viscosity than the linear polymer melt. The factor of 5 suggests that PB rings are even less able to form entanglements than PS rings, and again no satisfactory explanation for this has emerged. In summary, the rheology of ring polymers must be considered to be insufficiently understood at this time, at least for the purposes of shedding light on the significance of reptation. One final piece of experimental information comes from the measurement of D for PS rings into melts of high-molecular-weight linear PS.[118] In all cases the mobility of the rings was less than for linear polymers of identical molecular weight in the same matrix, with the difference increasing with M. Although a satisfying, quantitative interpretation of these results is not yet available, the results are at least consistent with the idea that rings should be less mobile than linear chains under identical entanglement conditions.

6. Unattached Chains in Networks

The dynamics of a linear chain trapped in a cross-linked matrix was the subject of de Gennes original paper on reptation,[1] and for this problem the reptation process seems to be the most reasonable possibility. Therefore a substantial amount of experimental effort has been expended on measuring either the diffusion or the relaxation of guest linear chains. As will emerge from the following discussion, the situation is rather unsatisfying in the sense that if one could establish unequivocally both the dominance and the detailed time scale of reptation in this circumstance, then the extension of reptation to uncross-linked systems would be on firmer ground. Alternatively, if it became apparent that trapped chains do not reptate, then the extension to melts would be untenable. However, to summarize the experimental picture in advance, the general features of trapped chain dynamics are not inconsistent with reptation, but the range and precision of the data are not great, and the production and generation of well-characterized model networks are major obstacles.

The situation we will consider is that of a small number of chains of length $M > M_e$ constrained by a network with an average molecular weight between cross-links M_x. The term network will also be used to include gels, that is, networks swollen with a small-molecular-weight solvent. To a first approximation one can envision two regimes. For $M_x > M_e$, the system approximates an entangled melt with $P = M$ when M_x is infinite, and with P infinite when $M_x = M_e$. (Note, however, that a network and a melt with $P = \infty$ are not exactly the same. First, in the former one has M_x as a variable, and second, even if $M_x = M_e$, the local mobility in the vicinity of a cross-link need not be the same as in the melt.) For $M_x < M_e$ one expects a reptation regime with a prefactor [i.e., k_D in Eq. (3.2)] which decreases with M_x because D decreases with tube diameter according to Eq. (3.4). However, the experimental difficulties in achieving a system in which these variables are the only ones that matter are substantial. Some of the salient hurdles are listed below.

1. How is the network prepared, such that the distribution of M_x is as narrow as possible? There are three generic routes to network formation: copolymerization of monomer with a polyfunctional comonomer, random cross-linking of preformed polymers, and end-linking of preformed telechelic "macromonomers." The chief drawback to the first is the inevitable production of microscopically inhomogeneous networks, with regions of high and low cross-link density. The second method probably produces a narrower distribution of M_x, but presents difficulties in getting the loose chains into the network. The third method in principle can give a narrow distribution for M_x, but is synthetically very challenging. Not only must the precursor be monodisperse, it should be almost completely telechelic (i.e., functionalized at both ends), and the cross-linking reaction must go readily to near completion.

2. How is the loose chain to be introduced? In the case of random copolymerization, it can be present in the reaction medium, but otherwise it must be introduced subsequently. The swelling of networks by loose chains is an extremely slow process, but, even more problematical, the network acts as a filter so that the more mobile, lower M chains are preferentially incorporated.

3. How is the network to be characterized? Values of M_x are usually inferred from either swelling or modulus measurements. Both ap-

proaches require applications of theoretical expressions which have been criticized, and the values determined by the two methods often disagree by a factor of 2 or more. Therefore one can only depend on relative values of M_x at best. Furthermore neither technique senses the distribution of M_x.

4. How is the inevitable chemical heterogeneity of the system accounted for? In some instances, the loose chains are chemically distinct from the network, raising the possibility of coil contraction, localization of loose chains in macropores, or even loose chain aggregation. Even when the two polymers are chemically identical, the cross-link points are not. This can influence the local free volume, for example, as T_g is a function of cross-link density. However, it is certainly not clear how to account for this in terms of the environment sensed by the loose chain.

These difficulties notwithstanding, a reasonable body of experimental data has been acquired, and some of the most relevant results are discussed below.

a. Diffusion of Unattached Chains.   Gent and Kaang have examined the diffusion of PI into PI networks by measuring the change in volume as the loose chains swell the network.[119] The results are reported to be quite consistent with an M^{-2} dependence and fall roughly a factor of 2–3 below those of Landry[81] and von Meerwall et al.[120] for PI melts. As in the melt case, the M^{-2} dependence persists for $M < M_e$, possibly for the same free-volume-based reasons. It should be noted that the factor of 2–3 brings these results into better agreement with the Doi–Edwards prediction, Eq. (3.4). However, some caution must be exercised before interpreting these data as strong evidence for reptation. First, the D values cited were those obtained for $M_x = M$. In other words, the matrix was different for each M. If D were compared at equal M_x, the M dependence would be significantly less. Second, D was consistently a decreasing function of M_x, implying that in the limit of zero cross-links, D would be less than in a tightly cross-linked system. This dependence on M_x is rather counterintuitive. An important point to emphasize is that the measured D is really a mutual diffusion coefficient, not a tracer diffusion coefficient, and that the sample was changing during the course of the experiment, being swollen by a factor of almost 2 by the time equilibrium swelling was achieved. The difficulty in achieving a quantitative interpretation of these data is therefore substantial. In an earlier study using poly (dimethyl siloxane) (PDMS) networks,[121] the apparent M dependence was approximately M^{-1}, but this result should also be viewed with caution. The diffusant was not monodisperse, and the experiment undoubtedly is heavily weighted toward the low end of the M distribution.

Garrido et al.[122] have used both NMR and swelling rates to examine D for PDMS chains in end-linked model networks. Again, D was less than for corresponding melts, and in this case D increased with M_x. However, the low values of M and the diffusant polydispersity effectively preclude a detailed analysis in terms of reptation; at best the crossover to entangled behavior would be examined.

The most revealing studies of D in networks are those of Antonietti and Sillescu,[123] where by virtue of the forced Rayleigh scattering technique tracer diffusion was examined. The polymer was PS, and the networks were prepared in a variety of ways. The most relevant series of measurements were made in randomly cross-linked (by Friedel–Crafts reaction) matrices, where exponents of -2 ± 0.2 and magnitudes about a factor of 2 below the melt case were observed. These data are shown in Fig. 3.24, and we believe they represent the strongest evidence for reptation in the experimental studies of loose chains in networks. Computer simulations of this problem, where ideal networks are easily generated, are also supportive of reptation, as discussed in Section III.A.7.

b.   Relaxation of Unattached Chains.   This process has been examined in two ways, by stress relaxation measurements on networks with and without guest chains,[107,124–128] and by dielectric relaxation on chains with dipole moments along the end-to-end vector.[129,130] In passing it seems to us that the measurement of infrared dichroism of deuterium-labeled chains, as mentioned in Section III.A.4, would be an excellent way to pursue this problem further, as neither of the other techniques is particularly well suited to the measurement

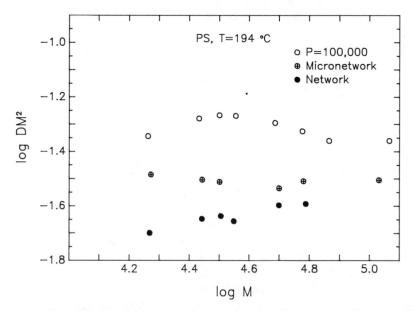

Figure 3.24.   Diffusion of unattached polystyrenes in melts, micronetworks, and networks.[123]

of a tracer component. The viscoelastic approach has been taken by Ferry and coworkers, while the dielectric studies have been carried out by Kotaka and coworkers. In the former, it was found that chains in networks relaxed more slowly than in melts, by a factor that could be as large as 10. An estimate of the relaxation time gave an M exponent in the range of 3–3.5, but the extraction of τ_1 was not sufficiently precise to make a strong statement. Three- and four-arm star polymers were found to relax even more slowly than linear polymers,[107] but a range of M_a sufficient to look for an exponential dependence was not examined. Chains which were anchored to the network at one end also relaxed more slowly than the unattached chains.[107] In the dielectric studies, the relaxation times were also longer than those in solutions and melts, and M exponents in the range from 3 to 4 were reported. However, on some systems both dielectric and viscoelastic measurements were made, and the results were always not in close agreement. This result underscores the difficulties inherent both in sample preparation and in the measurements themselves. In particular, the extraction of the loose chain relaxation requires some assumption about the appropriate blending law, which as discussed in Section III.A.4 is a far from trivial issue. Nevertheless, Fig. 3.25 presents data extracted

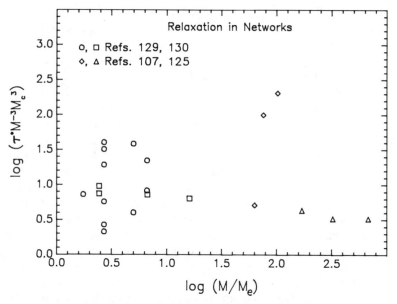

Figure 3.25. Longest relaxation time for unattached chains in networks. Ordinate is the measured longest time normalized to the longest time for a chain with $M = M_c$ in the melt, normalized by $(M_c/M)^3$. Abscissa is the molecular weight of the unattached chain normalized by the molecular weight between entanglements in the melt. Data from dielectric relaxation[129,130] and viscoelastic relaxation[107,125] for various chemical systems are included. Cross-link spacing in network is not considered in plot.

from the measurements of Ferry and coworkers and Kotaka et al. The vertical axis is the measured relaxation time, divided by M^3 and normalized to $\tau_1(M_c)$ in the melt. The results, although widely scattered, are consistent with $\tau_1 \sim M^3$. In summary, as a whole the results for loose chains in networks are quite consistent with the reptation hypothesis, but without being as clear-cut or as compelling as one might naively expect.

7. Computer Simulation

Computer simulations offer a unique opportunity to examine dynamics in a detailed and selective manner. They have proven quite informative in examining equilibrium as well as dynamic properties of polymer liquids. A variety of approaches have been adopted in order to apply the simulation approach to the entanglement phenomenon. Technically, while there are many difficulties to be overcome, the central problem is straightforward to understand. A realistic simulation must permit Brownian motion at the subchain or even monomer level, which maps an elementary computer step onto a submicrosecond time scale. The interesting dynamics occur over time scales at least a factor of 10^6 (and more commonly 10^9) longer, which places great demands on computer power. Equally importantly, a sample comprising a statistically significant number of chains must be used. The method of choice has been the Monte Carlo (MC) approach, usually on a lattice, but more recently molecular dynamics (MD) simulations have been successful in achieving an entangled system. In this section we consider only the simulation results for the diffusion coefficient and longest relaxation time; the more interesting monomer or subchain motions, which are much harder to examine experimentally, are examined in Section III.B.4. Even for the global chain dynamics discussed, however, the results of various workers are not always in agreement.

The simulations can be divided into two groups: those that examine single-chain dynamics in a fixed background, and those that examine a system of many mobile chains. The results for linear chains trapped in a regular array of obstacles or network are in most cases in agreement with strict reptation, at least at the level of exponents. Specifically, $D \sim N^{-2}$ and $\tau_1 \sim N^3$ are reported, within reasonable error bounds.[131-136] However, in a very recent paper, Deutsch and Madden report that the diffusion exponent is closer to -2.5,[33] in contrast to expectation. Their model was very similar to that used in the earlier work of Evans and Edwards,[132] but in the region where the simulations apparently overlap, the results are not identical. Clearly, there is some subtlety involved even in this case. Muthukumar and Baumgärtner have examined the very interesting situation of loose chains diffusing in irregular arrays of obstacles.[137,138] In this instance, stronger exponents are seen, roughly $D \sim N^{-3}$ and $\tau_1 \sim M^4$. This is ascribed to the entropic traps created by bottlenecks in the obstacle field. Apart from obvious applications to the properties of porous media, this work raises interesting questions about the domain of validity for reptation. For instance, it is possible that this system more closely imitates the chemical networks prepared for the study of un-attached chain dynamics than do perfect cages such as those employed by Evans and Edwards. Nevertheless, the evidence from the simulation of loose chains in networks is reasonably supportive of reptation, which is certainly not surprising. As a final note, in a study of three-arm stars diffusing through such a fixed network, strong evidence for the exponential dependence of D on arm length was obtained.[112]

In simulating a many-chain system, the crucial parameters are the chain length and the chain density in both the MC and MD approaches. It has proven quite difficult to examine a system that is thoroughly and un-ambiguously entangled, but certainly this limit is approached. The two most extensive studies have been those of Skolnick and coworkers[136, 139-141] using lattice MC, and Kremer et al.[142, 143] using MD. In the former, chains were observed for which the D and τ_1 exponents are close to those seen experi-mentally in entangled systems. However, examination of chain motion at the monomer level did not give evidence of reptation, which led the authors to suggest that reptation is not a correct picture of polymer melt dynamics, and to propose an alternative picture of their own, as discussed in Section II.D.2. In contrast, the MD studies also see the "correct" macroscopic exponents, but report monomer dynamics in better agreement with reptation. Because these two series of simulations apparently differ the most in the local motion domain, further discussion of this issue is deferred until Section III.B.4. Nevertheless, it has been established that computer simulations can achieve the macroscopic features associated with entangled polymer liquids, and undoubtedly further work will be revealing.

B.   Experiments for Times Shorter Than τ_1

A distinguishing characteristic of long, flexible chains is the breadth of time scale associated with relaxation processes, even in the unentangled state. If the longest relaxation time τ_1 is associated with the equilibration of the chain end-to-end vector, there may similarly be a sequence of shorter relaxation times associated with equilibration of vectors connecting any other pair of monomers. Representations of a flexible chain, such as in the original Rouse model, typically lead to a contracted set of relaxation times associated with the normal modes of the model chain. Although the concept of a discrete set of times has been criticized,[144] the resulting "relaxation spectrum" has been shown to be extremely successful in describing the frequency dependence of G^*, for example.[2] For the purposes of this chapter, the main point is that the reptation model leads directly to testable predictions for the effect of entanglement on some internal modes. Three different quantities will be considered. First, the dynamic structure factor $S(q,t)$, or its first cumulant $\Omega(q)$, can be examined by neutron spin echo (NSE). In the appropriate q and t regimes, it is possible, at least in principle, to examine the transition in monomer dynamics as the effects of entanglements are first felt. Second, by selective chemical labeling, the orientational relaxation of the chain ends relative to the chain middle can be determined. Third, in computer simulation it is possible to follow a variety of interesting spatial correlation functions for different monomers along the chain. Before proceeding to a comparison of these three classes of experiment with reptation predictions, it will be helpful to consider the corresponding results for isolated chains, following the original Rouse[29] and Zimm[30] models.

1. Internal Modes of Single Chains

a. $\Omega(q)$ for Isolated Chains. In either dynamic light scattering or NSE it is possible to examine the normalized (coherent) dynamic structure factor $\hat{S}(q,t)$ [or its cosine Fourier transform $\hat{S}(q,\omega)$]:

$$\hat{S}(q,t) \equiv \frac{S(q,t)}{S(q,0)} \qquad (3.18)$$

where

$$S(q,t) = \sum_{j} \sum_{k} \langle \exp\{i\mathbf{q}\cdot[\mathbf{r}_j(t) - \mathbf{r}_k(0)]\} \rangle \qquad (3.19)$$

and $\mathbf{r}_j$ is the position of monomer j. In general, it is quite difficult to calculate $S(q,t)$ explicitly, while experimentally the available range of q and t is limited. Consequently it is often more profitable to consider the first cumulant of $S(q,t)$ given by

$$\Omega(q) \equiv -\lim_{t \to 0} \left(\frac{d\hat{S}(q,t)}{dt} \right) \qquad (3.20)$$

Note that the limit $t \to 0$ actually means the limit of times shorter than the fastest process under consideration. In frequency–space measurements Ω corresponds to the line width of the scattering peak, while in the time domain Ω is the decay rate of the scattered field autocorrelation function (via the Wiener–Khintchine theorem). The magnitude of the scattering wave vector q sets the characteristic distance scale probed by a given technique, which is typically 5–200 Å for neutrons and 300–3000 Å for light. When the dimensionless product $qR_g \ll 1$, the polymer appears as a single scattering object, and no information on internal modes can be gained. In this regime it is well established that $\Omega \sim q^2$, which is really nothing more than a restatement of Fick's second law. On the other hand, when $qb > 1$, where b is the persistence length, specific motions with a subchain are probed; for a suitable time window, again $\Omega \sim q^2$, although with a different prefactor. The interesting intermediate regime defined by $R_g > q^{-1} > b$ is accessible to light scattering when M is very large, and to neutrons when b is relatively small. For a Rouse chain, $\Omega \sim q^4$ in this regime.[145] This exponent may be viewed in the following way. In the Rouse chain, disturbances travel from bead to bead only through the springs, and thus diffuse an average distance d along the chain in time $t^{0.5}$. Because of the random walk nature of the chain, however, the three-dimensional distance $x \sim d^{0.5} \sim t^{0.25}$. For the Zimm chain, where the beads also communicate via hydrodynamic interaction, the net result is $\Omega \sim q^3$.[145] This latter result has been amply verified in dilute solution,[146] although the exact value of the prefactor is still a subject of some debate. For low M melts, the Rouse result should be expected, but as M increases, a new length scale, the entanglement spacing, comes into the picture. This is discussed in Section III.B.2 in conjunction with the experimental results.

b. Subchain Relaxation in Dilute Solution. This issue may also be addressed via the Rouse–Zimm formalism, recognizing that while the Rouse model is mathematically more tractable and directly relevant to low M melts, the Zimm model is more appropriate to the isolated chain. The solution of the problem lies in identifying the eigenmodes of the chain [Eq. (1.7) for Rouse chains], and then considering the relaxation behavior of individual subchains. Although there are many possible ways to do this, one particularly revealing method has been developed by Man and coworkers.[147] Building on the Wang[148] extension to the Rouse–Zimm model for block copolymers, they have generated an algorithm to calculate the eigenmodes and relaxation times for a general triblock copolymer, for arbitrary degrees of hydrodynamic interaction. Furthermore, the programs have been developed to calculate the oscillatory flow birefringence properties as well as G^*. This proves particularly useful, because it is possible to set the optical anisotropy of all but a chosen bead–spring unit to zero. Then, only one subchain contributes a nonzero signal, and the chain position dependence of the dynamics can be examined. The detailed results of this kind of examination are well beyond the scope of this chapter, but the principal relevant result is this: chain ends relax considerably more rapidly than the middle sections, with the difference noticeably greater for the Rouse (relative to the Zimm) chain. An explicit expression for the relative orientation of the center of a Rouse chain is given in Section III.B.3.

c. Single-Bead Correlation Function. The correlation function of direct interest is

$$g_1(t) \equiv \langle [\mathbf{r}_i(t) - \mathbf{r}_i(0)]^2 \rangle \tag{3.21}$$

corresponding to the diffusion of monomer i in the laboratory frame, as discussed in Section II.A. This function is best examined by computer simulation, either via MC lattice methods, or more recently by MD. For a Rouse chain, $g_1(t)$ should exhibit three regimes. At very short times, $g_1 \sim t$, corresponding to diffusion over sufficiently short times/distances that an individual bead is not aware that it belongs to a chain. At very long times, g_1 also increases linearly with t and follows the center-of-mass diffusion of the chain. In the intermediate regime, $g_1(t) \sim t^{0.5}$ for the reason alluded to in the discussion of $S(q, t)$. The crossovers between these various regimes should be observed when $g_1 \approx b^2$ and R_g^2, respectively. Such scaling behavior has also been clearly observed in computer simulation of isolated chains. It is worth pointing out, however, that computer simulation of a chain experiencing hydrodynamic interaction is a much more complicated problem, and only recently has systematic investigation of this regime been attempted.

2. Neutron Spin Echo Measurements of $\hat{S}(q, t)$

NSE measurements on polymer systems are rather difficult to perform, and the most informative data have been obtained primarily from the spectrometer at the ILL, Grenoble. Neutron fluxes at the sample are many orders of magnitude less than photon fluxes obtainable with even weak light sources, leading to inherently small and uncertain scattering intensities. The monochromaticity of the neutron beam is also a significant factor, rendering q uncertain to within $\pm 10\%$ in the typical NSE case. Dynamic data are obtained by examining the decay in a nuclear spin echo as a function of (imposed) precession frequency. A magnetic field of varying strength is applied to the incident, polarized neutron beam, and the small change of energy which each scattered neutron experiences is encoded as a shift in precessional phase. Time resolution is limited by the maximum strength of the applied field to less than about 4×10^{-8} s, and the relevant spatial scale is constrained by the neutron wavelength and scattering geometry to between about 5 and 50 Å. In order to investigate the transition from unconstrained to constrained monomer motion, therefore, the choice of polymer system is crucial; only the most dynamically flexible chains with rapid local motions are suitable.

The reptation prediction for $\hat{S}(q, t)$ may be written as[149]

$$\hat{S}(q, t) = \left(1 - \frac{q^2 a^2}{36} \right) + \left(\frac{q^2 a^2}{36} \right) \exp(\omega_R t) \, \text{erfc}[(\omega_R t)^{0.5}] \tag{3.22}$$

where a is the entanglement length scale, that is, the tube diameter in the Doi–Edwards picture, and ω_R is a characteristic Rouse frequency given by $kTb^2q^4/12\zeta$. This expression for $\hat{S}(q,t)$ has two parts. The first is constant only for an infinite reptating chain, but otherwise would relax slowly according to Eq. (2.13a). The second represents the same reptative behavior as the $t^{0.25}$ regime discussed in Section III.B.4, but does not include the crossover from Rouse-like motions at shorter times and distances, and would also require modification at long times, as shown by Kremer and Binder.[150] Thus it cannot be expected to describe the NSE data in the crossover region. An alternative model has been proposed by Ronca,[18] as discussed in Section II.C.1. His result for $\hat{S}(q,t)$ covers a different range, showing a smooth crossover from Rouse-like to a time-independent term given by

$$\hat{S}(q, t \to \infty) = 1 - \frac{q^4 a^4}{496} \tag{3.23}$$

where again a represents the entanglement spacing. In either case, an effectively time-independent term is predicted.

Two experimental systems have been examined extensively in this regard: PDMS[151] and poly(tetrahydrofuran) (PTHF).[152] For PDMS, a calculated according to the Doi–Edwards theory $[a^2 = \frac{4}{5}(\langle h^2 \rangle/M)(\rho RT/G_N^0)]^{56}$ is approximately 50 Å, and is thus on the edge of the NSE spatial scale. For PTHF it is approximately 30 Å. The characteristic Rouse parameter was obtained from experiment to be $q^4/\omega_R \approx 2 \times 10^{-12}$ s/Å^4 for PDMS and $\approx 1.2 \times 10^{-12}$ s/Å for PTHF. Thus PTHF is also slightly more mobile at the local scale, and thus probably more appropriate for the issue at hand. For PDMS, although deviations from Rouse dynamics are clearly observed, neither reptation nor the Ronca model provides a good description of the data. It is also noteworthy that over the measured range, the Rouse scaling $(\omega_R t)^{0.5}$ was followed. This was interpreted as indicating that no new length scale, that is, the entanglement spacing, is necessary to describe the data. In contrast, for PTHF[152] closer agreement with the reptation prediction [Eq. (3.22)] was observed. In Fig. 3.26, reproduced from Higgins and Roots,[152] data for an unentangled and an entangled melt are compared. The abscissa is the applied magnetic field current in amperes, which is proportional to t for these experiments, while the ordinate is $\hat{S}(q,t)$. The local dynamics in the entangled case are clearly much slower than in the unentangled melt. While it was not possible to distinguish between the Ronca and the reptation predictions, the value of a obtained was 30 Å, in excellent agreement with the value based on viscoelastic measurements. There is thus some disagreement between the measurements on PDMS and PTHF. However, for the reasons cited, PTHF may be a much better candidate for investigating this crossover, and in that case the results are at least con-

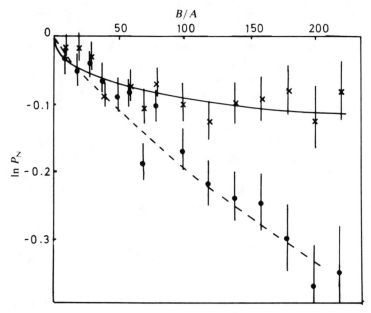

Figure 3.26. Neutron spin echo dynamic structure factor versus time for unentangled and entangled poly(tetrahydrofuran) melt. Dashed line corresponds to Rouse model; solid line is reptation or Ronca prediction with a entanglement spacing of 30 Å. From Higgins and Roots.[152]

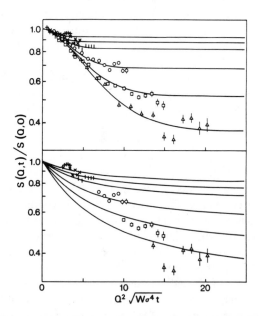

Figure 3.27. Neutron spin echo dynamic structure factor versus reduced time for poly-(ethylenepropylene) melts at various scattering vectors. Fits in upper plot are to Ronca model; lower curves are compared to limiting reptation prediction. From Richter et al.[153]

sistent with the reptation model. Very recently, a third polymer system, poly(ethylenepropylene) (PEP) has been examined.[153] Both the entanglement length and the local time scale for PEP are closer to those for PTHF than for PDMS, and for this polymer the effect of the entanglement length on the monomer dynamics was clearly observed, as illustrated in Fig. 3.27. Thus the NSE results as a whole are at least consistent with the onset of reptation. A stronger statement would definitely not be appropriate, as the difficulties inherent in interpreting crossover behavior in terms of asymptotic predictions are severe. More importantly, the existence of an intermediate length scale a, which affects even the microscopic dynamics, is confirmed. This is a central assumption of the tube models, for example.

3. Relaxation of Chain Segments

In principle, in a stress relaxation experiment on a concentrated solution or melt, it is possible to follow the relaxation of a single chain, or chain portion, by appropriate labeling. In this section some applications of flow birefringence, infrared dichroism, and small angle neutron scattering to chain relaxation are discussed. The first issue to be addressed is a comparison of the predictions of the Rouse and strict reptation models.

It is helpful to recall the Doi–Edwards strict reptation in a tube model and three characteristic times. The longest time τ_{rep} describes the time to escape completely from a given tube, that is, the reorientation time of the end-to-end vector, and is proportional to M^3. The Rouse time τ_{Rouse} describes the longest relaxation time for the internal modes of the chain in the tube and scales as M^2. The third time, τ_e, is essentially the Rouse time for a chain of length M_e and is therefore independent of M. These three times define four temporal regimes, of which the regime $\tau_{\text{Rouse}} < t < \tau_{\text{rep}}$ is of most interest here. The Doi–Edwards expression, Eq. (3.12), defines the relaxation function for a chain stretched at $t = 0$:

$$F(t) = \frac{8}{\pi^2} \sum_{j,\text{odd}} \left(\frac{1}{j^2}\right) \exp\left(-\frac{j^2 t}{\tau_{\text{rep}}}\right) \tag{3.24}$$

If a central section of the chain of fractional length ξ is labeled, such that the stress due to that section can be monitored, the result is[154,155]

$$F(t,\xi) = \frac{8}{\pi^2} \sum_{j,\text{odd}} \left(\frac{1}{j^2}\right) \exp\left(-\frac{j^2 t}{\tau_{\text{rep}}}\right) \cos\left[\frac{j\pi(1-\xi)}{2}\right] \tag{3.25}$$

If the ratio of the stress contributed by the central segment to that from the whole chain is computed, an increasing function of time is obtained, with the limiting values ξ at $t = 0$ and $\sin(\pi\xi/2)$ at $t = \infty$. This result is nothing more than an expression of the fact that the reptating chain relaxes from the ends in toward the middle. Although the net stress in any section is decreasing, the relative stress in the middle of the chain increases.

The same calculation has been performed for the Rouse chain, with the corresponding (approximate) result[154,156]

$$F_{Rouse}(t, \xi) = \sum_j \exp\left(-\frac{j^2 t}{\tau_{Rouse}}\right)\left[\xi + \frac{\sin[j\pi(1-\xi)]}{j\pi}\right] \qquad (3.26)$$

Again, the ratio of the stress in a central segment to that for the whole chain is an increasing function of time, with the limiting values ξ at $t = 0$ and $\xi + \sin(\pi\xi)/\pi$ at $t = \infty$. Interestingly, the two different models give qualitatively very similar results for the chain relaxation; a Rouse chain also relaxes more rapidly at the ends. This point has been overlooked on occasion; the observation of chain end relaxation preceding that for the chain center is not evidence of reptation per se. Indeed, the calculations of Man and coworkers[147] for Rouse and Zimm chains in dilute solution show qualitatively the same effect. Comparisons of the center segment to whole chain relaxation for $\xi = 0.333$ are shown in Fig. 3.28. In each case the abscissa is time reduced by the appropriate longest relaxation time, and thus the molecular-weight dependence is obscured. One can see that in the reptation case, the ends relax more rapidly relative to the chain middle than for a Rouse chain. The difference between the two models is more easily seen in the relaxation function for the middle block; the middle of the Rouse chain begins to relax immediately after the imposition of the strain, while the initial slope (dF/dt at $t = 0$) for the reptating chain is zero.

Osaki et al. have used flow birefringence to investigate these predictions[154,157] via the stress-optical relation discussed in Section III.A.4. They employed a concentrated solution of a poly(methyl methacrylate) (PMMA)–PS–PMMA triblock copolymer, where PS and PMMA have widely differing stress-optical coefficients. In that case it is clear that the stress-optical law will not be obeyed whenever the stress in the different blocks is not uniformly distributed, such as in either of the two cases in Fig. 3.28. In one study[154] the results were quite consistent with the predictions of the tube model, although the Rouse model could not be eliminated on the basis of the data as presented. However, in a subsequent publication on the same material, a quite different result was obtained, where the stress-optical law was valid over the measured range.[157] This result is very difficult to interpret except to say that the stress was uniformly distributed throughout the chain. This is in conflict with both Rouse and reptation models, and also extensive infrared dichroism measurements, some of which are discussed below. Thus at this stage this report remains an unexplained puzzle.

Two groups have used deuterium-labeled sections of PS to compare the

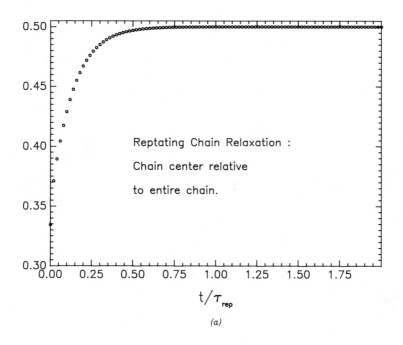

(a)

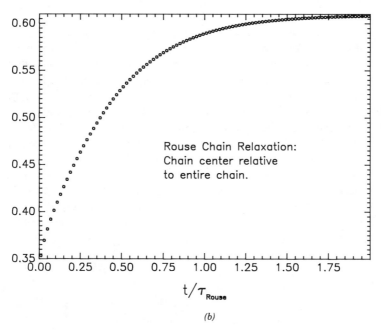

(b)

Figure 3.28. Calculated ratio of relaxation of middle third of a chain to that for entire chain. (a) For strict reptation. (b) For Rouse chain.

relaxation of chain ends and chain middle by infrared dichroism. Wool and Lee[158] saw clear evidence that the chain middle initially relaxed more slowly than the average for the entire chain after a uniaxial extension was applied. Although the test chain was relatively short, with $M = 75,000$, the matrix had a higher molecular weight, so it was reasonable to seek evidence of reptative motion. Estimates of the longest time from measurements at other M led to a scaling of approximately M^3 which, combined with the faster motion of the ends, was taken as evidence for reptation. However, the data were not compared explicitly with the equivalent of Eq. (3.25), and so this conclusion may not be robust. In other words, neither of the two observations cited requires reptation for their interpretation. Tassin et al. have also reported similar measurements, but during the course of the application of an elongation.[159] Again, the chain ends clearly were less oriented at early times than the chain middle, but the results were preliminary and a quantitative interpretation was not attempted. This group has also accumulated a vast body of chain orientation information, for various chain architectures and deformations, which have been interpreted in the context of tube models.[160] In general, the results can be successfully explained in this way, although it is clear that processes beyond those available to the Doi–Edwards chain are required. As with the results for $G(t)$ discussed in Section III.A.3, it remains to be established whether a given set of modifications to the strict reptation model can describe different experiments in a self-consistent manner.

Small-angle neutron scattering offers a means to examine the spatial monomer distribution function during the course of relaxation, in contrast to the orientational distribution discussed above. The strong difference in coherent scattering cross sections between hydrogen and deuterium provides a means for effective chain or chain segment labeling. Unfortunately the available q range restricts the distance scale over which the single-chain form factor $S(q)$ can be examined. In particular, when a chain of reasonably high M is extended, the component of R_g in the direction of extension $R_{g\parallel}$ is too large to be determined; conversely, $R_{g\perp}$ is amenable to study. The Doi–Edwards model has been applied to the problem of the temporal relaxation of $S(q)$ after imposition of a uniaxial elongation, with one particularly intriguing prediction.

Upon imposition of the deformation, the tube is assumed to deform affinely, and thus $R_{g\parallel}$ increases and $R_{g\perp}$ decreases. At this stage the chain is elongated along the tube axis. The next relaxation process is retraction of the chain to its equilibrium curvilinear length with a time constant $\tau_{\mathrm{Rouse}} \sim N^2$ in the still-deformed tube. This retraction is presumed to occur in all directions, resulting in a further contraction of $R_{g\perp}$. In the experiments of Boué et al. such an additional retraction was not observed,[161] and this observation has been construed as a significant failure of the tube model. However, from our perspective this prediction may be an unphysical artifact of taking the physical picture of a tube too literally, and is a manifestation of the difficulty of describing a material property with a single-chain model. In a sample with macroscopic dimensions held constant, it is difficult to envision a spontaneous, simultaneous contraction of every chain along all three directions. For $R_{g\perp}$ in particular, such a contraction would take each chain further away from its equilibrium dimensions.

4. Computer Simulation

The application of computer simulation to the study of entangled polymer liquids was discussed briefly in Section III.A.7. In this section we examine the results in more detail, with an emphasis on the evidence for the prediction of a $t^{0.25}$ regime for reptating monomers. As in Section III.A.7, we consider first the simulation of chains moving in a field of fixed obstacles, and then the more interesting case of a melt of chains.

The reptation model leads directly to the prediction that $g_1(t)$ should exhibit five distinct regimes, unlike the three for the Rouse chain discussed in Section III.B.1. In essence, the new feature is that the intermediate time $t^{0.5}$ regime for the Rouse chain splits into three, with $g_1(t)$ scaling as $t^{0.5}$, $t^{0.25}$, and $t^{0.5}$, respectively. A schematic representation of the predicted behavior of $g_1(t)$ is shown in Fig. 3.29. The first $t^{0.5}$ corresponds to the internal modes of a Rouse chain, which applies for chain segments less than M_e in size. It thus

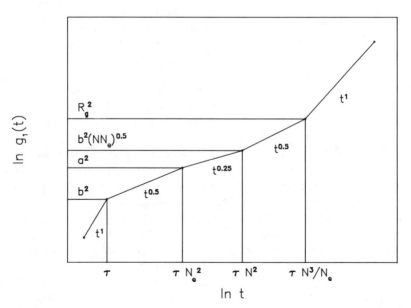

Figure 3.29. Schematic representation of time evolution of monomer correlation function $g_1(t)$ for reptating chain. Following Kremer and Grest.[143]

describes times shorter than the Rouse time for $M = M_e$, which scales as M_e^2. The $t^{0.25}$ regime, sometimes dubbed the "signature" of reptation, corresponds to the Rouse-like relaxation of chain segments which are confined to a tube. In essence, the 0.25 exponent reflects the product of a factor of 0.5 from the Rouse internal modes with a factor of 0.5 due to the confinement to a tube which is itself a random walk. The longest Rouse-like internal mode scales as M^2, which sets the upper limit for this regime. At longer times, but still less than the reptation time, the chain may be viewed as executing strict reptation, with the characteristic $t^{0.5}$ dependence [Eq. (2.7)]. The simulations of Evans and Edwards for chains in a rigid cage showed clearly the later $t^{0.25}$, $t^{0.5}$, and t^1 regimes,[132] but only when the obstacle spacing was sufficiently small; this was consistent with the length dependence of τ_1, which exhibited approximately N^3 scaling only for the same small spacings. Similarly, a $t^{0.25}$ regime was clearly evident in the simulations of Baumgärtner and Binder,[133] Richter et al.,[134] Kremer,[162] and Baumgärtner[163] when the environment was frozen. More recently, Kolinski et al.[136] report the same scaling under similar conditions. In short, the existence of this regime is amply confirmed for the case of a fixed environment.

The situation for a melt is not so clear. Although a great deal of work has been done in this area, we feel the essence of the problem is contained in the recent papers of Skolnick and coworkers[136,139-141] and Kremer et al.,[142,143] using lattice MC and MD methods, respectively. The former have presented extensive results, with two papers (involving a diamond lattice, I,[136] and a cubic lattice, II[140]) being particularly pertinent. In I, scaling of $D \sim N^{-2}$ and $\tau_1 \sim N^{3.4}$ was reported at a lattice occupancy $\phi = 0.75$. However, $g_1(t)$ scaled with $t^{0.5}$ in agreement with the Rouse model. In II, chains of up to $N = 800$ with $\phi = 0.5$ were examined. The cubic lattice has a smaller persistence length, so this N is effectively slightly larger than the same N on the diamond lattice. However, it was not possible to determine D and τ_1 unambiguously for this high an N. In the region $64 < N < 216$, the scaling laws $D \sim N^{-1.5}$ and $\tau_1 \sim N^{2.6}$ were observed. For $g_1(t)$, again no $t^{0.25}$ regime was observed, but for $N = 800$, a regime where $g_1(t) \sim t^{0.36}$ appeared. The authors remark that this should not be taken as evidence for reptation, because detailed examination of the monomer motion parallel and perpendicular to the "primitive path" shows, if anything, a preference for lateral motion. Their principal conclusion is that there is no evidence in their work for reptative motion or a confining tube, and that the motion of an individual chain is isotropic. The slowing down relative to Rouse chains reflects the cooperative constraints of surrounding chains, but does not require reptation.

The MD simulations of Kremer et al. were performed at a density $\rho = 0.85$, with chains of up to 200 in length.[142,143] The results for the global chain properties were consistent with other simulations and with experiment, but

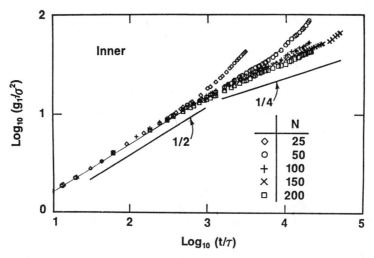

Figure 3.30.   Time evolution of $g_1(t)$ for central five monomers of various-length chains. From Kremer and Grest.[143]

the detailed analysis of monomer and chain motion was quite different from that of Skolnick and coworkers. There were four main differences in the results as presented. (1) The onset of a regime where $g_1^{(t)} \sim t^{0.25}$ is observed. (2) The "computer snapshots" of chain motion suggest the existence of a tube. (3) The mode index scaling of the Rouse modes is j^{-2} as predicted by the Rouse and reptation models, and not j^{-4} as predicted by generalized Rouse models (Section II.B.3). (4) A serious attempt is made to map the results onto real polymer systems. The results for $g_1(t)$ are shown in Fig. 3.30. The regime of approximately $t^{0.25}$ behavior is rather narrow and only evident for the largest chains. In addition, it must be emphasized that these results are averaged over the central five monomers on the chain. When the average is performed over the entire chain, the Rouse-like $t^{0.5}$ behavior obtains throughout.

The analysis of chain motion in terms of the existence/nonexistence of a confining tube was performed by constructing a coarse-grained representation of the actual chain, called a primitive chain (PC). It was then possible to examine the relative preference of monomer motion perpendicular and parallel to the PC, $g_{1\perp}(t)/g_{1\parallel}(t)$, for times much less than the longest relaxation time. This quantity is computed for each PC unit as follows. The subunit j at $t = 0$ is identified that is closest to the location of subunit i at time t. This distance defines $g_{1\perp}^i(t)$. At the same time, the distance along the PC proportional to $|i - j|$ defines $g_{1\parallel}^i(t)$. For a chain moving isotropically, this ratio should increase without bound as time progresses, except for an initial drop at short time, whereas for a chain undergoing strict reptation it should decrease steadily with time. For $N = 75$, the former behavior was observed, while for $N = 200$, $g_{1\perp}(t)/g_{1\parallel}(t)$ decreased monotonically. Furthermore, this evidence of confinement was much stronger for central monomers than for monomers near the chain ends.

Kremer et al. also examined the chain motion in terms of the Rouse modes, the normal modes of an unentangled chain. For short chains these modes were indeed orthogonal, decayed exponentially in time, with time constants scaling as mode number j^{-2}. Conversely, the generalized Rouse models, such as that of Noolandi, predict a scaling with j^{-4}. The simulation results distinguish very clearly between these two possibilities, and agree strongly with the j^{-2} scaling. In summary, the results of Kremer et al.'s simulations are quite consistent with a crossover to reptation at high N, although it is clear that the crossover is broad. For chains with $N/N_e \approx 5$, the largest studied in detail, reptation is not the dominant mode of motion for the whole chain. Nevertheless, this work represents the first simulation of an entangled melt which affords positive evidence for reptation.

The crucial issue when comparing simulation to experiment is to estimate the entanglement length N_e for the simulation, so that one has an idea of how entangled the chains may be. However, this not a trivial thing to do, because one does not have a plateau modulus to work with, for example. In I,[136] Skolnick and coworkers take the static screening length N_B, equal to N_e; N_B is obtained from the crossover from excluded volume to Gaussian statistics in R_g with increasing ϕ. For $\phi = 0.5$, $N_B = 18$, and for $\phi = 0.75$, $N_B = 12$. (Other estimates are provided, but all are lower, i.e., less than 18 monomers are required to produce an entanglement length for $\phi = 0.5$. However, none of these estimates comes from consideration of a dynamic property of the lattice melt.) These values imply that at $\phi = 0.5$, the chain with $N = 216$ has on average 12 entanglement lengths and is thus thoroughly entangled. However, we see no justification for taking the static screening length as equivalent to N_e, and in fact there is ample evidence that $N_B \ll N_e$. Furthermore, a closer examination of the results for D and τ_1 indicates that in some ways these lattice chains do not behave like flexible polymer melts.

The chain length exponents of Skolnick and coworkers for D and τ_1, $-\alpha$ and β, respectively, are plotted as a function of ϕ in Fig. 3.31. The following points are noteworthy. First, at $\phi = 0.5$, where most of the detailed analysis is reported, $\alpha = 1.6$ and $\beta = 2.7$. Neither correspond to entangled melt behavior. Second, the experimental values $\alpha = 2$ and $\beta = 3.4$ are matched near $\phi = 0.75$, so this is cited as evidence that the lattice chains are entangled. But α and β are changing rapidly with ϕ, whereas in real solutions, β is fixed near 3.4 from the onset of entangled behavior up to the melt. Thus in a concentrated solution one can vary ϕ by up to an order of magnitude (i.e., from 0.1 to 1) without affecting β significantly, whereas Fig. 3.31 implies that on the lattice

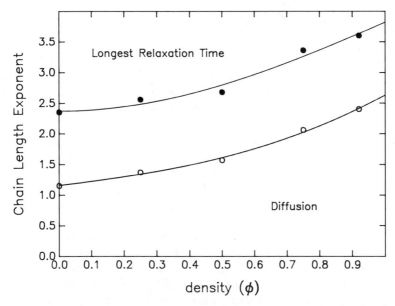

Figure 3.31.   Molecular weight exponents for diffusion α and longest relaxation time β as a function of lattice occupancy on a diamond lattice.[140] Smooth curves serve as guide to the eye.

the exponents vary smoothly over the entire range of ϕ. One must conclude that ϕ on the diamond lattice does not correspond to concentration in solution. Therefore the fact that there exists a particular ϕ where $\alpha = 2$ and $\beta = 3.4$ may have no real significance; it certainly does not guarantee that the lattice chains are entangled. The data in Fig. 3.31 extend to $\phi = 0.92$, which corresponds to a "computer" glass transition. It may well be that these exponents reflect more the onset of this glass transition than entangled dynamics. There are also other puzzling features in these results. Each value of α and β in Fig. 3.31 reflects the slope of a straight line drawn through from 4 to 7 points, extending from $N = 12$ up. In every case the straight line is an excellent representation of the data. Yet this does not correspond to what is observed in polymer solutions, because the range of N here extends from well below to well above N_B, and yet there is no change in the N dependence. In other words, the scaling of D and τ_1 with N is independent of whether $N < N_B$ or $N > N_B$, and thus clearly N_B does not parallel M_e determined experimentally in solutions or melts. Subsequently the authors suggests that $N_e \approx 120$, based on comparison with experiment. This estimate is much more reasonable. However, because most of the simulations were performed for $N = 216$, the chains were definitely not well entangled. This point is rather important, because these simulations are often cited as strong evidence that reptation does not occur. Note also that adopting $N_e \approx 120$ does not overcome all of the difficulties mentioned, in particular values of $\alpha = 1.6$ and $\beta = 2.7$ for chains that are $0.1N_e$ in length.

In II,[140] the simulations were restricted to $\phi = 0.5$, and again this did not correspond to the experimental exponents $\alpha = 2$ and $\beta = 3.4$ except possibly for the longest chains examined ($N = 216$ and $N = 800$). In this case $N_B \approx 16$, and again it is (initially) assumed that this corresponds to N_e. Equally clearly, however, this cannot be correct, because then α and β should adopt their characteristic values for $N > 80$ or so, which they do not; in the range $N = 64$ to $N = 216$, $\alpha = 1.5$ and $\beta = 2.6$. Later in the same paper, by comparing the data for D to a plausible functional form, an estimate of $N_e \approx 120$ is obtained, as mentioned. From this, one can see that none of the chains in I, and only the $N = 800$ chain in II, are likely to show any evidence of reptation. Indeed, it is only for this last chain that $g_1(t) \sim t^{0.36}$ is seen. As this chain is apparently about 6 or 7 entanglements long, one should not be surprised that a clear $t^{0.25}$ is not observed. Indeed, if the central five monomers are examined, as in the case of Kremer et al., a $t^{0.25}$ regime is evident. Nevertheless, even in this instance the consideration of the parallel and perpendicular components of $g_1(t)$ show no preference for longitudinal motion; the reasons for this contrast with the results of Kremer et al. are not yet apparent. Thus while these simulations do not show clear evidence of reptative behavior, we do not feel that they refute reptation either.

There are several important results which emerge from this extensive series of simulations, however. One is that a crossover from Rouse-like to fully entangled behavior, be it strict reptation or not, must necessarily be broad; this is in agreement with many experiments. Also, exponent values such as $\alpha = 2$ and $\beta = 3.4$ are not proof of reptation per se; they can be the result of other kinds of isotropic, constrained motion. Finally, the simulations were examined very carefully for the presence of long-lived two-chain contacts. A distribution of two-chain contact lifetimes was obtained, which decayed in a multiexponential manner, but which had a significant population at times greater than τ_1. It is definitely possible that simulations of this kind may provide crucial insights into the nature of the entanglement phenomenon.

Recently a different approach to computer simulation of entangled polymers has been attempted, using a density of 1 on a lattice.[164] In this case the only way to move a portion of a chain is by a series of simultaneous cooperative moves involving pieces of several chains. In a sense, then, this approach is closer to a real melt; in both the lattice MC and the MD approaches, void space is required for chain motion in quantities that are definitely not present in the melt. So far these results have also only achieved the crossover regime, with the appropriate exponents agreeing with experiment for the longest chains considered ($N = 256$ and 512). The behavior of $g_1(t)$ was also close to the predictions of the Rouse model, so no evidence of reptation was seen. One particularly interesting feature of these results is the appearance of KWW stretched exponential relaxation functions for the end-to-end vector. This is in agreement with rheological and dielectric relaxation measurements of the same quantity, and the predictions of the coupling model discussed in Section II.D.3. In summary, computer simulations have really only been able to achieve modest degrees of entanglement, and thus have not been able to provide definitive evidence on the issue of whether chains reptate. In the crossover regime, however, it is clear that the individual chains behave much more like Rouse chains than chains trapped in networks of fixed obstacles.

C. Related Phenomena

In this section we consider the extension and application of reptation-based models to physical processes related to, but also distinct from, the dynamics of entangled melts. Five different situations are considered. First, the dynamics of polymer solutions are examined, particularly in terms of the molecular weight and concentration dependence of diffusion. Second, treatments of the kinetics of polymer crystallization incorporating reptation are reviewed briefly. Third, the role of reptation in the gel electrophoresis of polyelectrolytes is considered. Fourth, polymer "welding," or the development of mechanical strength across a fracture surface in a polymer material, is discussed. Finally, the application of reptation to diffusion-controlled reactions of polymers is described. The common feature of these five quite disparate areas is the possibility that some form of reptative motion plays an important role. However, it should be emphasized from the outset that none of these processes really provides a critical test for the importance of reptation in melts. In other words, it is certainly conceivable that reptation could be the dominant mode of motion in highly entangled melts, but relatively unimportant in these other situations, or even vice versa. For this reason, the following discussion will be brief.

1. Dynamics of Polymer Solutions

a. General Comments. A quantitative description of chain dynamics in non-dilute polymer solutions is a much harder problem than in the melt or at infinite dilution. There are several reasons for this, the four most important of which follow.

1. In a good solvent, an isolated chain occupies a much larger volume than in the melt. The chains therefore contract with increasing concentration, and the relevant distribution functions (which are only approximately known) change with concentration.

2. In very dilute solution, hydrodynamic interaction plays a crucial role in all dynamic properties, whereas in the melt it is irrelevant. There is therefore a progressive diminution (sometimes called screening) of the hydrodynamic interaction with increasing concentration.

3. Just as intramolecular excluded volume and hydrodynamic interactions diminish, the effects of topological interactions, that is, entanglements, grow. The crossover from unentangled to entangled behavior with increasing concentration in solutions is at least as difficult to predict as the same crossover with increasing molecular weight in melts.

4. Implicit or explicit in all quantitative theories of polymer dynamics is a "microscopic jump frequency," or equivalently a subchain friction coefficient, which sets the time scale for local motions. In a homopolymer melt this is a function of temperature alone, but in solution it can be a strong function of concentration as well. For example, the most commonly studied polymer in dilute solutions is polystyrene, usually within 25 °C of room temperature. In this range the characteristic times for segmental motions are typically on the order of 1–10 ns. On the other hand, undiluted polystyrene is a glass at temperatures below $\approx 100\,°C$, which means that at room temperature the same segmental motions are almost completely frozen out. Changes in local friction alone can therefore result in orders of magnitude changes in the measured dynamic properties. As the subchain friction coefficient is a parameter and not a prediction of the model, it is customary to resort to empiricisms such as measuring the concentration dependence of the tracer diffusion of a small molecule in the polymer solution in order to estimate the change in this parameter, but this may not always be reliable.

These difficulties notwithstanding, a great deal of progress has been made toward understanding nondilute solution dynamics, and a large body of data exists. It is helpful to consider polymer solutions in terms of three concentration regimes. In *dilute* solutions, individual chains rarely encounter one another, and single-chain theories can be applied. In polymer volume fraction, dilute usually means $\phi < 0.001–0.01$, depending on molecular weight; dilute solutions are not considered further here. At the other extreme, in *concentrated* solutions the solvent acts only as a diluent, reducing the subchain friction coefficient and increasing the entanglement spacing. In these solutions the coils are assumed to have contracted to near their melt dimensions, so that Gaussian statistics apply; with a suitable renormalization of parameters it is possibly appropriate to apply reptation theories such as those of Doi and Edwards[7,8] or Curtiss and Bird[23,24] to concentrated solutions. The lower limit of this regime is typically taken to be $\phi \approx 0.1–0.2$. Intermediate concentrations are referred to as *semidilute*. The characteristic feature of this regime is that while individual coils interpenetrate, the solvent volume fraction is still much greater than the polymer volume fraction. It is also within this regime that the effects of entanglement first appear.

The demarcation between dilute and semidilute is called ϕ^*, or c^* in concentration units.[144] It may be estimated as

$$c^* \approx \frac{3M}{4\pi R_g^3 N_A} \tag{3.27}$$

that is, the total polymer concentration equals the monomer concentration inside a given coil. Following the scaling hypothesis, many properties X in the semidilute regime can be expressed as a power law in reduced concentration:

$$X(c) = X(0)\left(\frac{c}{c^*}\right)^x \tag{3.28}$$

It is important to note that c^* is not defined precisely, and that the crossover from dilute to semidilute behavior can extend over a substantial range of concentration. It is also true that Eq. (3.28) holds much better for "static" properties (R_g, osmotic pressure, correlation length) than for dynamic properties. There is some disagreement in the literature as to whether the concentration c_e at which entanglement effects become significant should be considered as distinct from c^*. The experimental situation is quite unambiguous; changes in the M dependence of η and J_e, and the appearance of a plateau in $G(t)$ or $G'(\omega)$, appear at concentrations typically a factor of 2–5 higher than c^* defined by Eq. (3.27), or by measurements of osmotic pressure.[2,165] One must therefore conclude either that a small semidilute but not entangled regime exists, or that c^* depends not only on M and R_g, but also on the property in question. The latter position is analogous to the different critical molecular weights for $G(t)$, η, and J_e^0 in melts, so the disagreement may be largely semantic. With the preceding comments as an introduction, we can now discuss some of the results for diffusion and viscosity for $c > c^*$.

b. *Diffusion.* The reptation hypothesis has been combined with the scaling hypothesis to yield scaling law predictions for the dependence of D on c, M, and P in the semidilute regime, for monodisperse, linear chains of molecular weight M, which move predominantly by reptation through a solution of polymers with molecular weight P. The principal results are[144]

$$D \sim M^{-2}c^{-1.75}P^0 \tag{3.29a}$$

in a good solvent and

$$D \sim M^{-2}c^{-3}P^0 \qquad\qquad (3.29b)$$

in a theta solvent. The M dependence is in fact assumed, because reptative motion is assumed; the P independence is also a direct consequence of this assumption. A great deal of data have been accumulated in various chemical systems and compared with Eqs. (3.29).[166-173] These results have been summarized elsewhere.[172] In no case have all three exponents been demonstrated simultaneously. In several cases, apparent M exponents below -2 and approaching -3 have been seen. Furthermore, for $P > M$, $D(M)$ decreases until $P \approx 3$–$5M$. The concentration dependence is rarely well described by a power law over any substantial range, and instantaneous slopes as low as -10 have been reported.

In their pioneering work Leger et al. observed the scaling laws $M^{-2}c^{-1.75}$, and this work is frequently cited as evidence for the importance of reptation in semidilute solutions.[166] This conclusion must now be regarded as incorrect, and the agreement between observed and predicted exponents fortuitous. First, the later studies revealed that D was not independent of P in these measurements. Second, the M and c exponents included data for solutions that could not possibly be entangled in the viscoelastic sense (e.g., $M = 78,000$ at $c = 10\%$, $M = 245,000$ at $c = 3\%$, given $M_e = 18,000$ in the melt, and assuming M_e increases approximately linearly with solvent fraction; see later). Third, using the tube concept of Doi and Edwards, or even a generic entanglement correlation length, one must conclude that only extremely high-molecular-weight polystyrenes would be sufficiently entangled in semidilute solutions to be constrained to reptate. In the melt the data indicate a tube diameter $a \approx 60$ Å, corresponding to R_g for a chain with $M \approx 40,000$, which is close to M_c. Recall also that $a^{-2} \sim G_N^0$. It is well established that G_N decreases with at least the second power of c, meaning that a decreases at least linearly with c, and thus the tube diameter for $c = 0.1$ is approximately 600 Å, corresponding to R_g for $M \approx 4 \times 10^6$. This rough calculation suggests that it is very unlikely that reptation is significant for linear polymers in the semidilute regime. Fourth, it is clear that exponents of approximately -2 for the M dependence of D are by no means unique to reptative models. Similarly, the concentration dependence of D for linear polymers is very similar to that for branched polymers and for latex spheres in polymer solutions, so there is nothing unique to reptation in this behavior. In melts in the crossover regime it is clear that reptation is not the only relaxation process available to linear chains, and there is no a priori reason why this should be different in solution. In other words, the requirement for reptation to be dominant in the melt, which one might argue is $M/M_c \approx 5$–10, might be even stronger in solution.

Taken as a whole, the results for $D(c, M, P)$ in solution appear to be rather confusing when analyzed in power law form. However, a considerable degree of universality is exhibited when the data are examined in more general ways.

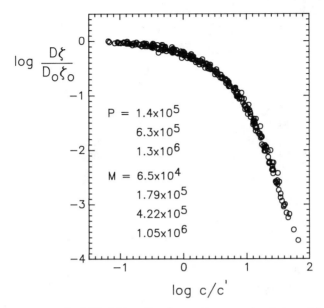

Figure 3.32. Normalized diffusivities of four linear polystyrenes in solutions of three different poly(vinyl methyl ether) solutions as a function of reduced matrix concentration. From Wheeler et al.[172]

This approach has been taken by a number of authors, and two examples are offered in Figs. 3.32 and 3.33. In the former, Wheeler et al. used dynamic light scattering to monitor the tracer diffusion of PS chains present in vanishing amounts in poly(vinyl methyl ether) (PVME) solutions.[172] The solvent and temperature were selected to match the solvent and PVME refractive indices exactly, so that the signal was due entirely to the PS molecules. Three different values of P and four different M were examined for concentrations up to 30%. The effect of changing local friction was also followed by NMR measurements of solvent diffusivity. The quantity plotted logarithmically in Fig. 3.32 is D/D_0 corrected for local friction, where D_0 is the infinite dilution value, against a reduced concentration c/c'. The reference concentration c' is selected to obtain the best superposition, but it correlates very well with the quantity $[c_{PS}^* c_{PVME}^*]^{0.5}$. The reduction to a master curve of such an extensive set of data is indicative of one dominant length scale in the solution dynamics. In Fig. 3.33 an even more extensive set of data are plotted in a similar format, involving NMR measurements of D for 17 different polymer–solvent systems.[173] In particular, the solvent quality ranged from theta to good, and some of the polymers were branched, suggesting that the apparent universality transcends the arguments leading to Eqs. (3.29).

In addition to the linear PS diffusion examined in Fig. 3.32, Lodge et al.

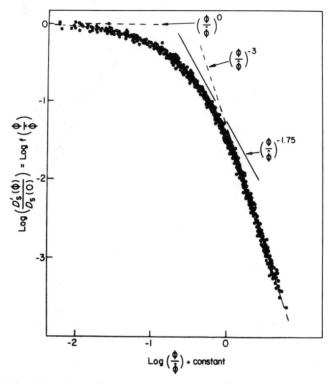

Figure 3.33. Normalized polymer diffusivities for 17 different polymer–solvent systems as a function of reduced concentration. From Skirda et al.[173]

have measured D for a series of three-arm and twelve-arm star PS in the same matrix solutions.[174] In their general character, the results are very similar to those for the linear polymers, but specific comparisons can also be made. For example, the ratio D_{star}/D_{linear} can be computed as a function of concentration. In the case of a diffusion mechanism dominated by hydrodynamic size, this ratio should not be a strong function of concentration, whereas if chain architecture becomes important, as is obviously the case in melts, the ratio should fall. The comparison can be made for star and linear polymers of equal arm length, or for equal total M. For $f = 3$, the two formats yield the same result; as c or P increases, the ratio drops by up to an order of magnitude, implying the onset of topological contributions to the diffusion mechanism, as illustrated in Fig. 3.34. For $f = 12$, at equal total M the ratio actually increases slightly, which is attributed to the earlier onset of entanglements for the less compact linear molecule. In summary, these results are at least consistent with a crossover to a reptation regime with increasing concentration.

Phillies and Peczak recently proposed a universal hydrodynamic scaling

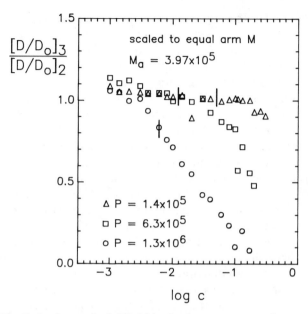

Figure 3.34. Ratio of normalized diffusivities for three-arm star to a linear polystyrene, at equal arm molecular weight, as a function of matrix concentration. Vertical lines indicate the entanglement concentrations for different matrices. From Lodge et al.[174]

function to describe $D(c, M, P)$:

$$\frac{D}{D_0} = \exp(-\alpha c^\nu) \qquad (3.30)$$

where α and ν are functions of M and P.[175] This functional form describes almost all of the available data quite well, including those for latex spheres, proteins, and branched polymers as well as linear ones. From this universality, they conclude that reptation cannot be significant in solution. The physical processes in the derivation of Eq. (3.30) are all hydrodynamic in origin, that is, with no topological component, so it is not immediately obvious how to reconcile this picture with well-established viscoelastic results, such as the expressions of a plateau in G' or $G(t)$ for entangled solutions. Therefore it is likely that while indeed there is a great deal of universality exhibited in $D(c, M, P)$, this may not have solely the origins suggested by Phillies and Peczak. Ngai and Lodge have shown that the stretched exponential concentration dependence of Eq. (3.30) can be viewed as the natural consequence of a progressive increase in the coupling model parameter n with increasing concentration.[53]

While there are many studies of $\eta(c, M)$ in the concentrated regime, there are very few for D. Recently Nemoto et al. have presented measurements of $D(M)$ and $\eta(M)$ over a wide range of M at $c = 40\%$ for PS in dibutyl phthalate.[176] Interestingly, $D \sim M^{-2.5}$ over more than a decade in M above $M_c \approx 10^5$, including data for $M = P$ and $M < P$. Indeed, values of M/M_c achieved were as high as those obtained in the melt, where M^{-2} is a reasonable description, as shown in Fig. 3.1. The authors suggest that "... a theory ... on the basis of a new physical concept (i.e., other than reptation) seems to be required." Just as we feel it was incorrect to take the initial studies of Leger et al.[166] as clear evidence of reptation in semidilute solution, it might be premature to reject reptation on the basis of this, the first extensive study of D in concentrated solution. The authors point out correctly that one cannot use the crossover arguments used heretofore[69,80] to rationalize the -2.5 exponent because it persists at least up to $M/M_c \approx 30$. However, it may still be a crossover effect, in the following sense. As mentioned, several studies have observed M dependences stronger than -2 in semidilute solution, presumably due to a complicated mixture of hydrodynamic and topological effects, while M^{-2} seems to be a good description in the melt. Therefore the apparent M exponent presumably will decrease in magnitude as concentration increases in this regime. The necessary next step must be to measure D for still higher concentrations, and with other solvents, to investigate this regime further. As a final comment on this study, it is interesting to note that the combination of exponents for D and η renders the paradox discussed in Section III.A.2 moot; the ratio of diffusional to relaxational time constants does not increase with increasing M. This raises the possibility that the high M crossover in the melt might occur in D and not in η, but this is purely speculative at this point.

c.   *Viscosity.*   The dependence of η on c and M in the concentrated regime has been much more extensively studied than that of D. In general, the results are describable as $\eta \sim (cM)^{3.4}$, for $M > M_c(c)$.[2,177,178] The 3.4 exponent for M is almost as well established as in the melt, suggesting strongly that the dominant modes of motion are similar to those in the melt, albeit with a renormalized local friction and M_c. Thus strict reptation is not in agreement with these results either. The concentration dependence is not as well established, but is certainly not far from that suggested. Again, the changing local friction is one factor that makes extraction of $\eta(c)$ less certain. In semidilute solutions there have been attempts to describe η as a power law in c based on reptation plus scaling, analogously to Eqs. (3.29), but so far the agreement with experiment is not compelling.[177-179]

2.   Polymer Crystallization from the Melt

Polymers with a high degree of stereoregularity may exhibit a rich array of crystalline morphologies, representing a complicated balance of kinetic and thermodynamic influences. Linear polyethylene is the paradigm of crystallizable polymers, and has been studied very extensively. It is well established, for example, that the preferred morphology is a lamellar structure with the individual chain backbones oriented perpendicular to the lamellar or crystal face. However, the thickness of a lamella is much less than the contour length of the chain, except for rather short chains, so each chain has the possibility of leaving and reentering the same crystal domain several times. For many years there was a great deal of controversy over the prevalence of "adjacent reentry" of chains, that is, whether or not the tail of a given chain immediately folded back to reenter the crystal domain.[180] A primary objection to adjacent reentry was the speed with which the crystal domains grew; it seemed implausible that individual chains would be able to arrange themselves as neatly as adjacent reentry required yet as rapidly as the growth kinetics implied. Conversely, "random reentry", with chains reentering the domain in a spatially random fashion, led to a density paradox. The resulting amorphous phase between lamellae would need to have a density considerably above that for the melt in order to accommodate all the strands emerging from the crystal.[181] However, it now appears that adjacent reentry is generally accepted to occur a significant fraction (perhaps two-thirds) of the time. One reason for this acceptance is the incorporation of the reptation concept into models for the crystallization kinetics.

The idea of incorporating reptation into the crystal growth mechanism is at least a decade old,[182] but it has most recently been discussed in detail by Hoffman and Miller.[183] They considered the molecular-weight dependence of the growth rate for PE melts of varying molecular weights cooled below the melting temperature T_m. In essence, the free energy of crystallization provides a force pulling on a chain from a segment attached to the crystal face, and the remainder of the chain is "reeled in." The frictional resistance for this motion should scale as M^{-1}, as discussed in Section II.A. This prediction is apparently in good agreement with the results, over at least a decade in molecular weight. At very high molecular weights, an individual chain has a significant possibility of being attached simultaneously to two growing faces, thus invalidating this treatment. This is also in accordance with the observation that the ultimate degree of crystallinity attainable begins to drop as molecular weight is increased.

From the perspective of the adjacent reentry debate, the reptation mechanism provides a means whereby segments can be deposited at the growth face at a much higher rate than heretofore estimated. Note also that this process is much more rapid than the equilibrium reptation discussed above. The reason is clear: rather than execute Brownian motion within a tube, possibly accompanied by substantial lateral excursions, a crystallizing chain moves steadily in one direction along the tube. (At reasonable degrees of undercooling, the reverse process, or chain melting, is unlikely to contribute significantly.) It is also appropriate to point out that there are other ways by which reptation could be incorporated into the crystallization process. For example, Klein and Ball consider the reeling in of chain "slack," rather than the whole chain.[182] The slack is equivalent to the defects considered in the original paper of de Gennes.[1] The retraction of these defects is essentially independent of molecular weight, so a different result is obtained. As a final comment, at this stage the possible contribution of reptation to the crystallization kinetics has no direct bearing on the melt problem, by virtue of the force imposed by the free energy of crystallization. (Hoffman and Miller estimate that for 16.5 °C undercooling in PE this force is equivalent to a pressure of 3–4 atm at the end of the tube.)[183] Nevertheless, it is an interesting application of the reptation concept and might prove to be a method for measuring the longitudinal chain friction coefficient directly.

3. *Electrophoresis*

Gel electrophoresis is currently the method of choice for the separation of biopolymers in general, and DNAs in particular. Nevertheless, the detailed mechanism is not well understood, nor are the optimum conditions for a given separation predictable a priori. Recently, however, a great deal of progress has been made via the incorporation of reptation ideas. The following naive picture suggests how "biased" reptation can quite simply account for two important features of the experiment: at low fields and/or low molecular weights, the electrophoretic mobility μ is proportional to M^{-1}, while at high fields and/or high molecular weights, μ becomes independent of M.

Consider a chain of N rigid, freely jointed links of length l, each carrying a charge q. The chain is a three-dimensional random walk in a gel. An electric field E is applied in the x direction, resulting in a steady-state drift of the chain center of mass $\langle v_x \rangle$. We can define

$$\mu \equiv \frac{\langle v_x \rangle}{E} \qquad (3.31)$$

and obtain $\langle v_x \rangle$ as $\langle F_x \rangle/\zeta$, where $\langle F_x \rangle$ is the average net force on the chain in the x direction, and ζ is the chain friction coefficient. If we assume that the chain is reptating through the gel, then $\zeta = kT/D_{\text{tube}} \sim M$. The net force is computed as

$$\langle F_x \rangle = qE \left\langle \sum_j \frac{l_{j,x}}{l} \right\rangle = QE \langle h_x \rangle L^{-1} \qquad (3.32)$$

where $l_{j,x}$ is the x component of the vector representing the jth link, h_x is the x component of the end-to-end vector, Q is the net charge ($= qN$), and L is the contour (tube) length ($= Nl$). The result is thus

$$\mu = \frac{Q}{\zeta}\left(\frac{\langle h_x \rangle}{L}\right) \qquad (3.33)$$

Superposed on this drift is the random Brownian motion of the chain (strict reptation) in the tube. Including this contribution modifies Eq. (3.33) only to

$$\mu_{\text{net}} = \frac{Q}{\zeta}\left(\frac{\langle h_x^2 \rangle}{L^2}\right) \qquad (3.34)$$

a result obtained by several authors.[184-186] In the low field limit, the presence of the field exerts a slight bias on the random walk executed by the end link as it escapes the tube, leading to a net drift in the x direction. However, as long as the chain approximates a Gaussian distribution, $\langle h_x^2 \rangle \sim M$ and so $\mu_{\text{net}} \sim M^{-1}$. When the field strength increases, the chains become extended in the x direction, until eventually $\langle h_x^2 \rangle \sim M^2$, and the mobility becomes independent of chain length. One method to overcome this difficulty is to alternate applying fields in the x and y directions. After a period of time in one direction such that saturation effects are seen, the change in direction of the field rerandomizes the chain distribution, more rapidly for shorter chains; this approach has been demonstrated experimentally.[187]

The preceding analysis is a vast oversimplification of the models that have been developed. At this stage, however, the field is in a state of rapid activity because the physical phenomena are much more complicated than even the most sophisticated models. Some of the additional factors to be considered are now mentioned.

1. The gels (usually agarose or polyacrylamide) are in general poorly characterized and difficult to reproduce from laboratory to laboratory. Clearly the mean mesh size is an important parameter, but the inhomogeneity of the gel, which for obvious reasons is not considered in tractable models, can be substantial. The simulations of Muthukumar and Baumgärtner[138] discussed in Section III.A.7 illustrate the kinds of influence this can have.

2. This picture assumes that the field effectively labels one chain end as the head, which the rest of the chain follows. However, in many chain conformations the head and the tail may have similar x coordinates, with the chain middle looped around the gel. In such a conformation the chain can be effectively trapped in a metastable state.

3. In general, the distribution function of the chain is not determined, nor are the charges along the chain assumed to have any influence other than affecting the persistence length of the chain and providing the

electrophoretic force. There is ample evidence, for example, that the chain changes its dimensions quite dramatically, and even in an oscillatory fashion, in the presence of a steady field.[188,189]

4. The effect of internal modes, in addition to the strict but biased reptation contribution, is often not included; it has recently been suggested that these modes may be essential to describe field-inversion (i.e., E applied alternately in $+x$ and $-x$ directions, albeit for different times) electrophoresis.[190,191]

5. A recent simulation of Deutsch and Madden[189] suggests the phenomenon of a "bunching instability," whereby a leading section of the chain develops a small kink which retards its motion. The trailing segments can then build up in one region of the gel, resulting in a locally enhanced friction, which in turn results in a substantial chain contraction. Eventually the chain unwinds and is reextended. This physical picture appears to be quite similar to that reported in direct fluorescence visualization of a DNA molecule undergoing electrophoresis.[188]

In summary, while the application of reptation ideas to the mechanism of gel electrophoresis has led directly to substantial advances in understanding, it is clear that the field is still in the early stages of development.

4. Polymer Welding

The process of polymer welding, or crack healing, refers to the buildup of mechanical strength across an interface between two polymer samples, due to chain interdiffusion. The interface may be created either by bringing two "fresh" surfaces together, or by annealing a fracture in a bulk sample. This phenomenon is clearly fundamental to polymer adhesion and reactive polymer blending, but it also offers the possibility of unique insight into the nature of entanglement. A more complete discussion of polymer welding can be found in recent reviews.[192,193]

A crucial observation is that the fracture energy for glassy and rubbery polymers is orders of magnitude greater than that associated with the surface tension of the newly created surfaces. The additional strength of the material is quite reasonably assumed to reflect the entanglement interaction. In support of this idea, it is also observed that the strength of the material increases with M up to a value comparable to M_e. Therefore in following the time evolution of mechanical strength across an interface, the dynamics of chain diffusion and the formation of entanglements can be investigated. In modeling this process it is necessary to select a characteristic property of the chains which determines the mechanical strength. As pointed out by Kausch and Tirrell,[193] this selection must also correspond to what is known about the mechanical strength of the bulk polymer. For example, the density of chains crossing the plane of the interface is not sufficient, because this number is independent of M for all but the lowest M in bulk polymers, while the strength is not. On the other hand, depth of penetration across the interface is not sufficient either, because this implies that strength increases indefinitely with M. The authors suggest that penetration of segments across the interface to a depth comparable to R_g for a chain of length M_e is a more appropriate quantity.

The last criterion suggests that most of the healing should be associated with interdiffusion on a length scale much shorter than the overall chain R_g, at least for very long chains, and therefore the quantity of interest may be $g_1(t)$, as discussed in Sections III.B.1 and 4. For simple interdiffusion of particles governed by Fick's law, one expects a $t^{0.5}$ growth of the interdiffusion layer. However, if one takes a reptation-based model, such as that of Prager and Tirrell,[194] the relevant regime is that where $g_1(t) \sim t^{0.5}$ for $\tau_{\text{Rouse}} < t < \tau_{\text{rep}}$, which leads to a $t^{0.25}$ dependence for the buildup of mechanical strength. This dependence has been observed experimentally,[195] but it should be emphasized that the experimental situation is far from clear, and that the experiments are not easy to perform. Among the experimental difficulties are the reproducible formation of the interface, the elimination of density fluctuations at the interface when two surfaces are brought into contact, and the need to use sharp molecular-weight fractions to avoid distortion of the time evolution due to fractionation of the chains by their mobilities at the interface. Nevertheless, the experimental data are in reasonable agreement with the $t^{0.25}$ dependence predicted on the basis of reptation. It remains to be seen whether other detailed models for the mechanism of chain diffusion can be similarly successful.

5. Diffusion-Controlled Reactions

The subject of diffusion-controlled reactions in polymer systems has been reviewed recently.[196] Intermolecular chemical reactions in entangled polymer liquids are generally diffusion-limited, for obvious reasons. The reptation process suggests a way to predict the dependence of rate constants on polymer molecular weight and concentration, and on time for reactions where at least one participant is a polymer. The most common example is the batch polymerization reaction, of either the step-growth (poly-condensation) or the chain-growth (free-radical) kind. The latter is of more interest here because much higher molecular weights are typically achieved, and because of the phenomenon of autoacceleration (the Trommsdorff effect). Under appropriate conditions it is observed that the rate of conversion of monomer to polymer increases dramatically when the extent of conversion is about 50%. As the addition of monomer to a growing polymer (i.e., propagation) is an exothermic step, and because the reaction temperature is usually below the glass transition temperature of the bulk polymer, the onset of autoacceleration has profound consequences from a reaction control perspective. The accepted interpretation of this phenomenon is that when the average polymer molecular weight and concentration are sufficiently high, polymer diffusion is severely retarded by the onset of entanglement. As a consequence, the termination reaction of two growing polymers combining to form one "dead" polymer is suppressed, leading to an increase in overall conversion rate. Tulig and Tirrell[197] have incorporated scaling predictions based on reptation for polymer diffusion, and via the Smoluchowski approach have modeled the behavior of the termination rate constant k_t:

$$k_t = 4\pi\rho(D_n + D_m) \tag{3.35}$$

where ρ is the intermolecular separation required for reaction, and D_n and D_m are the diffusivities of growing chains of length n and m, respectively. The analysis is reasonably successful in predicting the reaction time at which the fractional conversion begins to rise sharply and the concomitant increase in weight average molecular weight. At approximately 75% conversion, the reduction in monomer mobility as the glass transition is approached causes an abrupt decrease in the polymerization rate (so-called limiting conversion), which is beyond the scope of this treatment.

It is important to emphasize that a free-radical polymerization is a very complicated process, and much more work along these lines will be necessary. There are at least three factors that need to be addressed in utilizing reptative models for the polymer mobility. First, the system is inherently very poly-disperse. As discussed in Section III.A.4, this can make the analysis very complicated, as each molecular weight in the sample might be viewed as having a different reptative component to its overall mobility. Second, the chains whose mobility is of interest are growing in time as well as diffusing; a diffusion coefficient measured over a distance equal to many times the chain's instantaneous radius of gyration may not be the appropriate quantity to use in modeling the reaction medium. Third, it is not necessarily the polymer center-of-mass diffusivity that enters the problem. This depends on how much space the average reactive chain end needs to explore before encountering a potential reactant. This issue has been treated by de Gennes, with the following predictions emerging.[198]

1. For times much longer than τ_{rep}, where $g_1(t) \sim t$, the diffusion coefficient that matters is still that of the center of mass, but the effective ρ becomes the polymer R_g, and consequently $k_t \sim M^{-1.5}$, which differs from Eq. (3.35) if strict reptation were assumed and $D_n \sim M^{-2}$ were inserted into the Smoluchowski equation.

2. For intermediate times, $\tau_{rep} > t > \tau_{Rouse}$, where $g_1(t) \sim t^{0.5}$, $k_t \sim M^{-0.75}t^{-0.25}$.

3. For shorter times, where $t < \tau_{\text{Rouse}}$, the characteristic reptation prediction of $g_1(t) \sim t^{0.25}$ results in $k_t \sim M^0 t^{-0.375}$.

4. At still shorter times the effects of entanglement are not felt, and $g_1(t) \sim t^{0.5}$ leads to $k_t \sim t^{-0.25}$, as in 2, but independent of M, as in 3.

Bernard and Noolandi[199] have considered the end-to-end cyclization rate for an entangled, reptating polymer. For $t \ll \tau_{\text{rep}}$, they obtained $k_{\text{cyc}} \sim t^{-0.25}$, while for $t > \tau_{\text{rep}}$, $k_{\text{cyc}} \sim 1/\tau_{\text{rep}}$. On the whole, there have not been sufficient model studies to assess the reliability of these various predictions. This could certainly prove to be a fruitful area for further experimental investigation of the reptation concept.

In another application to diffusion-controlled reactions, Cates[200] has considered the effect of random chain scission and recombination in a polymer liquid, with potential application to surfactant organization, reactive "macromonomers," and "dynamic equilibrium" polymers, such as liquid sulfur. When the characteristic time constant for scission is less than that for chain reptation, the scission process can contribute significantly to stress relaxation, with a concomitant change in the character of $G(t)$.

IV. SUMMARY AND CONCLUSIONS

In this section we present our answers to the questions posed at the end of the Introduction, and some additional general discussion.

Do linear polymer chains execute primarily reptative motion over a nonvanishing region of parameter space?

We feel that the accumulated evidence argues strongly, but by no means conclusively, for the answer "yes." But there are probably only two experimental situations where linear chains might execute essentially only strict reptation: sufficiently long tracer chains surrounded by a matrix of much longer chains, and unattached chains in a regular network. Furthermore, the latter situation is very hard to achieve experimentally. In any entangled melt of linear chains, monodisperse or polydisperse, the cooperative nature of the problem requires consideration of other mechanisms, which can contribute substantially to the mobility of any chain. Thus while we believe reptation is an excellent first approximation to the behavior of individual chains, it is rarely sufficient for a quantitative description of the dynamics of polymer melts. Furthermore, it would definitely be premature to consider this issue closed.

What is the strongest evidence for the correctness of the reptation hypothesis?

Here we offer two general statements and three specific claims.

1. The reptation hypothesis, and models built thereon, give an extremely good qualitative description of a broad range of experimental observations; in many cases the agreement is quantitative. Furthermore, in many cases the predictions were made prior to the experimental observations, so it was not a case of generating an explanation a posteriori. If one adopts the position that it is more important for a good model to explain a wide variety of phenomena reasonably well than to be in precise agreement with a restricted set of observations, then reptation-based models must be viewed as extremely successful.

2. At least as significant to our view is the fact that there are no experimental observations of which we are aware that contradict the reptation hypothesis. There are abundant illustrations of the fact that strict reptation alone cannot describe the dynamics of polymer melts, but that is distinct from proving that chains do not reptate.

3. To our knowledge, no molecular model has been advanced other than reptation which can predict the dramatic retardation of branched polymer relaxation and diffusion, relative to linear polymers of comparable size. In particular, this difference increases apparently exponentially with increasing arm length. To explain these observations without invoking a snakelike escape out of entanglements seems to us to remain a substantial challenge.

4. In computer simulation via molecular dynamics, the observation of a time regime where the mean square displacement of the center monomers follows a time scaling of approximately $t^{0.25}$, combined with evidence for a preference for longitudinal motion, is supportive of the microscopic basis of reptation.

5. Neutron spin echo measurements on three polymer systems, once viewed as contradictory, now seem to provide clear evidence of an entanglement length scale, reflected in the random motion of monomers. While this is not evidence of reptation per se, it is certainly consistent with the basic physical picture incorporated in the Doi–Edwards model.

What phenomena, if any, can reptation theory not describe, and what are its main weaknesses?

It is important to recall what reptation is, and what it is not. It is a postulate for the long-time motion of linear chains in entangled polymer systems, whereby the inherently many-body problem is reduced to that of a single chain in an effective field. It does not explain what entanglements are, or when they come about, a feature shared by other models. Nor can it really explain, by itself, any of the collective properties of a polymer material. Only single-chain scaling laws, such as $D \sim M^{-2}$ and $\tau_1 \sim M^3$, are direct consequences of the strict reptation hypothesis. When additional assumptions are built on, such as the tube ansatz, it is possible to develop predictions for material properties and constitutive equations. The Doi–Edwards and Curtiss–Bird theories are the two most prominent of this kind, and both have remarkable success in describing some material properties as well as significant deficiencies with others. The similarities and differences between these two approaches continue to be the subject of intense discussion, and are most evident in the domain of nonlinear viscoelasticity.[45,46] However, in our view the most significant result is that two models which are developed in substantially different ways arrive at constitutive equations that are almost equivalent. The main similarities between the two models are that they are single-chain-in-a-mean-field models which assume a preference (absolute in the DE case) for longitudinal chain motion. This may reflect a fundamental robustness in the reptation hypothesis which transcends the formalism built around it. Alternatively it may reflect the inevitability of getting $\tau_1 \sim M^3$ for a system with long-range correlations in the interaction potentials.

The main weaknesses of reptation and reptation-based models are three. First, the reptation hypothesis is just that, and has no real justification in molecular theory. Second, reptation assumes the phenomenon of entanglement, and therefore offers little insight into what entanglements are, or under what molecular conditions they become effective. Third, as a single-chain-in-a-mean-field postulate, it is difficult to begin with reptation and build in the effects of many-chain interactions in anything but an arbitrary fashion, no matter how plausibly it may be done. For example, it has recently been suggested that none of the reptation models can predict the phenomenon of elastic recovery.[201] One possible advantage of the Curtiss–Bird approach is that it establishes a formalism independent of a reptation assumption, and it is possible, at least in principle, to explore the results of other postulates in a systematic fashion.

What sets the limits of applicability of reptation, and what other processes need to be considered?

A detailed answer to this question would depend on the physical property of interest. For example, in the case of viscosity, the strict reptation scaling has never been observed, while for tracer diffusion with $M \ll P$ it almost certainly has. The basic requirement for reptation to contribute significantly is the effective suppression of large-scale lateral motion, which requires the formation of an entanglement network with a low enough disentanglement frequency relative to the reptation time, and a correlation length (or tube diameter) substantially less than R_g for the chain in question. Given the entanglement molecular weight M_e as the fundamental (and as yet unexplained) parameter, this condition translates into something like $M \approx 5-10 M_e$ for monodisperse melts. Furthermore, the extension to solution suggests that for a concentration of 10%, only very high-molecular-weight chains could exhibit reptation, and that reptation is almost certainly not important in the semidilute regime. The crossover from Rouse-like to reptation-like behavior with increasing M is a region that is very hard to model and extends over an order of magnitude in M. In our opinion the reptation picture has been applied too optimistically in this regime, and a more thorough theoretical analysis is required.

Clearly, processes other than reptation are significant in polymer melts, even at the highest realizable molecular weights. Several that have been proposed in various forms have been discussed in the text. All, either implicity or obviously, are attempts to modify the effective mean field experienced by a test chain. One, the tube fluctuation correction of Doi, may be able to account for the discrepancy in exponents for the viscosity. "Constraint release" and "tube dilation" are terms that describe attempts to correct for the mobility of neighboring chains which strict reptation neglects. It is clear from experiment that a contribution from something we might call constraint release is required to bring reptation theory into line with the data, but this could be done in a variety of ways. Furthermore, it is very difficult to test these ideas independently. Two features of constraint release models serve to illustrate some of these difficulties. First, it was argued in the Introduction, and assumed by the tube model, that the phenomenon of entanglement does not reflect a specific, local interaction between two chains, but an effective, smeared interaction with a correlation length that might be taken as the diameter of a constraining tube. Within the experimentally inferred tube diameter, or between effective entanglements, there are many other chains in close contact with the test chain. Yet, constraint release is often modeled as a special, *single* chain which is acting as an entanglement and which moves away. Second, the motion of this special chain is always assumed to be purely reptative which, as argued, is only likely to be correct in a restricted range of circumstances. The P^3 dependence ascribed to the relaxation of these constraining chains has been seen only very approximately in one study and not in several others.

The so-called orientational coupling is a more recent addition to the set of other processes to be considered. In this case the strict reptation process is modified by biasing the random walk of the end segments of the chain leaving the tube, when the surrounding polymers are in an oriented state. This notion has some direct molecular support, in the surprising degree of orientation that can be induced in small molecules in the presence of oriented polymers, but it remains to be seen how successful this additional contribution will be in describing experimental results.

How much further progress can be made by modifications to the basic reptation picture, or will a new approach ultimately be necessary?

The answer to this question depends on the interpretation of the word "progress." Clearly, the reptation model serves as an excellent template upon which to build a quantitative description of polymer melt dynamics. However, to a skeptic, the grafting of additional processes onto strict reptation often has an ad hoc flavor. The price for increased agreement with experiment is the expanding number of terms in the equations, and the concomitant difficulty in testing the appropriateness of these terms independently. We believe a new approach is already necessary, in that the reptation hypothesis needs microscopic theoretical justification in addition to the extensive experimental support it already has. Simply stated, the reptation model with or without modifications is basically a phenomenological theory, which while possibly physically appealing and relatively straightforward to use, provides only a partial picture of a bulk polymer at the molecular level.

Some Additional Comments

There are a variety of arguments that may be raised against reptation models; some have been discussed in the course of this chapter. We offer comments on a few of these below.

1. *Strict reptation, or the tube model extension, does not describe the experimentally observed viscosity of polymer melts, either in terms of power-law behavior or in absolute magnitude.* This is true. However, when the reptation hypothesis is viewed as a major step toward a final answer, and not the "last word," it actually does remarkably well. A facter of 2–3 in a quantity like η, which varies over at least 10 orders of magnitude, is not so bad for a model with no freely adjustable parameters (i.e., in Eqs. (3.4) and (3.6)). Most importantly, strict reptation over-estimates the viscosity (and underestimates the diffusivity), which implies that additional relaxation processes are present in the liquid. This is consistent with the neglect of all cooperative chain motions in the strict reptation assumption. If reptation underestimated η, it would be difficult to explain. Although it requires more than strict reptation to explain the behavior of melts, this does not by itself invalidate the hypothesis. Certainly it is difficult to conceive of a conceptually simple approach that will do better; the $M^{3.4}$ scaling of η undoubtedly reflects complicated physics, or at least a wide transition zone.

2. *Strict reptation also fails to describe a variety of other properties.* This is also true. For example, the tube model underestimates the diffusivity, and the breadth of the relaxations in the terminal regime, as reflected, for example, in the product $G_N^0 J_e^0$. However, this situation is similar to that discussed above. The reptation hypothesis describes the dominant mode of motion, and should not be expected to be the only mode available. It is also true that the more detailed the examination of the data, the more likely it is that some discrepancy

will emerge between experiment and reptation theory. But this is true of almost any theory for complex physical processes. We have attempted in this chapter to take a consistently broader view, and not be immersed in the fine details. Every experiment is subject to error, and every theory contains approximations, so too much weight should not be assigned to any single experiment or class of experiments. If there were a rival model that made as many testable predictions as reptation, then a more detailed analysis might be necessary; however, at this time such a rival model does not exist.

3. *The scaling laws $D \sim M^{-2}$ and $\tau_1 \sim M^3$ are not unique to reptation, and experimental observation of this behavior proves nothing.* This is a very important point. It seems clear that the M^3 scaling for the longest relaxation time can be derived from a number of different viewpoints, and the -2 exponent is a direct result. Thus observation of either scaling law indeed proves nothing. Particularly, $D \sim M^{-2}$ has too often been taken as being synonymous with reptation, for example, in semidilute solution. Apart from the inherent difficulties in extracting reliable power-law exponents from experiment, an apparent M^{-2} dependence can result from a number of other factors. If there were no other evidence for reptation than measurements of D and η, it would be very difficult to make a strong case in its favor. The three crucial additional pieces of information are the retardation of branched polymer mobility, the $t^{0.25}$ scaling of the simulated monomer correlation function, and the appearance of an entanglement length scale in the experimental dynamic structure factor. All three are completely independent of $\tau_1 \sim M^3$, in the sense that one could have the latter without predicting anything about star or monomer motion. Furthermore, they are observations that as yet have no other single explanation, as far as we are aware. Thus the reptation models make testable predictions for a range of independent observables, for molecules of different architectures, and for time scales much longer and much shorter than the longest relaxation time. No other single model so far presented can match the breadth of this range.

4. *The behavior of ring polymers conflicts with reptation.* As discussed in Section III.A.5b we do not believe this to be established. Certainly for the available samples, η is less for rings than for linear polymers of equal molecular weight. But the basic parameter of the reptation model is the entanglement length, which for rings appears to be at least twice that for linear chains. When compared at an equal number of entanglements, ring melts have comparable or higher viscosities than linear chains. At this stage of development we do not feel that the results on cyclic polymers are sufficiently well understood to be employed as an argument for or against reptation.

5. *The reptation model is invoked more frequently than is appropriate.* This is actually not a criticism of the model itself, but of its application. It is a view

that we endorse. As discussed, there are a very limited number of circumstances where strict reptation is sufficient to give a quantitative description of polymer dynamics, and too many examples of a naive or overoptimistic invocation of reptation when other processes may well be dominant. Furthermore, for many authors it is apparent that the issue is settled, that is, reptation is established as the correct underlying relaxation mechanism. We do not feel that the experimental evidence supports this rigid a stance.

6. *Too much of the appeal of reptation models lies in their (apparent) physical and mathematical simplicity.* This point has been raised previously. It is a constant dilemma in describing complicated physical phenomena. Models that are very elaborate may be largely ignored, while simplicity, however appealing, is no guarantee of correctness. Nevertheless, progress often hinges on bold simplification, and we feel that reptation has been a tremendous success in this regard, even if in no other. One particular advantage of this simplicity has been the ability to make reptation-based predictions for a variety of observables before the appearance of extensive data; in general, these predictions have later been shown to be at least reasonable and often excellent. In contrast, for example, the "sparse constraints" models discussed in Section II.D have almost without exception been developed to describe established results, and have had very little predictive success.

What further experiments, if any, are required to resolve the first question of this summary beyond any reasonable doubt?

Obviously it is a simple matter to call for more and better experiments. However, we believe that there are a number of areas where more experimental work would be revealing. These are listed below in no particular order of priority.

1. *Dynamic structure factor for labeled chain sections.* The neutron spin echo experiment has been demonstrated to provide a unique experimental window on monomer motions, but the results to date have not been able to establish or refute reptation clearly. One promising way to do this would be to examine triblock copolymers in which the center block is isotopically labeled. This would be the experimental analog of the computer simulations in which the correlation functions are examined for monomers at different locations along the chain. In addition, it would be very interesting to examine the longer time behavior of the dynamic structure factor, to investigate the second $t^{0.5}$ regime in Fig. 3.29. Clearly, this is beyond the capability of neutron spin echo, but light scattering techniques may be applicable.

2. *Concentration dependence.* As argued in the text, it is unlikely that reptation is significant in the semidilute regime, whereas in entangled melts it is often the dominant mode. There has not been a thorough examination of

the concentration regime of 10–100% polymer, particularly for diffusion measurements. An extensive series of measurements of D_t for M chains in a P matrix over this composition range could be very revealing. In particular, one would expect the M^{-2} law to be established at some c, which decreases with increasing P, and then persist for all higher c. Based on the extensive semidilute solution measurements to date, it is quite possible that the M dependence would be stronger than M^{-2} before converging to that value, as suggested by recent results. It would be essential to have viscoelastic measurements over the same regime for the same chemical system. The polymer would need to be selected carefully to minimize the problem of changing local friction. However, it would also need to be chemically robust because it can take months to equilibrate concentrated polymer solutions; hydrogenated polyisoprenes and butadienes may be the best candidates.

3. *Loose chains in networks.* The difficulties associated with making measurements on this kind of system have been discussed, but such experiments are still well worth the effort. One possibility would be to use dichroism to examine the relaxation of single chains in networks. In addition, in concert with the concentration dependence studies mentioned, it would be useful to examine diffusion and relaxation in gels; some preliminary results have already been reported. One additonal variable is the regularity of the network. As suggested by computer simulation, the degree of regularity may have a substantial influence on the importance of reptation. In all cases, characterization of the network or gel structure is essential.

4. *Higher-molecular-weight branched polymers and rings.* Branched polymers have provided some of the strongest evidence in favor of the reptation model, but it is still not clear that recent models for branched polymer dynamics are adequate. Longer arms and different architectures are important variables for examining these issues. Larger rings would of course be most interesting, but synthetically they may be too demanding. The behavior of rings and branched polymers in networks and gels is a fascinating topic which has not yet received the attention it deserves.

5. *Molecular-weight blends.* This approach has proven to be very profitable in terms of examining processes in addition to strict reptation, and will continue to be so. In addition, it is not clear that extension of a model that is successful on a binary blend to a blend with a large number or continuous distribution of chain lengths will be simple or successful. However, this will ultimately be essential for describing industrial polymer samples.

6. *Temperature dependence of diffusion and viscoelasticity.* This is an often neglected aspect of polymer dynamics. Essentially, it is assumed (by reptation or most other theories) that all chain dynamic properties are factorizable into a structural term that depends on M, c, and chain architecture, and a local

friction term, that depends on T and c, but not M for large M. Thus all dynamic properties are presumed to have the same T dependence through the local friction or microscopic "jump frequency." However, there is evidence that this is not always true. For example, the apparent activation energies for diffusion and viscosity are different in some systems, while in other cases different frequency regions of the viscoelastic response exhibit distinct time-temperature superposition shift factors.[202] This has been rationalized in terms of the coupling model, but at this point it appears that a much more thorough examination of this problem is called for. Unfortunately this will require tedious and extensive measurements of different properties as a function of T. It will be important to establish whether the reptation hypothesis can be reconciled with a different T dependence for different observables.

7. *Single-chain structure factor under deformation.* A variety of techniques, including neutron scattering, flow birefringence, and infrared dichroism, have all been employed to some extent to follow the time evolution of single-chain distribution functions, either in steady flow or after imposition of a deformation. However, in most cases these studies must be regarded as preliminary or inconclusive. The possibility of examining directly the response of single chains or chain segments to external stresses is attractive and should receive more attention.

The preceding list consists of experiments designed to examine in more detail the domain of applicability of the reptation concept. For many purposes these issues may be considered sufficiently settled that other problems are more deserving of attention. For example, we have not addressed in any detail the ability of various reptation-based constitutive equations to describe non-linear rheological behavior. The reason for this is that we have been trying to assess the importance of reptation itself, and not the merits and weaknesses of constitutive equations per se. Yet it is this kind of problem that is of the most practical importance, and much interesting and difficult experimental work is necessary. It is beyond the scope of this chapter to examine these issues, however.

In what direction should future theoretical efforts be aimed?

In this concluding section we identify issues that we feel are important and insufficiently resolved.

1. As pointed out several times, the nature of an entanglement is not established, nor are the criteria for the onset of entanglement effects understood.
2. In general, the most successful models to date are of the single chain in an effective field variety. At the least, a more self-consistent approach

would be desirable, including the viscoelastic nature of the environment and the effect of the test chain on the field. The extended Langevin equation developments are promising in this regard.

3. The question of whether the viscosity–molecular-weight relationship should approach an M^3 dependence at high M, or whether an indefinite dependence of approximately $M^{3.5}$ is acceptable, has not been resolved. It seems to us that this question could be addressed on general grounds, without necessarily specifying a particular mechanism.

4. The crossover region from unentangled to entangled behavior is worthy of more investigation, both in melts with increasing M and in solution with increasing c; successful models should also incorporate this transition.

5. More consideration should be given to the "nonlinear" viscoelastic properties of polymer liquids, although we have not discussed them here. These phenomena are arguably even more characteristic of entangled polymers than the linear regime, and ultimately will provide even stronger and more discriminating tests for molecular models. As mentioned previously, the differences between the Doi–Edwards and Curtiss–Bird approaches are more evident here.

6. The issue of whether reptation-based models, or any model treating a single chain in an effective mean field, can predict the phenomenon of elastic recovery in polymer liquids deserves further examination. If not, as recently suggested,[201] then it is even more essential that point 2 above be addressed.

7. For reptation-based approaches, the two outstanding issues are to try and establish a microscopic justification of the reptation hypothesis, and to establish one self-consistent, small set of modifications which account for the additional degrees of freedom.

8. For the generalized Langevin equation approach, the main developments are too recent to be judged. However, it is clear that the main challenge is to carry through the development of the model to the level of testable predictions for experimental observables.

9. For the sparse constraints models, the challenges are twofold. First, the models themselves need to be more clearly codified so that it is possible to develop predictions for observables in advance of the experiments. Second, they need to address the claim that they are not able to provide an adequate description of the behavior of branched polymers.

10. For network models, which have not been discussed in detail here, the crucial next step would be to introduce molecular weight dependence through the relaxation function for a single chain.

11. For the Curtiss-Bird reptation model, further progress will require more information about polymer-polymer, link-pair, and momentum space distribution functions. In addition, implementation of the model for single chain observables, such as D and τ_1, would be useful.

In this chapter we have described and assessed the application of the reptation concept to the dynamics of entangled polymer liquids. Over the past 15 years there has been a tremendous resurgence of interest in a molecular understanding of these phenomena. This resurgence is due predominantly to the introduction of the reptation hypothesis. Even if this approach is eventually supplanted, its contribution to the field is therefore secure.

Acknowledgment

The authors appreciate helpful discussions with, or comments from, a number of colleagues in the field, including R. B. Bird, J. des Cloizeaux, S. F. Edwards, J. D. Ferry, H. Fujita, W. W. Graessley, E. J. Kramer, K. Kremer, A. S. Lodge, G. B. McKenna, M. Muthukumar, K. L. Ngai, J. Noolandi, D. S. Pearson, G. D. J. Phillies, K. S. Schweizer, H. Sillescu, J. Skolnick, and M. Tirrell. Acknowledgment is also made to the donors of the Petroleum Research Fund, administered by the American Chemical Society, and to the National Science Foundation (DMR-8715391) for support of this work.

References

1. P. G. de Gennes, *J. Chem. Phys.* **55**, 572 (1971).
2. J. D. Ferry, *Viscoelastic Properties of Polymers*, 3rd ed., Wiley, New York, 1980.
3. L. R. G. Treloar, *The Physics of Rubber Elasticity*, 3rd ed., Clarendon, Oxford, UK, 1975.
4. J. Meissner, *Rheol. Acta* **10**, 230 (1971).
5. S. Onogi, T. Masuda, and K. Kitagawa, *Macromolecules* **3**, 109 (1970).
6. G. C. Berry and T. G. Fox, *Adv. Polym. Sci.* **5**, 261 (1968).
7. M. Doi and S. F. Edwards, *J. Chem. Soc., Faraday Trans. 2*, **74**, 1789; 1802; 1818 (1978); **75**, 38 (1978).
8. M. Doi and S. F. Edwards, *The Theory of Polymer Dynamics*, Oxford Univ. Press, Oxford, UK, 1986.
9. M. S. Green and A. V. Tobolsky, *J. Chem. Phys.* **14**, 80 (1946).
10. M. Yamamoto, *J. Phys. Soc J.* **11**, 413 (1956); **12**, 1148 (1957); **13**, 1200 (1958).
11. A. S. Lodge, *Trans. Faraday Soc.* **52**, 120 (1956); *Rheol. Acta* **7**, 379 (1968).
12. F. Bueche, *J. Chem. Phys.* **20**, 1959 (1952); *Physical Properties of Polymers*, Interscience, New York, 1962.
13. M. Shen, D. R. Hansen, and M. C. Williams, *Macromolecules* **9**, 345 (1976).
14. W. W. Graessley, *J. Chem. Phys.* **43**, 2696 (1965); **54**, 5143 (1971).
15. H. Fujita and Y. Einaga, *Polym. J. (Tokyo)* **17**, 1131 (1985).
16. A. Kolinski, J. Skolnick, and R. Yaris, *J. Chem. Phys.* **88**, 1407; 1418 (1988).
17. K. L. Ngai, R. W. Rendell, A. K. Rajagopal, and S. Teitler, *Ann. NY Acad. Sci.* **484**, 150 (1985).
18. G. Ronca, *J. Chem. Phys.* **79**, 1031 (1983).
19. M. Fixman, *J. Chem. Phys.* **89**, 3892, 3912 (1988).
20. K. S. Schweizer, *J. Chem. Phys.* **91**, 5802; 5822 (1989).
21. W. Hess, *Macromolecules* **19**, 1395 (1986); **20**, 2587 (1987); **21**, 2620 (1988).

128   T. P. LODGE, N. A. ROTSTEIN, AND S. PRAGER

22. J. Noolandi, G. W. Slater, and G. Allegra, *Makromol. Chem., Rapid Commun.* **8**, 51 (1987); T. A. Kavassalis and J. Noolandi, *Phys. Rev. Lett.* **59**, 2674 (1987); *Macromolecules* **21**, 2869 (1988); **22**, 2709 (1989).
23. R. B. Bird and C. F. Curtiss, *J. Chem. Phys.* **74**, 2016; 2026 (1981).
24. R. B. Bird, O. Hassager, R. C. Armstrong, and C. F. Curtiss, *Dynamics of Polymeric Liquids*, vol. 2: *Kinetic Theory*, Wiley, New York, 1987.
25. B. J. Geurts and R. J. J. Jongschaap, *J. Rheol.* **32**, 353 (1988); B. J. Geurts, *J. Non-Newtonian Fluid Mech.* **31**, 27 (1989).
26. J. C. Le Guillou and J. Zinn-Justin, *Phys. Rev. Lett.* **39**, 95 (1977).
27. P. J. Flory, *Principles of Polymer Chemistry*, Cornell Univ. Press, Ithaca, NY, 1953.
28. R. G. Kirste, W. A. Kruse, and J. Schelten, *Makromol. Chem.* **162**, 299 (1973); D. G. Ballard, J. Schelten, and G. D. Wignall, *Eur. Polym. J.* **9**, 965 (1973); J. P. Cotton, D. Decker, H. Benoit, B. Farnoux, J. Higgins, G. Jannink, R. Ober, C. Picot, and J. des Cloizeaux, *Macromolecules* **7**, 863 (1974).
29. P. E. Rouse, *J. Chem. Phys.* **21**, 1272 (1953).
30. B. H. Zimm, *J. Chem. Phys.* **24**, 269 (1956).
31. M. Doi, *J. Polym. Sci., Polym. Lett.* **19**, 265, (1981); *J. Polym. Sci., Polym. Phys. Ed.* **21**, 667 (1983).
32. W. W. Graessley, *Adv. Polym. Sci.* **47**, 68 (1982).
33. J. M. Deutsch and T. L. Madden, *J. Chem. Phys.* **91**, 3252 (1989).
34. M. Rubinstein, *Phys. Rev. Lett.* **59**, 1946 (1987).
35. J. M. Deutsch, Phys. Rev. Lett. **54**, 56 (1985); H. Scher and M. F. Shlesinger, *J. Chem. Phys.* **84**, 5922 (1984).
36. M. G. Brereton and A. Rusli, *Chem. Phys.* **26**, 23 (1979).
37. M. Daoud and P. G. de Gennes, *J. Polym. Sci., Polym. Phys. Ed.* **17**, 1971 (1979).
38. J. Klein, *Macromolecules* **11**, 852 (1978).
39. G. Marrucci, *J. Polym. Sci., Polym. Phys. Ed.* **23**, 159 (1985).
40. J. L. Viovy, *J. Phys. (Les Ulis)* **46**, 847 (1985).
41. H. Watanabe and M. Tirrell, *Macromolecules* **22**, 927 (1989).
42. E. Helfand and D. S. Pearson, *J. Chem. Phys.* **79**, 2054 (1983).
43. J. des Cloizeaux, *Europhys. Lett.* **5**, 437; **6**, 475 (1988).
44. M. Doi, W. W. Graessley, E. Helfand, and D. S. Pearson *Macromolecules* **20**, 1900 (1987).
45. G. Marrucci, *A. I. Ch. E. Journal* **35**, 1399 (1989); R. B. Bird, C. F. Curtiss, R. C. Armstrong, and O. Hassager, *A. I. Ch. E. Journal* **35**, 1400 (1989); T. C. B. McLeish, *J. Non-Newtonian Fluid Mech.* **33**, 107 (1989); R. B. Bird, C. F. Curtiss, R. C. Armstrong, and O. Hassager, *J. Non-Newtonian Fluid Mech.* **33**, 111 (1989).
46. R. B. Bird, H. H. Saab, and C. F. Curtiss, *J. Chem. Phys.* **77**, 4747 (1982); H. H. Saab, R. B. Bird, and C. F. Curtiss, *J. Chem. Phys.* **77**, 4758 (1982); R. B. Bird, H. H. Saab, and C. F. Curtiss, *J. Phys. Chem.* **86**, 1102 (1982); J. D. Schieber, *J. Chem. Phys.* **87**, 4917; 4928 (1987).
47. W. W. Graessley, *Adv. Polym. Sci.* **16**, 1 (1974).
48. K. L. Ngai, A. K. Rajagopal, and S. Teitler, *J. Chem. Phys.* **88**, 5086 (1988).
49. R. W. Rendell, K. L. Ngai, and G. G. McKenna, *Macromolecules* **20**, 2250 (1987).
50. G. B. McKenna, K. L Ngai, and D. J. Plazek, *Polymer* **26**, 1651 (1985).
51. W. W. Graessley, personal communication.
52. C. R. Bartels, B. Crist, and W. W. Graessley, *Macromolecules* **17**, 2702 (1984).
53. K. L. Ngai and T. P. Lodge, manuscript in preparation.
54. T. P. Lodge, "Techniques for Measuring Polymer Diffusion," in *Applied Polymer Analysis and Characterization*, vol. 2, J. Mitchell, Jr., Ed., Hanser, Munich, W. Germany, in press.
55. D. A. Bernard and J. Noolandi, *Macromolecules* **16**, 548 (1983).
56. W. W. Graessley, *J. Polym. Sci., Polym. Phys. Ed.* **18**, 27 (1980).

57. F. Bueche, *J. Chem. Phys.* **48**, 1410 (1968).
58. Y. Kumagai, H. Watanabe, K. Miyasaka, and T. Hata, *J. Chem. Eng. Jpn.* **12**, 1 (1979).
59. R. Kimmich and R. Bachus, *Coll. & Polym. Sci.* **260**, 911 (1982).
60. R. Bachus and R. Kimmich, *Polymer* **24**, 964 (1983).
61. P. F. Green, P. J. Mills, C. J. Palmstrom, J. W. Mayer, and E. J. Kramer, *Phys. Rev. Lett.* **53**, 2145 (1984).
62. P. J. Mills, P. F. Green, C, J. Palmstrom. J. W. Mayer, and E. J. Kramer, *Appl. Phys. Lett.* **45**, 958 (1984).
63. P. F. Green, C. J. Palmstrom, J. W. Mayer, and E. J. Kramer, *Macromolecules* **18**, 501 (1985).
64. P. F. Green, P. J. Mills, and E. J. Kramer, *Polymer* **27**, 1063 (1986).
65. P. F. Green and E. J. Kramer, *Macromolecules* **19**, 1108 (1986).
66. P. F. Green and E. J. Kramer, *J. Mater. Res.* **1**, 202 (1986).
67. M. Antonietti, J. Coutandin, R. Grutter, and H. Sillescu, *Macromolecules* **17**, 798 (1984).
68. M. Antonietti, J. Coutandin, and H. Sillescu, *Macromolecules* **19**, 793 (1986).
69. M. Antonietti, K. J. Folsch, and H. Sillescu, *Makromol. Chem.* **188**, 2217 (1987).
70. G. Fleischer, *Makromol. Chem., Rapid Commun.* **6**, 403 (1985).
71. G. Fleischer, *Coll. & Polym. Sci* **265**, 89 (1987).
72. G. Fleischer, *Polym. Bull.* **9**, 152 (1983).
73. G. Fleischer, *Polym. Bull.* **11**, 75 (1984).
74. D. W. McCall, D. C. Douglass, and E. W. Anderson, *J. Chem. Phys.* **30**, 771 (1959).
75. J. Klein and B. J. Briscoe, *Proc. R. Soc. Lond. A.* **365**, 53 (1979).
76. J. Klein, D. Fletcher, and L. J. Fetters, *Nature* **304**, 526 (1983).
77. A. Peterlin, *Makromol. Chem.* **184**, 2377 (1983).
78. I. Zupancic, G. Lahajnar, R. Blinc, D. H. Reneker, and D. L. Vanderhart, *J. Polym. Sci., Polym. Phys. Ed.* **23**, 387 (1985).
79. D. S. Pearson, G. Ver Strate, E. von Meerwall, and F. C. Schilling, *Macromolecules* **20**, 1133 (1987).
80. H. Watanabe and T. Kotaka, *Macromolecules* **20**, 530 (1987).
81. M. R. Landry, Ph. D. Thesis, University of Wisconsin, 1985.
82. N. Nemoto, M. Moriwaki, H. Odani, and M. Kurata, *Macromolecules* **4**, 215 (1971).
83. G. V. Vinogradov, A. Y. Malkin, Y. G. Yanovskii, E. K. Borisenkova, B. V. Yarlinkov, and G. V. Berezhnaya, *J. Polym. Sci. A2* **10**, 1061 (1972).
84. A. Rudin and K. K. Chee, *Macromolecules* **6**, 613 (1973).
85. T. Fujimoto, N. Ozaki, and M. Nagasawa, *J. Polym. Sci. A2* **6**, 129 (1974).
86. V. R. Raju, G. G. Smith, G. Marin, J. R. Knox, and W. W. Graessley, *J. Polym. Sci., Polym. Phys. Ed.* **17**, 1183 (1979).
87. V. R. Raju, H. Rachapudy, and W. W. Graessley, *J. Polym. Sci., Polym. Phys. Ed.* **17**, 1223 (1979).
88. J. T. Gotro and W. W. Graessley, *Macromolecules* **17**, 2767 (1984).
89. M. J. Struglinski and W. W. Graessley, *Macromolecules* **18**, 2630 (1986).
90. G. B. McKenna, G Hadziiaounnou, P. Lutz, G. Hild, C. Strazielle, C. Straupe, P. Rempp, and A. J. Kovacs, *Macromolecules* **20**, 498 (1987).
91. R. H. Colby, L. J. Fetters, and W. W. Graessley, *Macromolecules* **20**, 2226 (1987).
92. W. W. Graessley and S. F. Edwards, *Polymer* **22**, 1329 (1981).
93. K. Osaki, K. Nishizawa, and M. Kurata, *Macromolecules* **15**, 1068 (1982).
94. Y.-H. Lin, *Macromolecules* **17**, 2846 (1984); **19**, 159; 168 (1986).
95. M. Rubinstein, E. Helfand, and D. S. Pearson, *Macromolecules* **20**, 822 (1987).
96. M. Rubinstein and R. H. Colby, *J. Chem. Phys.* **89**, 5291 (1988).
97. H. Janeschitz-Kriegl, Polymer Melt Rheology and Flow Birefringence, Springer, Berlin, W. Germany, 1983.

98. J. A. Kornfield, G. G. Fuller, and D. S. Pearson, *Macromolecules* **22**, 1334 (1989).
99. W. W. Merrill, M. Tirrell, J. F. Tassin, and L. Monnerie, *Macromolecules* **22**, 896 (1989).
100. M. Doi, D. S. Pearson, J. A. Kornfield, and G. G. Fuller, *Macromolecules* **22**, 1488 (1989).
101. P. G. de Gennes, *J Phys. (Paris)* **36**, 1199 (1975).
102. M. Doi and N. Y. Kuzuu, *J. Polym. Sci., Polym. Lett. Ed.* **18**, 775 (1980).
103. J. Klein, *Macromolecules* **19**, 105 (1086).
104. C. R. Bartels, B. Crist, Jr., L. J. Fetters, and W. W. Graessley, *Macromolecules* **19**, 785 (1986).
105. D. S. Pearson and E. Helfand, *Macromolecules* **17**, 888 (1984).
106. V. C. Long, G. C. Berry, and L. M. Hobbs, *Polymer* **5**, 517 (1964).
107. H. C. Kan, J. D. Ferry, and L. J. Fetters, *Macromolecules* **13**, 1571 (1981).
108. M. Antonietti and H. Sillescu, *Macromolecules* **19**, 798 (1986).
109. K. R. Shull, E. J. Kramer, G. Hadziiaounnou, M. Antonietti, and H. Sillescu, *Macromolecules* **21**, 2578 (1988).
110. W. W. Graessley, *Acct. Chem. Res.* **10**, 332 (1977).
111. D. S. Pearson and L. J. Fetters, manuscript in preparation.
112. R. J. Needs and S. F. Edwards, *Macromolecules* **16**, 1492 (1983).
113. R. C. Ball and T. C. B. McLeish, *Macromolecules* **22**, 1911 (1989).
114. P. Lutz, G. B. McKenna, P. Rempp, and C. Strazielle, *Makromol. Chem., Rapid Commun.* **7**, 599 (1986).
115. J. Roovers *Macromolecules* **18**, 1359 (1985).
116. G. B. McKenna, B. J. Hostetter, N. Hadjichristidis, L. J. Fetters, and D. J. Plazek, *Macromolecules* **22**, 1834 (1989).
117. J. Roovers, *Macromolecules* **21**, 1517 (1988).
118. P. J. Mills, J. E. Mayer, E J. Kramer, G. Hadziiaounnou, P. Lutz, C. Strazielle, P. Rempp, and A. J. Kovacs, *Macromolecules* **20**, 513 (1987).
119. A. N. Gent and S. Y. Kaang, *J. Polym. Sci., Polym. Phys. Ed.* **27**, 893 (1989).
120. E. von Meerwall, J. Grigsby, D. Tomich, and R. Van Antwerp, *J. Polym. Sci., Polym. Phys. Ed.* **20**, 1037 (1982).
121. A. N. Gent and R. H. Tobias, *J. Polym. Sci., Polym. Phys. Ed.* **20**, 2317 (1982).
122. L. Garrido, J. E. Mark, S. J. Clarson, and J. A. Semlyen, *Polym. Commun.* **25**, 218 (1984); L. Garrido and J. E. Mark, *J. Polym. Sci., Polym. Phys. Ed.* **23**, 1933 (1985); L. Garrido, J. E. Mark, J. L. Ackerman, and R. A. Kinsey, *J. Polym. Sci., Polym. Phys. Ed.* **26**, 2367 (1988).
123. M. Antonietti and H. Sillescu, *Macromolecules* **18**, 1162 (1985).
124. O. Kramer, R. Greco, R. A. Neira, and J. D. Ferry, *J. Polym. Sci., Polym. Phys. Ed.* **12**, 2361 (1974); O. Kramer, R. Greco, J. D. Ferry, and E. T. McDonel, *J. Polym. Sci., Polym. Phys. Ed.* **13**, 1675 (1975).
125. G. W. Nelb, S. Pedersen, C. R. Taylor, and J. D. Ferry, *J. Polym. Sci., Polym. Phys. Ed.* **18**, 645 (1980); C. R. Taylor, H. C. Kan, G. W. Nelb, and J. D. Ferry, *J. Rheol.* **25**, 507 (1981).
126. S. Granick, S. Pedersen, G. W. Nelb, J. D. Ferry, and C. W. Macosko, *J. Polym. Sci., Polym. Phys. Ed.* **19**, 1745 (1981).
127. G. W. Kamykowski, J. D. Ferry, and L. J. Fetters, *J. Polym. Sci., Polym. Phys. Ed.* **20**, 2125 (1982).
128. G. Kraus and K. W. Rollmann, *J. Polym. Sci., Polym. Phys. Ed.* **15**, 385 (1977).
129. B. T. Poh, K. Adachi, and T. Kotaka, *Macromolecules* **20**, 2563; 2569; 2574 (1987).
130. K. Adachi, T. Nakamoto, and T. Kotaka, *Macromolecules* **22**, 3106; 3111 (1989).
131. M. Doi, *Polymer J.* **5**, 288 (1973).
132. K. E. Evans and S. F. Edwards, *J. Chem. Soc., Faraday Trans. 2* **77**, 1891; 1921; 1929 (1981).
133. A. Baumgärtner and K. Binder, *J. Chem. Phys.* **75**, 2994 (1981).
134. D. Richter, A. Baumgärtner, K. Binder, B. Ewen, and J. B. Hayter, *Phys. Rev. Lett.* **47**, 109 (1981).

135. A. Baumgärtner, in *Polymer Motion in Dense Systems*, D. Richter and T. Springer, Eds., Springer, Berlin, W. Germany, 1987.
136. A. Kolinski, J. Skolnick, and R. Yaris, *J. Chem. Phys.* **86**, 1567 (1987).
137. A. Baumgärtner and M. Muthukumar, *J. Chem. Phys.* **87**, 3082 (1987).
138. M. Muthukumar and A. Baumgärtner, *Macromolecules* **22**, 1937; 1941 (1989).
139. A. Kolinski, J. Skolnick, and R. Yaris, *J. Chem. Phys.* **84**, 1922 (1986).
140. A. Kolinski, J. Skolnick, and R. Yaris, *J. Chem. Phys.* **86**, 7164 (1987).
141. A. Kolinski, J. Skolnick, and R. Yaris, *J. Chem. Phys.* **86**, 7174 (1987).
142. K. Kremer, G. S. Grest, and I. Carmesin, *Phys. Rev. Lett.* **61**, 566 (1988).
143. K. Kremer and G. S. Grest, *J. Chem. Phys.* **92**, 5057 (1990).
144. P. G. de Gennes, *Scaling Concepts in Polymer Physics*, Cornell Univ. Press, Ithaca, NY, 1979.
145. P. G. de Gennes, *Physics* **3**, 37 (1967); E. Dubois-Violette and P. G. de Gennes, *Physics* **3**, 181 (1967).
146. M. Adam and M. Delsanti, *Macromolecules* **10**, 1229 (1977); T. Nose and B. Chu, *Macromolecules* **12**, 1122 (1979); C. C. Han and A. Z. Akcasu, *Macromolecules* **14**, 1080 (1981).
147. V. F Man, Ph. D. Thesis, University of Wisconsin, 1984; V. F. Man, J. L. Schrag, and T. P. Lodge, *Macromolecules*, submitted.
148. F. W. Wang, *Macromolecules* **8**, 364 (1975); **11**, 1198 (1978).
149. P. G. de Gennes, *J. Phys. (Paris).* **42**, 735 (1981).
150. K. Kremer and K. Binder, *J. Chem. Phys.* **81**, 6381 (1984).
151. D. Richter, B. Ewen, B. Farago, and T. Wagner, *Phys. Rev. Lett.* **62**, 2140 (1989).
152. J. S. Higgins and J. E. Roots, *J. Chem. Soc., Faraday Trans. 2*, **81**, 757 (1985).
153. D. Richter, B. Farago, C. Lartigue, L. J. Fetters, J. S. Huang, and B. Ewen, *Phys. Rev. Lett.*, in press.
154. K. Osaki, E. Takatori, M. Kurata, H. Ohnuma, and T. Kotaka, *Polym. J. (Jpn)* **18**, 947 (1986).
155. B. D. Lawrey, R. K. Prud'homme, and J. T. Koberstein, *J. Polym. Sci., Polym. Phys. Ed.* **24**, 203 (1984).
156. D. S. Pearson, personal communication.
157. K. Osaki, E. Takatori, M. Ueda, M. Kurata, T. Kotaka, and H. Ohmuna, *Macromolecules* **22**, 2457 (1989).
158. A. Lee and R. P. Wool, *Macromolecules* **19**, 1063 (1986); **20**, 1924 (1987).
159. J. F. Tassin, L. Monnerie, and L. J. Fetters, *Polym. Bull.* **15**, 165 (1986).
160. J. F. Tassin and L. Monnerie, *Macromolecules* **21**, 1846 (1988); J. F. Tassin, L. Monnerie, and L. J. Fetters, *Macromolecules* **21**, 2404 (1988); C. W. Lantman, J. F. Tassin, L. Monnerie, L. J. Fetters, E. Helfand, and D. S. Pearson, *Macromolecules* **22**, 1184 (1989).
161. F. Boué, M. Nierlich, G. Jannink, and R. Ball, *J. Phys. (Paris)* **43**, 137 (1982); F. Boué, K. Osaki, and R. C. Ball, *J. Polym. Sci., Polym. Phys. Ed.* **23**, 833 (1985).
162. K. Kremer, *Macromolecules* **16**, 1632 (1983).
163. A. Baumgärtner, *Ann. Rev. Phys. Chem.* **35**, 419 (1984).
164. T. Pakula, *Macromolecules* **20**, 679 (1987); T. Pakula and S. Geyler, *Macromolecules* **20**, 2909 (1987); **21**, 1665 (1988).
165. Y. Takahashi, I. Noda, and M. Nagasawa, *Macromolecules* **18**, 2220 (1985).
166. L. Leger, H. Hervet, and F. Rondelez, *Macromolecules* **14**, 1732 (1981); H. Deschamps and L. Leger, *Macromolecules* **19**, 2760 (1986); M. F. Marmonier and L. Leger, *Phys. Rev. Lett.* **55**, 1078 (1985).
167. P. T. Callaghan and D. N. Pinder, *Macromolecules* **13**, 1085 (1980); **14**, 1334 (1981); **17**, 431 (1982).
168. J. A. Wesson, I. Noh, T. Kitano, and H. Yu, *Macromolecules* **17**, 431 (1984); H. Kim, T. Chang, J. M. Yohanan, L. Wang, and H. Yu, *Macromolecules* **19**, 2737 (1986).
169. E. D. von Meerwall, E. J. Amis, and J. D. Ferry, *Macromolecules* **18**, 260 (1985).

170. N. Numasawa, T. Hamada, and T. Nose, *J. Polym. Sci., Polym. Phys. Ed.* **24**, 19 (1986); N. Numasawa, K. Kuwamoto, and T. Nose, *Macromolecules* **19**, 2593 (1986).

171. N. Nemoto, S. Okada, T. Inoue, and M. Kurata, *Macromolecules* **21**, 1502; 1509 (1988).

172. L. M. Wheeler, T. P. Lodge, B. Hanley, and M. Tirrell, *Macromolecules* **20**, 1120 (1987); L. M. Wheeler and T. P. Lodge, *Macromolecules* **22**, 3399 (1989).

173. V. D. Skirda, V. I. Sundukov, A. I. Maklakov, O. E. Zgadzai, I. R. Gafurov, and G. I. Vasiljev, *Polymer*, **29**, 1294 (1988).

174. T. P. Lodge, P. Markland, and L. M. Wheeler, *Macromolecules* **22**, 3409 (1989).

175. G. D. J. Phillies, *Macromolecules* **19**, 2367 (1986); **20**, 558 (1987); G. D. J. Phillies and P. Peczak, *Macromolecules* **21**, 214 (1988).

176. N. Nemoto, T. Kojima, T. Inoue, M. Kishine, T. Hirayama, and M. Kurata, *Macromolecules* **22**, 3793 (1989).

177. W. W. Graessley, in *Physical Properties of Polymers*, Eds., J. E. Mark, A. Eisenberg, W. W. Graessley, L. Mandelkern, and J. L. Koenig, ACS Publ., Washington, DC, 1984.

178. D. S. Pearson, *Rubber Chem. Tech.* **60**, 439 (1987).

179. M. Adam and M. Delsanti, *J. Phys. (Paris)* **44**, 1185 (1983); **45**, 1513 (1984); Y. Takahashi, Y. Isono, I. Noda, and M. Nagasawa, *Macromolecules* **18**, 1002 (1985).

180. F. C. Frank, *Faraday Disc. Chem. Soc.* **68**, 7 (1979).

181. C. M. Guttman, J. D. Hoffman, and E. A. DiMarzio, *Faraday Disc. Chem. Soc.* **68**, 297 (1979).

182. J. D. Hoffman, C. M. Guttman, and E. A. DiMarzio, *Faraday Disc. Chem. Soc.* **68**, 177 (1979); J. Klein and E. Ball, *Faraday Disc. Chem. Soc.* **68**, 198 (1979); J. D. Hoffman, *Polymer* **23**, 656 (1982).

183. J. D. Hoffman and R. L. Miller, *Macromolecules* **21**, 3083 (1988).

184. L. S. Lerman and H. C. Frisch, *Biopolymers* **21**, 955 (1982).

185. O. J. Lumpkin and B. H. Zimm, *Biopolymers* **21**, 315 (1982); O. J. Lumpkin, P. Dejardin, and B. H. Zimm, *Biopolymers* **24**, 1573 (1985).

186. G. W. Slater and J. Noolandi, *Biopolymers* **25**, 431 (1986).

187. D. C. Schwartz and C. R. Cantor, *Cell* **37**, 67 (1984).

188. D. C. Schwartz and M. Koval, *Nature* **338**, 520 (1989).

189. J. M. Deutsch and T. L. Madden, *J. Chem. Phys.* **90**, 2476 (1989).

190. M. Lalande, J. Noolandi, C. Turmel, J. Rousseau, and G. W. Slater, *Proc. Nat. Acad. Sci. USA* **84**, 8011 (1987).

191. J. L. Viovy, *Phys. Rev. Lett.* **60**, 855 (1988).

192. H. H. Kausch, *Polymer Fracture*, 2nd ed., Springer, Berlin, W. Germany, 1987.

193. H. H. Kausch and M. Tirrell, *Ann. Rev. Mater. Sci.* **19**, 341 (1989).

194. S. Prager and M. Tirrell, *J. Chem. Phys.* **75**, 5194 (1981).

195. K. Jud, H. H. Kausch, and J. G. Williams, *J. Mater. Sci.* **16**, 204 (1981).

196. I. Mita and K. Horie, *Rev. Macromol. Chem. Phys.* **C27**, 91 (1987).

197. T. J. Tulig and M. Tirrell, *Macromolecules* **14**, 1501 (1981).

198. P. G. de Gennes, *J. Chem. Phys.* **76**, 3316; 3322 (1982).

199. D. A. Bernard and J. Noolandi, *Phys. Rev. Lett.* **50**, 253 (1983).

200. M. E. Cates, *Macromolecules* **20**, 2289, (1987).

201. A. S. Lodge, *Rheol. Acta* **28**, 351 (1989).

202. K. L. Ngai and D. J. Plazek, *J. Polym. Sci., Polym. Phys. Ed.* **23**, 2159 (1985); **24**, 619 (1986).

ULTRAFAST PROCESSES AND TRANSITION-STATE SPECTROSCOPY

S. H. LIN, B. FAIN*, and N. HAMER†

*Department of Chemistry
and
Center for the Study of Early Events in Photosynthesis,
Arizona State University, Tempe, AZ 85287-1604*

CONTENTS

* *Permanent address:* School of Chemistry, Tel-Aviv University, Ramat Aviv, 69978, Tel-Aviv, ISRAEL
† *Permanent address:* Group A-6, Mail Stop: M997, Los Alamos National Laboratory, Los Alamos, New Mexico 87545

Advances in Chemical Physics, Volume LXXIX, Edited by I. Prigogine and Stuart A. Rice.
ISBN 0-471-52768-8   © 1990 John Wiley & Sons, Inc.

INTRODUCTION

An accurate interpretation of the spectra of short-lived species requires a careful analysis of the detailed dynamics of the absorption of radiation by the molecular system. The usual picture of instantaneous photon absorption is incomplete whenever there are molecular processes competing on the time scale of the initial excitation processes. For example, if intramolecular relaxation or vibrational predissociation are fast compared to the initial excitation time, then the initially prepared state might be poorly characterized; interpretation of the associated spectrum is thus clouded. In particular, the precise nature of the transition that produced the absorption or photochemical event is no longer obvious. In this chapter we discuss the theoretical treatment of three commonly used experimental techniques for studying ultrafast processes, that is, pump-probe experiments, time-resolved emission spectroscopy, and impulsive Raman scattering. We shall restrict ourselves to the quantum theoretical treatment of these experiments (especially to the density matrix method).

Optical absorption of an ultrashort pulse into an electronic state which is unstable with respect to dissociation (or other chemical change) may result in coherent wave-packet propagation from the initially excited geometry to the photodissociated state. This type of phase-coherent chemistry can be time-resolved to reveal dissociation dynamics and information about the structures of the unstable species formed between reactant and product. In some cases, time-resolved observation of coherent wave-packet propagation over a local potential maximum or saddle point and characterization of the corresponding activated complex (i.e., transition state) should prove possible. The influence of the environment, such as a solvent, on transition-state structure and evolution could also be explored.

Recently there has been much interest in probing the dynamics and potential energy surfaces of chemically reactive species. For small molecules in the gas phase, dissociation dynamics may be determined uniquely by the potential energy surfaces of the lowest energy electronic states. In condensed phases, interactions between the reacting species and its environment cause dissipation and dephasing, which destroy the one-to-one correspondence between reactive potential energy surface and reaction dynamics. Recent advances in both time- and frequency-domain spectroscopies have raised hopes for rather complete experimental determination of potential energy surfaces and molecular dynamics. Femtosecond spectroscopy techniques have been used to make direct time-resolved observations of coherent wave-packet propagation. Individual oscillation cycles of phase-coherent molecular vibrational motion have been resolved, as have chemical bond breakage and formation events, which in the simplest cases may be thought of as half-cycles of oscillation.

The time-domain experiments are carried out with pulses whose duration is short compared to the time required for any significant nuclear motion in the direction of chemical change. These experiments are facilitated by the fact that phase-coherent nuclear motion (i.e., phase-coherent wave-packet propagation) is initiated whenever a sufficiently short laser pulse passes through most materials. In this chapter we also review the transition-state spectroscopy. Both time-resolved and steady-state transition-state spectra are discussed.

I. ULTRAFAST PROCESSES

A. Pump-Probe Experiments

Conventionally absorption is described by a frequency-dependent absorption coefficient. This coefficient is connected with the imaginary part of the susceptibility in the isotropic case, or with its anti-Hermitian part in the anisotropic case. Thus if $f_a(t)$ is the ath component of the force acting on the system (e.g., an electric field component in the case of interaction with the electric field), then the susceptibility $\chi_{ab}(\omega)$ is defined by the relation

$$\langle x_a(t) \rangle = \sum_b [\chi_{ab}(\omega)f_b(\omega)e^{-i\omega t} + \chi_{ab}^*(\omega)f_b^*(\omega)e^{i\omega t}] \tag{1.1}$$

where $f_b(\omega)$ denotes the complex amplitude of $f_b(t)$. Here x_a is some physical quantity of the system (e.g., the dipole moment or polarization operator for the electromagnetic interaction case), and it is assumed that the external force depends harmonically on time. In the case of an arbitrary dependence on time, a summation or integration over frequencies ω has to be performed. The absorption can in general be characterized by the rate of energy dissipation. Then, assuming that the interaction with the system is described by the interaction energy

$$V = -\sum_a f_a(t)x_a \tag{1.2}$$

one gets for the dissipation of the energy per unit time the following expression[1]

$$Q = i\omega \sum_{a,b} [\chi_{ab}^*(\omega) - \chi_{ba}(\omega)]f_a^*(\omega)f_b(\omega) \tag{1.3}$$

In the isotropic case $\chi_{ab}(\omega)$ is diagonal, and the expression for the energy dissipation takes the form

$$Q = 2\omega \sum_a \chi_{aa}''(\omega)|f_a(\omega)|^2 \tag{1.4}$$

In the quantum case when forces f_a are described by the corresponding operators, expressions (1.3) and (1.4) have been generalized by Fain.[2] It turns out that even in the case where f_a is a nonclassical quantum quantity, the dissipation can still be described by the frequency-dependent susceptibility $\chi_{ab}(\omega)$.

The basic underlying assumption of the linear response theory is the stationarity of the system (medium). Recent development of the femtosecond (fs) technique has made it necessary to revise this assumption. The aim of this section is the generalization of the conventional theory in order to describe real-time femtosecond experiments.[3-6]

1. Generalized Linear Response Theory

Considering the general linear response theory, we have in mind femtosecond experiments in which a laser pulse, lasting about 10–100 fs, excites the system to certain electronic states while a second probe femtosecond laser pulse is used to measure absorption from the excited state to another electronic state. Since the duration of the first pulse, 10^{-14}–10^{-13} s, is of the same order or shorter than a typical vibration period, it creates a nonstationary wave packet. Therefore the probe pulse measures the absorption in the nonstationary system.

To find a linear response in this case, we start from the von Neumann equation in the interaction representation (see Fig. 1.1)

$$i\hbar \frac{\partial \rho}{\partial t} = [V(t), \rho] \tag{1.5}$$

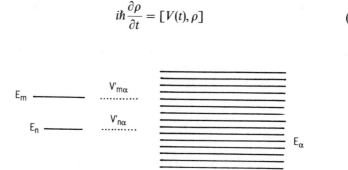

Figure 1.1.   Energy-level scheme for coupling between discrete states and a continuum.

where V is given by Eq. (1.2) and ρ is the density matrix of the system subjected to the external forces $f_a(t)$. A mean value of some operator O of the system takes the form

$$\langle O(t) \rangle = \mathrm{Tr}(\rho O(t)) = \mathrm{Tr}(\rho_i O(t)) + \frac{1}{i\hbar} \int_{t_i}^{t} \mathrm{Tr}(\rho_i [O(t), V(t_1)]) \, dt_1 \quad (1.6)$$

We choose the operator O to be the operator x_a (it can be the dipole moment of the molecule) whose mean value $\mathrm{Tr}(\rho_i x_a)$ is zero or lies in the low-frequency range and does not affect further results. In this case we get

$$\langle x_a(t) \rangle = \frac{i}{\hbar} \int_{t_i}^{t} \sum_b \langle [x_a(t), x_b(t_1)] \rangle f_b(t_1) \, dt_1 \quad (1.7)$$

where t_i denotes the time of creation of the coherent state ρ_i by the initial laser pulse. It should be stressed that averaging $\langle \cdots \rangle$ in the right-hand side of Eq. (1.7) is performed over the unperturbed state ρ_i, and time dependence $[x_a(t), x_b(t_1)]$ in the right-hand side is determined by the unperturbed Hamiltonian [without the forces $f_a(t)$].

Introducing the response function

$$\phi_{ab}(t, t_1) = \frac{i}{\hbar} \langle [x_a(t), x_b(t_1)] \rangle \quad (1.8)$$

we can rewrite Eq. (1.7) as

$$\langle x_a(t) \rangle = \sum_b \int_{t_i}^{t} \phi_{ab}(t, t_1) f_b(t_1) \, dt_1 \quad (1.9)$$

In the frequency representation we expand $f_b(t_1)$ into the Fourier integral

$$f_b(t_1) = \int f_b(\omega_1) e^{-i\omega_1 t_1} \, d\omega_1 \quad (1.10)$$

and obtain

$$\langle x_a(t) \rangle = \sum_b \iint \chi_{ab}(\omega, \omega_1) e^{-i\omega t} f_b(\omega_1) \, d\omega_1 \, d\omega \quad (1.11)$$

or

$$\langle x_a(\omega_1) \rangle = \sum_b \int \chi_{ab}(\omega, \omega_1) f_b(\omega_1) \, d\omega_1 \qquad (1.12)$$

Here

$$\chi_{ab}(\omega, \omega_1) = \frac{1}{2\pi} \int_{-\infty}^{\infty} e^{i\omega t} \, dt \int_{t_i}^{t} \phi_{ab}(t, t_1) e^{-i\omega_1 t_1} \, dt_1$$

$$= \frac{1}{2\pi} \int_{-\infty}^{\infty} e^{i(\omega - \omega_1)t} \, dt \int_{0}^{t-t_i} \phi_{ab}(t, t - \tau) e^{i\omega_1 \tau} \, d\tau \qquad (1.13)$$

Thus $\phi_{ab}(t, t_1)$ may be called the two-time, or time–time, representation of the response function; $\chi_{ab}(\omega, \omega_1)$ may be called the two-frequency, or frequency–frequency, representation of the susceptibility. It is important to notice that in the nonstationary case, the two times t and t_1 and the two frequencies ω and ω_1 are independent variables.

Now, we also introduce the time–frequency representation of the susceptibility. Having in mind the femtosecond experiments mentioned, we will be more specific about the time dependence of the laser pulse. We assume that the time dependence of the probe laser pulse may be presented in the form

$$f_a(t_1) = [\tilde{f}_a(\omega) e^{-i\omega t_1} + \tilde{f}_a(-\omega) e^{i\omega t_1}] L(t_1) \qquad (1.14)$$

Here $L(t_1)$ is the dimensionless shape function. The latter is defined to be equal to 1 at the maximum $t_1 = t_p$, the probing time, and decaying when

$$|t_1 - t_p| \gtrsim T \qquad (1.15)$$

Here T may be called the (effective) time of pulse duration. We can present here two simple forms of $L(t_1)$

$$L(t_1) = \begin{cases} 1, & -\dfrac{T}{2} \leq t_1 - t_p \leq \dfrac{T}{2} \\[2mm] 0, & |t_1 - t_p| > \dfrac{T}{2} \end{cases} \qquad (1.16)$$

or

$$L(t_1) = e^{-2|t_1 - t_p|/T} \qquad (1.17)$$

Notice that the integral of both functions is

$$\int_{-\infty}^{\infty} L(t_1)\, dt_1 = T \tag{1.18}$$

This may serve as a justification of calling T the pulse duration time. Substituting Eq. (1.10) into Eq. (1.9) we obtain time–frequency representation of the susceptibility

$$\langle x_a(t)\rangle = \sum_b \int \chi_{ab}(\omega, t) f_b(\omega) e^{-i\omega t}\, d\omega \tag{1.19}$$

where

$$\chi_{ab}(\omega, t) = \int_0^{t - t_i} \phi_{ab}(t, t - \tau) e^{i\omega\tau}\, d\tau \tag{1.20}$$

Notice that there is a relation between $\chi_{ab}(\omega, \omega_1)$ and $\chi_{ab}(\omega, t)$,

$$\chi_{ab}(\omega, \omega_1) = \frac{1}{2\pi} \int_{-\infty}^{\infty} e^{i(\omega - \omega_1)t} \chi_{ab}(\omega_1, t)\, dt \tag{1.21}$$

It may also be convenient to use another definition of time–frequency susceptibility, defining it as a response to the amplitudes $\tilde{f}_a(\omega)$ given by Eq. (1.14),

$$\langle x_a(t)\rangle = \sum_b \tilde{\chi}_{ab}(\bar{\omega}, t) \tilde{f}_b(\bar{\omega}) e^{-i\bar{\omega} t} + \text{c.c.} \tag{1.22}$$

where

$$\tilde{\chi}_{ab}(\bar{\omega}, t) = \int_0^{t - t_i} \phi_{ab}(t, t - \tau) L(t - \tau) e^{i\bar{\omega}\tau}\, d\tau \tag{1.23}$$

Strictly speaking $\tilde{\chi}_{ab}(\omega, t)$ cannot be considered as the susceptibility characterizing property of the matter only. It contains in its definition, Eq. (1.23), the property of the electromagnetic field: the laser pulse shape $L(t)$. However, $\tilde{\chi}_{ab}(\omega, t)$ is useful because it is a directly observable property. There exists a relation between $\tilde{\chi}_{ab}(\omega, t)$ and $\chi_{ab}(\omega, t)$ following from Eqs. (1.23) and (1.20),

$$\tilde{\chi}_{ab}(\bar{\omega}, t) = \int_{-\infty}^{\infty} \chi_{ab}(\omega, t) \hat{L}(\omega - \bar{\omega}) e^{-i(\omega - \bar{\omega})t}\, d\omega \tag{1.24}$$

where $\hat{L}(\omega)$ is a Fourier transform of $L(t)$. Notice that amplitudes $f_a(\omega)$ in Eq.

(1.19) and $\tilde{f}_a(\bar{\omega})$ in Eq. (1.22) are connected by a relation following from Eq. (1.14),

$$f_a(\omega) = \frac{1}{2\pi} \int_{-\infty}^{\infty} e^{i\omega t} f_a(t)\, dt = \tilde{f}_a(\bar{\omega})\hat{L}(\omega - \bar{\omega}) + \tilde{f}_a(-\bar{\omega})\hat{L}(\omega + \bar{\omega}) \quad (1.25)$$

Similar to expression (1.3) there exist relations connecting susceptibilities $\chi_{ab}(\omega, \omega_1)$, $\chi_{ab}(\omega, t)$, and $\tilde{\chi}_{ab}(\omega, t)$ with the dissipation of the energy per unit time Q. The derivation of these relations is given in Appendix A. Here are the results. The relation between Q and $\tilde{\chi}_{ab}(\omega, t)$ has the form

$$Q = i\omega \sum_{a,b} [\tilde{\chi}_{ab}^*(\omega, t) - \tilde{\chi}_{ab}(\omega, t)]\tilde{f}_a(\omega)\tilde{f}_b^*(\omega)L(t) \quad (1.26)$$

and in the isotropic case

$$Q = 2\omega\tilde{\chi}''(\omega, t)|\tilde{f}(\omega)|^2 L(t) \quad (1.27)$$

where

$$|\tilde{f}(\omega)|^2 = \sum_a |\tilde{f}_a(\omega)|^2 \quad (1.28)$$

The two-frequency susceptibility $\chi_{ab}(\omega_1, \omega_2)$ determines the Fourier transform of the time-dependent rate of the dissipation energy,

$$\begin{aligned}
\hat{Q}(\Omega) &= \frac{1}{2\pi} \int_{-\infty}^{\infty} Q(t)e^{i\Omega t}\, dt \\
&= i \sum_{a,b} \int\int \omega_1 \chi_{ab}(\Omega - \omega_1, \omega_2) f_a(\omega_a) f_b(\omega_2)\, d\omega_2\, d\omega_1 \quad (1.29)
\end{aligned}$$

For the monochromatic probe pulse, $\hat{L}(\omega) = \delta(\omega)$, the Fourier transform of the energy dissipation rate takes the form

$$\hat{Q}(\Omega) = i\omega \sum_{a,b} [\chi_{ab}(\Omega - \omega, -\omega) - \chi_{ba}(\Omega + \omega, \omega)] f_a(\omega) f_b(-\omega) \quad (1.30)$$

while $Q(t)$ can be expressed through $\chi_{ab}(\omega, t)$ as

$$Q(t) = \sum_{a,b} [\chi_{ab}^*(\omega, t) - \chi_{ba}(\omega, t)] f_a(\omega) f_b(-\omega) \quad (1.31)$$

Generalized susceptibilities $\chi_{ab}(\omega_1, \omega_2)$, $\chi_{ab}(\omega, t)$, and $\tilde{\chi}_{ab}(\omega, t)$ can be expressed through the density matrix ρ_{mn} of the unperturbed (without the external field) system (Appendix A).

2. Nonstationary Absorption in the Born–Oppenheimer Approximation

We now analyze the expression for $\tilde{\chi}_{ab}(\omega, t)$ or $\chi(\omega, t)$ given by Eq. (1.23) in the Born–Oppenheimer (or adiabatic) approximation,

$$\psi_{\alpha v}(q, \xi) = \Phi_\alpha(q, \xi)\Theta_{\alpha v}(\xi) \tag{1.32}$$

where the molecular wave function $\psi_{\alpha v}$ is written as a product of electronic wave function Φ_α and nuclear wave function $\Theta_{\alpha v}$.

Using Eq. (1.32) we obtain

$$\langle [x_a(t), x_b(t - \tau)] \rangle = \sum_{vv'} \sum_{\beta u} (\rho_i)_{\alpha v, \alpha v'} e^{-it\omega_{\alpha v, \alpha v'}} [(x_a)_{\alpha v', \beta u}(x_b)_{\beta u, \alpha v} e^{-i\tau\omega_{\beta u, \alpha v}}$$
$$- (x_b)_{\alpha v', \beta u}(x_a)_{\beta u, \alpha v} e^{i\tau\omega_{\beta u, \alpha v'}}] \tag{1.33}$$

Here it is assumed that the pump laser prepares the system coherently in the electronic state α as described by $(\hat{\rho}_i)_{\alpha v, \alpha v'}$. Choosing $L(t)$ as

$$L(t) = \exp\left(-\frac{2}{T}|t - t_p|\right) \tag{1.34}$$

we assume that $L(t)$ has characteristic duration length T [see Eqs. (1.16) and (1.17)], which satisfies the condition

$$\omega_{vv'}^{-1} \gg T \gg \omega_{el}^{-1} \tag{1.35}$$

This means that the duration of the probe laser pulse is much larger than the period of the electronic motion, but much smaller than the period of the vibrational motion. Notice that condition (1.35) does not enable us to neglect τ in the commutator in Eq. (1.23) since it contains time dependence corresponding to both electronic and vibrational motions.

Using Eq. (1.32), the generalized linear susceptibility function becomes

$$\tilde{\chi}_{ab}(\omega, t_p) = \frac{i}{\hbar} \sum_{vv'} \sum_{\beta u} (\rho_0)_{\alpha v, \alpha v'} \exp[-i(t_p - t_i)\omega_{\alpha v, \alpha v'}] \Big\{ (x_a)_{\alpha v', \beta u}(x_b)_{\beta u, \alpha v}$$
$$\times \frac{\exp[(t_p - t_i)(i\omega - 2/T - i\omega_{\beta u, \alpha v})] - 1}{i(\omega - \omega_{\beta u, \alpha v}) - 2/T} - (x_b)_{\alpha v', \beta u}(x_a)_{\beta u, \alpha v}$$
$$\times \frac{\exp[(t_p - t_i)(i\omega - 2/T + i\omega_{\beta u, \alpha v'})] - 1}{i(\omega + \omega_{\beta u, \alpha v'}) - 2/T} \Big\} \tag{1.36}$$

where ρ_0 denotes the original density matrix at $t = t_i$. Here for simplicity we set $t = t_p$, that is, at the probing time. In particular, if $2(t_p - t_i)/T \gg 1$, then Eq. (1.36) reduces to

$$\tilde{\chi}_{ab}(\omega, t_p) = \frac{i}{\hbar} \sum_{vv'} \sum_{\beta u} (\rho_0)_{av, av'} \exp[-i(t_p - t_i)\omega_{av, av'}]$$

$$\times \left[(x_a)_{av', \beta u}(x_b)_{\beta u, av} \frac{i(\omega - \omega_{\beta u, av}) + 2/T}{(\omega - \omega_{\beta u, av})^2 + (2/T)^2} \right.$$

$$\left. - (x_b)_{av', \beta u}(x_a)_{\beta u, av} \frac{i(\omega + \omega_{\beta u, av'}) + 2/T}{(\omega + \omega_{\beta u, av'})^2 + (2/T)^2} \right] \qquad (1.37)$$

The second term on the right-hand side is usually negligible compared with the first term. The conventional expression for $\tilde{\chi}''_{ab}(\omega, t_p)$ is recovered by using the relation $(\rho_0)_{av, av'} = \delta_{vv'}(\rho_0)_{av, av}$ with $(\rho_0)_{av, av}$ being the Maxwell–Boltzmann distribution function,

$$\tilde{\chi}''_{ab}(\omega, t_p) = \frac{1}{\hbar} \sum_{v} \sum_{\beta u} (\rho_0)_{av, av}(x_a)_{av, \beta u}(x_b)_{\beta u, av} \frac{2/T}{(\omega - \omega_{\beta u, av})^2 + (2/T)^2} \qquad (1.38)$$

Notice that the nuclear wave functions $\Theta_{\beta u}$ and Θ_{av} oscillate rapidly except near the classical turning points where $E_{\beta u} - E_{av} = U_\beta(\xi) - U_\alpha(\xi)$, with U_β and U_α representing the potential energy functions of electronic states β and α, respectively. In the rotating wave approximation we obtain

$$\tilde{\chi}_{ab}(\omega, t) = \frac{i}{\hbar} \int_0^{t-t_i} d\tau \sum_v \left\langle \Theta_{av} \left| (\rho_i)_{\alpha\alpha}(x_a)_{\alpha\beta}(x_b)_{\beta\alpha} \right.\right.$$

$$\times \left. \exp\left[\frac{it}{\hbar} \{U_\alpha(\xi) - U_\beta(\xi)\} \right] \right| \Theta_{av} \right\rangle e^{it\omega} L(t - \tau) \qquad (1.39)$$

The vibrational initial state (prepared by the first laser) is assumed to be a certain wave packet. Therefore averaging over vibrational coordinates ξ in $x_a(\xi)$, $x_b(\xi)$, and $\omega_{\alpha\beta}(\xi)$ leads to time dependence of these quantities in the averaged quantities. If we assume that at the probing time

$$t = t_p \qquad (1.40)$$

the probe pulse is resonant with the electronic terms difference

$$\omega = \omega_{\beta\alpha}(\xi) = \frac{1}{\hbar}[U_\beta(\xi) - U_\alpha(\xi)] \qquad (1.41)$$

we get for $\tilde{\chi}_{ab}$ the expression

$$
\tilde{\chi}_{ab}(\omega, t_p) = \frac{i}{\hbar} \sum_v \left\langle \Theta_{av} \left| (\rho_i)_{\alpha\alpha}(x_a)_{\alpha\beta}(x_b)_{\beta\alpha} \right. \right.
$$

$$
\times \left. \frac{\exp\{(t_p - t_i)[i(\omega - \omega_{\beta\alpha}) - 2/T]\} - 1}{i(\omega - \omega_{\beta\alpha}) - 2/T} \right| \Theta_{av} \right\rangle \quad (1.42)
$$

For the case $(2/T)(t_p - t_i) \gg 1$, Eq. (1.42) reduces to

$$
\tilde{\chi}_{ab}(\omega, t_p) = \frac{i}{\hbar} \sum_v \left\langle \Theta_{av} \left| \frac{(\rho_i)_{\alpha\alpha}(x_a)_{\alpha\beta}(x_b)_{\beta\alpha}}{2/T - i(\omega - \omega_{\beta\alpha})} \right| \Theta_{av} \right\rangle \quad (1.43)
$$

For the isotropic case, the imaginary part of $\chi(\omega, t_p)$ is given by

$$
\tilde{\chi}''(\omega, t_p) = \frac{1}{\hbar} \sum_v \left\langle \Theta_{av} \left| \frac{(2/T)(\rho_i)_{\alpha\alpha}(x_a)_{\alpha\beta}(x_a)_{\beta\alpha}}{(2/T)^2 + (\omega - \omega_{\beta\alpha})^2} \right| \Theta_{av} \right\rangle \quad (1.44)
$$

This expression has been postulated by Bersohn and Zewail[6] in their analysis of the experimental femtosecond transition spectra of the photodissociation of ICN.

Only in the case of a very narrow wave packet where one can neglect the dispersion

$$
\langle \xi(t)^2 - \overline{\xi(t)^2} \rangle \ll \overline{\xi(t)^2} \quad (1.45)
$$

is it possible to use

$$
\zeta = \overline{\xi(t)} \quad (1.46)
$$

in Eq. (1.39), that is,

$$
(x_a)_{\alpha\beta} = x_a(\overline{\xi})_{\alpha\beta} \quad (1.47)
$$

3. Discussion

It should be noted that the theoretical results presented in the previous sections are based on the use of molecular eigenstates. However, for a complicated system (or a large molecule) it is usually more convenient to separate the total system into the system (or the degrees of freedom) under observation and the heat bath (or the irrelevant degrees of freedom). The Liouville equation for this type of problem has been derived previously [16,17] and can approximately be expressed as

$$\frac{d\rho_t}{dt} = -iL_0\rho_t - \frac{1}{\hbar}[V,\rho_t] - R\rho_t \tag{1.48}$$

where R represents the damping operator due to the coupling to the heat bath and/or dissociation continuum; it provides the description of both relaxation and/or dissociation and dephasing. In Eq. (1.5) the $iL_0\rho$ term has been eliminated because of the interaction representation.

Using Eq. (1.48) and repeating the derivation as shown in Section I.A.1 we obtain

$$\tilde{\chi}_{ab}(\omega,t) = \frac{i}{\hbar}\int_0^{t-t_i} \text{Tr}\left\{x_a \exp\left[-i\tau(L_0-iR)\right][x_b,\rho_t(t-\tau)]\right\}e^{i\tau\omega}L(t-\tau)\,d\tau \tag{1.49}$$

In the adiabatic approximation (see Section I.A.2) and at $t = t_p$, $\tilde{\chi}_{ab}(\omega,t)$ can be written as

$$\tilde{\chi}_{ab}(\omega,t_p) = \sum_v \sum_{v'} (\rho_0)_{\alpha v, \alpha v'} \exp\left[-i(t_p-t_i)(\omega_{\alpha v,\alpha v'} - iR_{\alpha v,\alpha v'})\right]F_{\alpha v,\alpha v'}(\omega)_{ab} \tag{1.50}$$

and
$$F_{\alpha v,\alpha v'}(\omega)_{ab} = \frac{i}{\hbar}\sum_u \frac{(x_a)_{\alpha v',\beta u}(x_b)_{\beta u,\alpha v}}{(2/T + R_{\beta u,\alpha v'}) - i(\omega - \omega_{\beta u,\alpha v'})} \tag{1.51}$$

where, for example, the dephasing constant $R_{\alpha v,\alpha v'}$ is related to the damping constants $R_{\alpha v}$ and $R_{\alpha v'}$ of αv and $\alpha v'$,

$$R_{\alpha v,\alpha v'} = \tfrac{1}{2}(R_{\alpha v} + R_{\alpha v'}) \tag{1.52}$$

From Eq. (1.51) we can see that the absorption line-shape function is determined by the imaginary part of $F_{\alpha v,\alpha v'}(\omega)_{ab}$,

$$\text{Im}\left[F_{\alpha v,\alpha v'}(\omega)_{ab}\right] = \frac{1}{\hbar}\sum_\mu \frac{(2/T)(x_a)_{\alpha v',\beta u}(x_b)_{\beta u,\alpha v}}{(\omega - \omega_{\beta u,\alpha v})^2 + (2/T)^2} \tag{1.53}$$

Here we neglected $R_{\beta u,\alpha v'}$ in comparison with $2/T$, with T being in the femtosecond region. The nuclear wave functions $\Theta_{\beta u}$ and $\Theta_{\alpha v}$ oscillate rapidly, except near the classical turning points where $E_{\beta u} - E_{\alpha v} = U_\beta(\xi) - U_\alpha(\xi)$, with U_β and U_α representing the potential energy functions of electronic states β and α. In this approximation, the line-shape function becomes

$$\text{Im}\left[F_{\alpha v,\alpha v'}(\omega)_{ab}\right] = \frac{1}{\hbar}\frac{(2/T)\langle\Theta_{\alpha v'}|(x_a)_{\alpha\beta}(x_b)_{\beta\alpha}|\Theta_{\alpha v}\rangle}{[\omega - (1/\hbar)(U_\beta - U_\alpha)]^2 + (2/T)^2} \tag{1.54}$$

From Eq. (1.50) we can see that for the case of αv and $\alpha v'$ being bound states the imaginary part of $\chi_{ab}(\omega, t_p)$ plotted against t_p exhibits an oscillatory decay provided the system is coherently prepared to a group of states (i.e., wave packet). The oscillation is due to the term $\exp[-i(t_p - t_i)\omega_{\alpha v, \alpha v'}]$, while the decay is due to $\exp[-(t_p - t_i)R_{\alpha v, \alpha v'}]$. The coherence term $(\rho_0)_{\alpha v, \alpha v'}$ is determined by the first laser (i.e., the pump laser), its intensity, pulse duration, and pulse width. The exact determination of $(\hat{\rho}_0)_{\alpha v, \alpha v'}$ requires the solution of the generalized master equations resulting from Eq. (1.48). To obtain an estimate of the coherence, one can consider a simplified model for the pumping laser pulse which has a constant amplitude with duration T. In this case we find (see the Appendix B and Fain et al.[18])

$$(\rho_0)_{\alpha v, \alpha v'} = \frac{1}{\hbar^2} V_{\alpha v, \gamma 0} V_{\gamma 0, \alpha v'} T^2 \tag{1.55}$$

where $V_{\alpha v, \gamma 0}$ and $V_{\gamma 0, \alpha v'}$ denote the interaction matrix elements involved in the pumping laser.

Now we apply the theoretical results developed in the previous sections to the photodissociation of NaI and NaBr. The real-time femtosecond transition-state (FTS) spectra of the dissociation of these systems have been reported by Zewail et al.[4-6] The idealized energy level scheme for this system is shown in Fig. 1.2. It should be noted that in this case the initially prepared wave packet or a group of coherently prepared states are coupled to the dissociation continuum of the lower electronic state. In other words, in this case R in Eq. (1.48) denotes the damping due to the coupling with the dissociation continuum. From Eq. (1.50) we can see that $\tilde{\chi}_{ab}(\omega, t_p)$ consists of two parts, one due to the time-dependent behavior of the coherence and the other due to the band-shape function of the probing laser. In the simplest case, if only two vibrational levels are involved, then $\tilde{\chi}''_{ab}(\omega, t_p)$ will behave as

$$\tilde{\chi}''_{ab}(\omega, t_p) \propto \left(\frac{T}{2}\right) |\langle \Theta_{\alpha v} | \Theta_{\beta u} \rangle|^2_{av} \{|\langle \Theta_{\alpha v} | \Theta_{\gamma 0} \rangle|^2 e^{-(t_p - t_i)R_{\alpha v, \alpha v}}$$

$$+ |\langle \Theta_{\alpha v'} | \Theta_{\gamma 0} \rangle|^2 e^{-(t_p - t_i)R_{\alpha v', \alpha v'}}$$

$$+ 2D \langle \Theta_{\alpha v} | \Theta_{\gamma 0} \rangle \langle \Theta_{\gamma 0} | \Theta_{\alpha v'} \rangle e^{-(t_p - t_i)R_{\alpha v, \alpha v'}} \cos[(t_p - t_i)\omega_{\alpha v, \alpha v'}]\} \tag{1.56}$$

where

$$D = \frac{\sum_u \langle \Theta_{\alpha v'} | \Theta_{\beta u} \rangle \langle \Theta_{\beta u} | \Theta_{\alpha v} \rangle}{|\langle \Theta_{\alpha v} | \Theta_{\beta u} \rangle|^2_{av}} \tag{1.57}$$

The expression of $\chi''_{ab}(\omega, t_p)$ for the three-level case can be obtained similarly.

In carrying out the numerical solution of the reduced response function, we used the data specified in the papers by Zewail's group[4] and Schaefer et

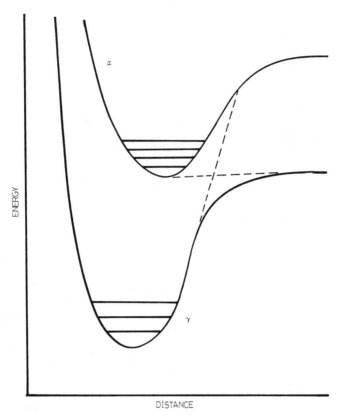

Figure 1.2. Energy-level schemes for FTS of dissociation of molecules. α and γ refer to the excited and ground electronic states, respectively. Another excited electronic state β due to probing excitation from α is not shown.

al.[19] for NaI. In the experimental work on NaI the system was usually pumped at 300–328 nm, which according to the analysis of Schaefer et al. corresponds to vibrational levels in the range of $v = 140$–240. The choice of the equilibrium distance for the excited state was somewhat difficult. Berry[20] used 5.29 Å while an average of turning points used by Schaefer et al. gave a value of 4.72 Å at $v = 200$. In Fig. 1.3 we show a comparison of the two- and three-level cases by choosing the values of 4.72 Å, $v = 200$, $\omega = 5.28 \times 10^{12}$ s^{-1}, and $R = 1.00 \times 10^{12}$ s^{-1}. As can be seen from Fig. 1.3, the effect of the number of levels is not significant.

In fact, these calculations have been carried out for a four-level system, but we did not include them in Fig. 1.3 since the four-level system is only barely distinguishable from the three-level system. One can assume that systems

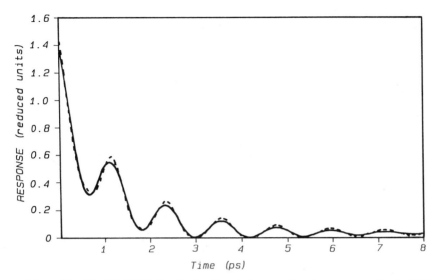

Figure 1.3. Simplified FTS of molecular dissociation using best known parameters for NaI. Solid line—two-level case; dashed line—case for three pumped levels.

with greater than four levels would be almost completely indistinguishable from the four-level system. In Fig. 1.4 we fitted the experimental data of the pumping wave length of 310 nm by using the two-level case; the agreement between experiment and theory is reasonable.

In Fig. 1.4 we only used the two-level system for the reasons stated above. It is interesting to note that to achieve the residual continuous signal for longer times, as seen in the experimental results of Zewail et al., it was necessary to assume that upper levels decay on a time scale that is long compared to the time of observation. We have not given figures for NaBr as the parameters available for a meaningful analysis are even more sketchy than those available for NaI. From the expressions obtained here one can easily produce curves that are similar to the experimental results for NaBr by using decay parameters which are shorter than the frequency of the oscillation of the wave packet in the pumped state. This is to say that the time for decay from the pumped state in NaBr to the continuum is equal to or shorter than the period of oscillation of the prepared wave packet.

In summary, in Section I.A we discuss the generalized linear response theory which can be used in the femtosecond time range. We show how to employ this formalism to treat the FTS of elementary chemical reactions. It should be noted that this theory can be applied to systems in the collision-free condition, the gas phase and dense media. The advantage of the theoretical treatment presented in this section over the conventional perturbation theory

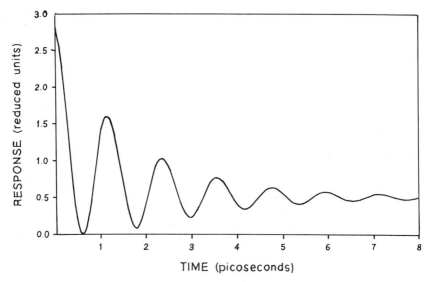

Figure 1.4. Best fit FTS corresponding to results of Zewail's group[4] for NaI. See text for discussion.

for transient processes[21-23] is that the generalized linear response theory properly treats the dynamics of both population and coherence of the system in the femtosecond region.

B. Time-Resolved Emission Spectroscopy

Recently picosecond and femtosecond time-resolved emission spectroscopy has been applied to probe the solvation dynamics of polar solvents and study the mechanism of the solvation effect on processes such as electron transfer and proton transfer.[24-30] Typically a fluorescent probe molecule is excited electronically and the fluorescence spectrum is monitored as a function of time. Relaxation of the solvent polarization about the newly created excited-state dipole leads to a red shift of the fluorescence spectrum. The microscopic solvation relaxation is conventionally probed by monitoring the spectral shift as a function of time. In this section we present the theoretical treatment of the time-resolved emission spectroscopy in the ultrashort time range. In this case it is important to note that when an ultrashort laser pulse is used to pump a molecule system, in general more than one state is coherently pumped and the resulting time-resolved emission spectra will in general consist of the contribution from the evolution of both population and coherence. In other words, the band-shape functions associated with both population and coherence are required in order to deconvolute an observed ultrafast time-resolved spectrum.

1. Theory

We consider a total system consisting of a molecular system and a radiation field. The Hamiltonian of the total system is given by $H = H_s + H_r + V$, where H_s and H_r denote the Hamiltonians of the molecular system and the radiation field, respectively, and V represents the interaction between the molecular system and the radiation field. V is given by[31,32]

$$V = -\sum_k \frac{e_k}{m_k c} \vec{P_k} \cdot \vec{A}(\vec{r}_k) \tag{1.58}$$

where $\vec{P_k}$ denotes the linear momentum of the kth particle and $\vec{A}$ is the vector potential of the radiation field. For a molecular system, in the dipole approximation V can be written as $V = -\vec{X} \cdot \vec{Y}$, where $\vec{X} = (1/c)\vec{A}$ and $\vec{Y} = (e/m)\vec{P}$. Here e and m represent the charge and mass of the electron and $\vec{P} = \sum_k \vec{P_k}$. For the case in which the field is quantized $\vec{A}$ is given by

$$\vec{A} = \sum_r \left(\frac{2\pi\hbar c^2}{\omega_r U}\right)^{1/2} \vec{e}_r (a_r + a_r^*) \tag{1.59}$$

where a_r and a_r^* are boson operators, $\vec{e}_r$ denotes the unit polarization vector, and U is the volume of the system. If follows that $\vec{X} = \sum_r (\vec{x}_r + \vec{x}_r^*)$, where $\vec{x}_r = \vec{e}_r (2\pi\hbar/U\omega_r)^{1/2} a_r$ and $\vec{x}_r^* = \vec{e}_r (2\pi\hbar/U\omega_r)^{1/2} a_r^*$.

Next we consider the evolution of the density matrix ρ of the total system,

$$i\hbar \frac{d\rho}{dt} = [H_0, \rho] + [V, \rho] \tag{1.60}$$

where $H_0 = H_r + H_s$. In the interaction representation we have

$$\rho = \exp\left(-\frac{it}{\hbar} H_0\right) \sigma \exp\left(\frac{it}{\hbar} H_0\right) \tag{1.61}$$

and
$$i\hbar \frac{d\sigma}{dt} = [V(t), \sigma] \tag{1.62}$$

where
$$V(t) = \exp\left(\frac{it}{\hbar} H_0\right) V \exp\left(-\frac{it}{\hbar} H_0\right) = -\vec{X}(t) \cdot \vec{Y}(t) \tag{1.63}$$

To the first-order approximation, Eq. (1.62) can be integrated as

$$\sigma(t) = \sigma_i - \frac{i}{\hbar} \int_{t_i}^{t} [V(\tau), \sigma_i] \, d\tau = \sigma_i + \frac{i}{\hbar} \int_{t_i}^{t} [\vec{X}(\tau) \cdot \vec{Y}(\tau), \sigma_i] \, d\tau \quad (1.64)$$

where σ_i denotes the density matrix σ at $t = t_i$.

In this section we are concerned with the time-resolved emission process of the system, which is described by

$$\left\langle \frac{dH_r}{dt} \right\rangle = \mathrm{Tr}(\rho \dot{H}_r) = \langle \dot{H}_r \rangle = \frac{i}{\hbar} \mathrm{Tr}(\rho[H, H_r]) = -\frac{i}{\hbar} \mathrm{Tr}(\sigma[\vec{X}(t) \cdot \vec{Y}(t), H_r(t)])$$

$$(1.65)$$

Using the relation

$$\frac{d}{dt} \vec{X}(t) = \dot{\vec{X}}(t) = \frac{i}{\hbar} [H_r(t), \vec{X}(t)] \quad (1.66)$$

we can rewrite Eq. (1.65) as

$$\langle \dot{H}_r \rangle = \frac{i}{\hbar} \int_{t_i}^{t} \mathrm{Tr}([\vec{X}(\tau) \cdot \vec{Y}(\tau), \sigma_i] \vec{Y}(t) \cdot \dot{\vec{X}}(t)) \, d\tau \quad (1.67)$$

which can be written as

$$\langle \dot{H}_r \rangle = \frac{i}{\hbar} \int_{0}^{t-t_i} [\langle \dot{\vec{X}}(t) \vec{X}(t - \tau) \rangle : \langle \vec{Y}(t) \vec{Y}(t - \tau) \rangle$$

$$- \langle \vec{X}(t - \tau) \dot{\vec{X}}(t) \rangle : \langle \vec{Y}(t - \tau) \vec{Y}(t) \rangle \} \, d\tau \quad (1.68)$$

where, for example, $\langle \vec{Y}(t) \vec{Y}(t - \tau) \rangle = \mathrm{Tr}(\sigma_i^{(s)} \vec{Y}(t) \vec{Y}(t - \tau))$, $\sigma_i^{(s)}$ being the density matrix of the molecular system.

Notice that

$$\dot{\vec{X}}(t) = i \sum_r \omega_r [-\vec{x}_r(t) + \vec{x}_r^*(t)] \quad (1.69)$$

For spontaneous emission, the radiation field is initially in the vacuum state. In this case we have

$$\langle \dot{\vec{X}}(t) \vec{X}(\tau) \rangle = -i \sum_r \omega_r \langle \vec{x}_r(t) \vec{x}_r^*(\tau) \rangle_0 \quad (1.70)$$

A similar expression for $\langle \vec{X}(\tau)\dot{\vec{X}}(t)\rangle$ can be obtained. Substituting Eq. (1.70) into Eq. (1.68) we obtain

$$\langle \dot{H}_r \rangle = \frac{1}{\hbar}\sum_r \omega_r \int_0^{t-t_i} \langle \vec{x}_r \vec{x}_r^* \rangle_0 : [e^{-it\omega_r}\langle \vec{Y}(t)\vec{Y}(t-\tau)\rangle$$
$$+ e^{it\omega_r}\langle \vec{Y}(t-\tau)\vec{Y}(t)\rangle]\, d\tau \qquad (1.71)$$

In obtaining Eqs. (1.70) and (1.71) it has been assumed that the density matrix σ_i can be written as a product of the density matrix of the molecular system and that of the radiation field.

To simplify the expressions $\langle \vec{Y}(t)\vec{Y}(t-\tau)\rangle$ and $\langle \vec{Y}(t-\tau)\vec{Y}(t)\rangle$ for the molecular system, we use the adiabatic approximation as a basis set. In particular, we use the notation (βv) for the vibronic states of the excited electronic state β and the notation (αw) for the vibronic states of the lower electronic state α. In this case we have

$$\langle \vec{Y}(t)\vec{Y}(t-\tau)\rangle = \sum_{vv'w} (\sigma_i^{(s)})_{\beta v,\beta v'}(\vec{Y}_{\beta v',\alpha w}\vec{Y}_{\alpha w,\beta v})\exp{(it\omega_{\beta v',\beta v} - i\tau\omega_{\alpha w,\beta v})} \quad (1.72)$$

and

$$\langle \dot{H}_r \rangle = \frac{1}{\hbar}\sum_r \sum_{vv'w} \omega_r [\langle \vec{x}_r \vec{x}_r^* \rangle_0 : (\vec{Y}_{\beta v',\alpha w}\vec{Y}_{\alpha w,\beta v})](\sigma_i^{(s)})_{\beta v,\beta v'}\exp{(it\omega_{\beta v',\beta v})}$$
$$\times \int_0^{t-t_i} [\exp\{-i\tau(\omega_r + \omega_{\alpha w,\beta v})\} + \exp\{i\tau(\omega_r - \omega_{\beta v',\alpha w})\}]\, d\tau$$
$$\qquad (1.73)$$

For $\omega^*|t - t_i| \gg 1$, where ω^* is the effective bandwidth of photon modes r, Eq. (1.73) reduces to

$$\langle \dot{H}_r \rangle = \frac{\pi}{\hbar}\sum_r \sum_{vv'w} (\rho_0^{(s)})_{\beta v,\beta v'}\exp{[i(t-t_i)\omega_{\beta v',\beta v}]}\left(\frac{2\pi\hbar}{V}\right)(\vec{e}_r \cdot \vec{Y}_{\beta v',\alpha w})(\vec{e}_r \cdot \vec{Y}_{\alpha w,\beta v})$$
$$\times \left[\delta(\omega_r + \omega_{\alpha w,\beta v}) + \delta(\omega_r - \omega_{\beta v',\alpha w})\right.$$
$$\left. + \frac{i}{\pi}P\frac{\omega_{\beta v',\beta v}}{(\omega_r - \omega_{\beta v',\alpha w})(\omega_r + \omega_{\alpha w,\beta v})}\right] \qquad (1.74)$$

Here the relation $(\sigma_i^{(s)})_{\beta v,\beta v'} = (\rho_0^{(s)})_{\beta v,\beta v'}\exp{(-it_i\omega_{\beta v',\beta v})}$ has been used.
    Notice that

$$\sum_r \rightarrow \frac{V}{4\pi^3 c^3} \int_0^\infty \omega^2 \, d\omega \int_\Omega d\Omega \qquad (1.75)$$

Here a factor of 2 has been introduced for polarizations. Using Eq. (1.75), we can rewrite Eq. (1.74) as

$$\langle \dot{H}_r \rangle = \frac{2\omega^2}{c^3} \sum_{vv'w} (\rho_0^{(s)})_{\beta v, \beta v'} \exp\left[i(t - t_i)\omega_{\beta v', \beta v}\right](\vec{e} \cdot \vec{Y}_{\beta v', \alpha w})(\vec{e} \cdot \vec{Y}_{\alpha w, \beta v})$$

$$\times \left[\delta(\omega + \omega_{\alpha w, \beta v}) + \delta(\omega - \omega_{\beta v', \alpha w}) \right.$$

$$\left. + \frac{i}{\pi} P \frac{\omega_{\beta v', \beta v}}{(\omega - \omega_{\beta v', \alpha w})(\omega + \omega_{\alpha w, \beta v})} \right] \qquad (1.76)$$

The time-resolved emission spectra measured in terms of the number of emitted photons per unit time is described by

$$P(\omega, t) = \frac{1}{\hbar \omega} \langle \dot{H}_r \rangle$$

$$= \frac{2\omega}{\hbar c^3} \sum_{vv'w} (\rho_0^{(s)})_{\beta v, \beta v'} \exp\left[i(t - t_i)\omega_{\beta v', \beta v}\right](\vec{e} \cdot \vec{Y}_{\beta v', \alpha w})(\vec{e} \cdot \vec{Y}_{\alpha w, \beta v})$$

$$\times \left[\delta(\omega + \omega_{\alpha w, \beta v}) + \delta(\omega - \omega_{\beta v', \alpha w}) \right.$$

$$\left. + \frac{i}{\hbar} P \frac{\omega_{\beta v', \beta v}}{(\omega - \omega_{\beta v', \alpha w})(\omega + \omega_{\alpha w, \beta v})} \right] \qquad (1.77)$$

If molecules are randomly oriented, then Eq. (1.77) becomes

$$P(\omega, t) = \frac{2\omega}{3\hbar c^3} \sum_{vv'w} (\rho_0^{(s)})_{\beta v, \beta v'} \exp\left[i(t - t_i)\omega_{\beta v', \beta v}\right](\vec{Y}_{\beta v', \alpha w} \cdot \vec{Y}_{\alpha w, \beta v})$$

$$\times \left[\delta(\omega + \omega_{\alpha w, \beta v}) + \delta(\omega - \omega_{\beta v', \alpha w}) \right.$$

$$\left. + \frac{i}{\pi} P \frac{\omega_{\beta v', \beta v}}{(\omega - \omega_{\beta v', \alpha w})(\omega + \omega_{\alpha w, \beta v})} \right] \qquad (1.78)$$

The Einstein A coefficient (i.e., the spontaneous emission rate constant) can be obtained from Eq. (1.78) by setting $(\rho_0^{(s)})_{\beta v, \beta v'} = \delta_{vv'}(\rho_0^{(s)})_{\beta v, \beta v}$ and integrating $P(\omega, t)$ over ω,

$$A = \frac{4}{3\hbar c^3} \sum_{vw} (\rho_0^{(s)})_{\beta v, \beta v} \omega_{\beta v, \alpha w} |\vec{Y}_{\beta v, \alpha w}|^2$$

$$= \frac{4}{3\hbar c^3} \sum_{vw} (\rho_0^{(s)})_{\beta v, \beta v} \omega_{\beta v, \alpha w}^3 |\vec{\mu}_{\beta v, \alpha w}|^2 \qquad (1.79)$$

Here the relation $\vec{Y}_{\alpha w, \beta v} = i\omega_{\alpha w, \beta v} \vec{\mu}_{\alpha w, \beta v}$ has been used; $\vec{\mu}_{\beta v, \alpha w}$ represents the transition moment. $P(\omega, t)$ given by Eq. (1.77) or Eq. (1.78) is a central result of this section; it can provide the emission spectra of the molecular system as a function of time.

Next we show the application of the theory of ultrashort time-resolved emission spectroscopy presented in this section. $P(\omega, t)$ given by Eqs. (1.77) or (1.78) is derived in terms of molecular eigenstates. In practice it is often more convenient to separate the total system into the molecular system and the heat bath. It should be noted that for the case in which the ultrashort laser pulse is used for the excitation of a molecular system embedded in a heat bath, the initially prepared wave packet or a group of coherently prepared states are coupled to the continuum (or quasi-continuum) of the heat bath, and the evolution of these states can be described by the stochastic Liouville equation. The Liouville equation for this type of problem has been derived and can be expressed as[31]

$$\frac{d\rho}{dt} = -\frac{i}{\hbar}[H \cdot \rho] - R\rho \qquad (1.80)$$

where R represents the damping operator due to the coupling of the system to the heat bath; it provides the description of the relaxation and dephasing (i.e., the dynamics) of the molecular system.

It should be noted that in Eq. (1.80) the damping operator has been introduced to describe the relaxation processes due to the presence of the heat bath. The validity of this approximation is based on the assumption that the relaxation time scale is much longer than the characteristic correlation time in the heat bath. In a liquid solution it should mean that the relaxation rate characterized by R is much smaller than the rate of the regression of fluctuations in the solvent. Clearly if one can prepare an excited molecular state having a rapid enough evolution time, this separation of time scales breaks down. In this case the non-Markovian effect has to be considered. This non-Markovian effect has been treated by Fain, Lin, and Wu.[18]

Using Eq. (1.80) and repeating the derivative presented in Section I.B.1, we obtain, for a system of randomly oriented molecules (see Appendix C),

$$P(\omega, t) = \frac{2\omega}{3\hbar c^3} \sum_{vv'w} \{\exp[-i(t - t_i)(L_s - iR)]\rho_0^{(s)}\}_{\beta v, \beta v'}(\vec{Y}_{\beta v', \alpha w} \cdot \vec{Y}_{\alpha w, \beta v})$$

$$\times \left[\delta(\omega + \omega_{\alpha w, \beta v}) + \delta(\omega - \omega_{\beta v', \alpha w}) \right.$$

$$\left. + \frac{i}{\pi} P \frac{\omega_{\beta v', \beta v}}{(\omega - \omega_{\beta v', \alpha w})(\omega + \omega_{\alpha w, \beta v})} \right] \tag{1.81}$$

where $R_{\beta v, \beta v'}$ represents the dephasing rate constant and is related to the total decay rates $R_{\beta v}$ and $R_{\beta v'}$ by $R_{\beta v, \beta v'} = \frac{1}{2}(R_{\beta v} + R_{\beta v'}) + R_{\beta v, \beta v'}^{(d)}$. Here $R_{\beta v, \beta v'}^{(d)}$ denotes the pure dephasing.

2. Time-Resolved Emission Spectroscopy of Electronic Transitions

We now use the expression for $P(\omega, t)$ given in the Born–Oppenheimer adiabatic approximation. The adiabatic approximation is applicable to low electronic states of molecules. For this purpose we let (βv) denote the vibronic states in the excited electronic state manifold, while (αw) denotes the vibronic states in the ground electronic state manifold. In this case Eq. (1.81) can be written as

$$P(\omega, t) = \frac{2\omega}{3\hbar c^3} \sum_{vv'w} \rho^{(s)}(\Delta t)_{\beta v, \beta v'}(\vec{Y}_{\beta v' \alpha w} \cdot \vec{Y}_{\alpha w, \beta v})$$

$$\times \left[\delta(\omega - \omega_{\beta v', \alpha w}) + \delta(\omega + \omega_{\alpha w, \beta v}) \right.$$

$$\left. + \frac{i}{\pi} P \frac{\omega_{\beta v', \beta v}}{(\omega + \omega_{\alpha w, \beta v})(\omega - \omega_{\beta v', \alpha w})} \right] \tag{1.82}$$

where $\rho^{(s)}(\Delta t) = G^{(s)}(\Delta t)\rho_0^{(s)}$ and $\Delta t = t - t_i$. From Eq. (1.82) we can see that $P(\omega, t)$ consists of the incoherent contribution $P(\omega, t)_n$ and the coherent contribution $P(\omega, t)_c$, that is,

$$P(\omega, t) = P(\omega, t)_n + P(\omega, t)_c \tag{1.83}$$

where

$$P(\omega, t)_n = \frac{4\omega}{3\hbar c^3} \sum_{vw} \rho^{(s)}(\Delta t)_{\beta v, \beta v} |\vec{Y}_{\alpha w, \beta v}|^2 \delta(\omega - \omega_{\beta v, \alpha w}) \tag{1.84}$$

or

$$P(\omega, t)_n = \frac{4\omega^3}{3\hbar c^3} \sum_{vw} \rho^{(s)}(\Delta t)_{\beta v, \beta v} |\vec{\mu}_{\alpha w, \beta v}|^2 \delta(\omega - \omega_{\beta v, \alpha w})$$

$$= \sum_{v} \rho^{(s)}(\Delta t)_{\beta v, \beta v} F_{\beta v, \beta v}(\omega) \tag{1.85}$$

and

$$P(\omega, t)_c = \frac{2\omega}{3\hbar c^3} \sum_{v}^{v \neq v'} \sum_{v'}' \sum_{w} \rho^{(s)}(\Delta t)_{\beta v, \beta v'} (\vec{Y}_{\beta v', \alpha w} \cdot \vec{Y}_{\alpha w, \beta v})$$

$$\times \left[\delta(\omega - \omega_{\beta v', \alpha w}) + \delta(\omega + \omega_{\alpha w, \beta v}) + \frac{i}{\pi} P \frac{\omega_{\beta v', \beta v}}{(\omega + \omega_{\alpha w, \beta v})(\omega - \omega_{\beta v', \alpha w})} \right]$$

$$= \sum_{v}^{v \neq v'} \sum_{v'}' \rho^{(s)}(\Delta t)_{\beta v, \beta v'} F_{\beta v, \beta v'}(\omega) \qquad (1.86)$$

where $F_{\beta v, \beta v}(\omega)$ and $F_{\beta v, \beta v'}(\omega)$ represent the band-shape functions,

$$F_{\beta v, \beta v}(\omega) = \frac{4\omega^3}{3\hbar c^3} \sum_{w} |\vec{\mu}_{\alpha w, \beta v}|^2 \delta(\omega - \omega_{\beta v, \alpha w}) \qquad (1.87)$$

$$F_{\beta v, \beta v'}(\omega) = \frac{2\omega}{3\hbar c^3} \sum_{w} (\vec{Y}_{\beta v', \alpha w} \cdot \vec{Y}_{\alpha w, \beta v})$$

$$\times \left[\delta(\omega - \omega_{\beta v', \alpha w}) + \delta(\omega + \omega_{\alpha w, \beta v}) \right.$$

$$\left. + \frac{i}{\pi} P \frac{\omega_{\beta v', \beta v}}{(\omega + \omega_{\alpha w, \beta v})(\omega - \omega_{\beta v', \alpha w})} \right] \qquad (1.88)$$

From the expression of $P(\omega, t)_n$ we can see that it consists of the dynamics of the population $\rho^{(s)}(\Delta t)_{\beta v, \beta v}$ and the associated band-shape function $F_{\beta v, \beta v}(\omega)$. Similarly, $P(\omega, t)_c$ can be separated into the coherence $\rho^{(s)}(t)_{\beta v, \beta v'}$ and the associated band-shape function $F_{\beta v, \beta v'}(\omega)$. For the case in which the Condon approximation applies (i.e., the electronic transition is an allowed transition), the band-shape functions $F_{\beta v, \beta v}(\omega)$ and $F_{\beta v, \beta v'}(\omega)$ become

$$F_{\beta v, \beta v}(\omega) = \frac{4\omega^3}{3\hbar c^3} |\vec{\mu}_{\alpha \beta}|^2 \sum_{w} |\langle \Theta_{\alpha w} | \Theta_{\beta v} \rangle|^2 \delta(\omega - \omega_{\beta v, \alpha w}) \qquad (1.89)$$

$$F_{\beta v, \beta v'}(\omega) = \frac{2\omega}{3\hbar c^3} |\vec{\mu}_{\alpha \beta}|^2 \sum_{w} \omega_{\beta v', \alpha w} \omega_{\beta v, \alpha w} \langle \Theta_{\beta v'} | \Theta_{\alpha w} \rangle \langle \Theta_{\alpha w} | \Theta_{\beta v} \rangle$$

$$\times \left[\delta(\omega - \omega_{\beta v', \alpha w}) + \delta(\omega + \omega_{\alpha w, \beta v}) \right.$$

$$\left. + \frac{i}{\pi} P \frac{\omega_{\beta v', \beta v}}{(\omega + \omega_{\alpha w, \beta v})(\omega - \omega_{\beta v', \alpha w})} \right] \qquad (1.90)$$

where $\Theta_{\alpha w}$, $\Theta_{\beta v'}$, and $\Theta_{\beta v}$ represent the nuclear wave functions and $\vec{\mu}_{\alpha \beta}$ denotes the electronic transition moment.

It should be noted that when the dephasing is much faster than the relaxation, in the time scale of molecular relaxation $\rho^{(s)}(\Delta t)_{\beta v, \beta v'}$ vanishes and we have $P(\omega, t) = P(\omega, t)_n$.

3.   Band-shape Functions

In Section I.B.2 we have shown that the time-resolved emission spectra $P(\omega, t)$ can be separated into the product of the population or coherence and the associated band-shape functions. In this section we show how to obtain the band-shape functions $F_{\beta v, \beta v}(\omega)$ and $F_{\beta v, \beta v'}(\omega)$. For this purpose we consider only the case where one degree of freedom is nonstationary and other degrees of freedom are in thermal equilibrium. In other words, only one optical mode is coherently pumped and the other modes maintain or reach equilibrium in the time scale under consideration. In this case,

$$F_{\beta v, \beta v}(\omega) = \frac{4\omega^3}{3\hbar c^3} |\bar{\mu}_{\alpha\beta}|^2 \sum_v \sum_w |\langle X_{\beta v_0} | X_{\alpha w_0} \rangle|^2 \prod_i{}' P_{\beta v_i} |\langle X_{\beta v_i} | X_{\alpha w_i} \rangle|^2$$
$$\times \, \delta(\omega_{\beta v, \alpha w} - \omega) \tag{1.91}$$

where $(X_{\beta v_0}, X_{\alpha w_0})$ and $(X_{\beta v_i}, X_{\alpha w_i})$ represent the wave functions of the optically pumped mode and the remaining other modes and $P_{\beta v_i}$ denotes the Boltzmann factor. For the case of harmonic oscillators, $F_{\beta v, \beta v}(\omega)$ can be written as[33]

$$F_{\beta v, \beta v}(\omega) = \frac{2\omega^3}{3\pi\hbar c^3} |\bar{\mu}_{\alpha\beta}|^2 \int_{-\infty}^{\infty} dt \exp\left[it(\omega - \omega_{\beta\alpha})\right] K_{v_0}(t) \prod_i{}' G_i(t) \tag{1.92}$$

where

$$K_{v_0}(t) = \sum_{w_0} |\langle X_{\beta v_0} | X_{\alpha w_0} \rangle|^2 \exp\left\{it\left[(w_0 + \tfrac{1}{2})\omega_0' - (v_0 + \tfrac{1}{2})\omega_0\right]\right\} \tag{1.93}$$

and

$$G_i(t) = \sum_{v_i} \sum_{w_i} P_{\beta v_i} |\langle X_{\beta v_i} | X_{\alpha w_i} \rangle|^2 \exp\left\{it\left[(w_i + \tfrac{1}{2})\omega_i' - (v_i + \tfrac{1}{2})\omega_i\right]\right\} \tag{1.94}$$

For the case of a harmonic oscillator, $G_i(t)$ has been evaluated[33]

$$G_i(t) = 2\beta_i \beta_i' \sinh\frac{\hbar\omega_i}{2kT} \exp\left[-\frac{k_i(t)}{(h_i(t))^{1/2}}\right] \tag{1.95}$$

where

$$k_i(t) = \frac{\beta_i^2 \beta_i'^2 d_i^2}{\beta_i'^2 \coth(\lambda_i/2) + \beta_i^2 \coth(\mu_i'/2)} \tag{1.96}$$

$$h_i(t) = \sinh \lambda_i \sinh \mu_i' \left(\beta_i'^2 \coth\frac{\mu_i'}{2} + \beta_i^2 \coth\frac{\lambda_i}{2} \right) \left(\beta_i'^2 \tanh\frac{\mu_i'}{2} + \beta_i^2 \tanh\frac{\lambda_i}{2} \right) \tag{1.97}$$

and $\mu_i' = -it\omega_i'$ and $\lambda_i = it\omega_i + \hbar\omega_i/kT$. Similarly $K_{v_0}(t)$ is given by

$$K_{v_0}(t) = K_{0_0}(t)\Delta K_{v_0}(t) \tag{1.98}$$

where

$$K_{0_0}(t) = \exp\left[-\frac{\beta_0^2 d_0^2}{2}(1 - e^{-\mu_0}) \right] \tag{1.99}$$

and

$$\Delta K_{v_0}(t) = \sum_{n_0=0}^{v_0} \frac{v_0!}{(n_0!)^2(v_0 - n_0)!} \left[\frac{\beta_0^2 d_0^2}{2}(e^{\mu_0/2} - e^{-\mu_0/2})^2 \right]^{n_0} \tag{1.100}$$

Here it is assumed that the oscillator is displaced but not distorted. The case of a displaced and distorted oscillator is shown in Appendix E. It should be noted that $K_{0_0}(t)$ given by Eq. (1.98) is equal to $G_{0_0}(t)$ at $T = 0$.

Substituting Eqs. (1.94) and (1.98) into Eq. (1.92) and using the short-time approximation (i.e., the strong coupling case), $F_{\beta v, \beta v}(\omega)$ becomes

$$F_{\beta v, \beta v}(\omega) = \frac{2\omega^3}{3\hbar c^3} |\vec{\mu}_{\alpha\beta}|^2 \Delta K_{v_0}(t^*) \sqrt{\frac{\pi}{D}} \exp\left[-\frac{(\omega_{\beta\alpha} - \omega_s - \omega)^2}{4D} \right] \tag{1.101}$$

where

$$D = \sum_i' \frac{\omega_i^2}{4}\left[\rho_i^2\left(1 - \frac{\rho_i}{2}\right)^2 \left(\coth\frac{\hbar\omega_i}{2kT}\right)^2 + \beta_i^2 d_i^2(1 - \rho_i)^4 \coth\frac{\hbar\omega_i}{2kT} \right]$$
$$+ \frac{\omega_0^2}{4}\beta_0^2 d_0^2 \tag{1.102}$$

$$\omega_s = \sum_i' \frac{\omega_i}{2}\left[\beta_i^2 d_i^2(1 - \rho_i)^2 - \rho_i\left(1 - \frac{\rho_i}{2}\right)\coth\frac{\hbar\omega_i}{2kT} \right] + \frac{\omega_0}{2}\beta_0^2 d_0^2 \tag{1.103}$$

and

$$t^* = \frac{i(\omega - \omega_{\beta\alpha} + \omega_s)}{2D}, \qquad \omega_i' = \omega_i(1 - \rho_i) \tag{1.104}$$

Due to the term $\Delta K_{v_0}(t^*)$, $F_{\beta v, \beta v}(\omega)$ takes the modified Gaussian form. Notice that at $v_0 = 0$, $\Delta K_{v_0}(t^*) = 1$. In other words, only for this case would $F_{\beta v, \beta v}(\omega)$ take the Gaussian form.

Next we consider $F_{\beta v, \beta v'}(\omega)$, which can be rewritten as

$$F_{\beta v, \beta v'}(\omega) = \frac{2\omega}{3\pi hc^3} |\vec{\mu}_{\alpha\beta}|^2 \sum_w {\sum_v}' P_{\beta v} \omega_{\beta v', \alpha w} \omega_{\beta v, \alpha w} \langle \Theta_{\beta v} | \Theta_{\alpha w} \rangle \langle \Theta_{\alpha w} | \Theta_{\beta v'} \rangle$$

$$\times \int_0^\infty \{ \exp[it(\omega - \omega_{\beta v', \alpha w})] + \exp[-it(\omega + \omega_{\alpha w, \beta v})] \} \, dt$$

$$\tag{1.105}$$

$F_{\beta v, \beta v'}(\omega)$ consists of two terms,

$$F_{\beta v, \beta v'}(\omega) = F_{\beta v, \beta v'}(\omega)_1 + F_{\beta v, \beta v'}(\omega)_2 \tag{1.106}$$

where

$$F_{\beta v, \beta v'}(\omega)_1 = \frac{2\omega}{3\pi hc^3} |\vec{\mu}_{\alpha\beta}|^2 \int_0^\infty \exp[it(\omega - \Delta v_0 \omega_0)]$$

$$\times \left(\frac{\partial^2}{\partial \theta^2} - \Delta v_0 \omega_0 \frac{\partial}{\partial \theta} \right) I_{vv'}(-\theta) \, dt \tag{1.107}$$

$$F_{\beta v, \beta v'}(\omega)_2 = \frac{2\omega}{3\pi hc^3} |\vec{\mu}_{\alpha\beta}|^2 \int_0^\infty e^{-it\omega} \left(\frac{\partial^2}{\partial \theta^2} + \Delta v_0 \omega_0 \frac{\partial}{\partial \theta} \right) I_{vv'}(\theta) \, dt \tag{1.108}$$

$$I_{vv'}(\theta) = \sum_w {\sum_v}' P_{\beta v} \langle \Theta_{\beta v'} | \Theta_{\alpha w} \rangle \langle \Theta_{\alpha w} | \Theta_{\beta v} \rangle e^{\theta \omega_{\beta v, \alpha w}} \tag{1.109}$$

Here $\theta = it$ and $v_0' = v_0 + \Delta v_0$. Eqs. (1.105)–(1.109) indicate that to evaluate $F_{\beta v, \beta v'}(\omega)$, it is necessary to evaluate $I_{vv'}(\theta)$. Notice that $I_{vv'}(\theta)$ can be expressed as

$$I_{vv'}(\theta) = e^{\theta \omega_{\beta \alpha}} K_{v_0 v_0'}(\theta) {\prod_i}' G_i(\theta) \tag{1.110}$$

where

$$K_{v_0 v_0'}(\theta) = \sum_{w_0} \langle X_{\beta v_0'} | X_{\alpha w_0} \rangle \langle X_{\alpha w_0} | X_{\beta v_0} \rangle \exp\{ \theta[(v_0 + \tfrac{1}{2})\omega_0 - (w_0 + \tfrac{1}{2})\omega_0'] \} \tag{1.111}$$

and

$$G_i(\theta) = \sum_{v_i} \sum_{w_i} P_{\beta v_i} |\langle X_{\beta v_i} | X_{\alpha w_i} \rangle|^2 \exp\{ \theta[(v_i + \tfrac{1}{2})\omega_i - (w_i + \tfrac{1}{2})\omega_i'] \} \tag{1.112}$$

$G_i(\theta)$ is exactly the same as $G_i(t)$ given by Eq. (1.94) by replacing $t = i\theta$. We now calculate $K_{v_0, v_0'}(\theta)$, which can be accomplished in the same way as for $K_{v_0}(t)$. We obtain, for the displaced oscillator case,

$$K_{v_0, v_0'}(\theta) = K_{0_0, 0_0}(\theta)\Delta K_{v_0, v_0'}(\theta) \tag{1.113}$$

where

$$K_{0_0, 0_0}(\theta) = \exp\left[\frac{\beta_0^2 d_0^2}{2}(1 - e^{-\omega_0\theta})\right] \tag{1.114}$$

$$\Delta K_{v_0, v_0'}(\theta) = \sum_{n_0=0}^{v_0} \frac{(v_0! v_0'!)^{1/2}}{n_0!(v_0-n_0)!(v_0'-n_0)!} e^{(v_0-n_0)\omega_0\theta}\left[\frac{\beta_0 d_0}{\sqrt{2}}(1 - e^{-\omega_0\theta})\right]^{v_0+v_0'-2n_0} \tag{1.115}$$

for $v_0' > v_0$. Similarly for the displaced oscillator case, $G_i(\theta)$ is given by

$$G_i(\theta) = \exp\left\{-\frac{\beta_i^2 d_i^2}{2}\left[\coth\frac{\hbar\omega_i}{2kT} - \operatorname{csch}\frac{\hbar\omega_i}{2kT}\cosh\left(\omega_i\theta - \frac{\hbar\omega_i}{2kT}\right)\right]\right\} \tag{1.116}$$

In this case we have

$$I_{vv'}(\theta) = \exp\left\{-\frac{\beta_0^2 d_0^2}{2}(1 - e^{-\omega_0\theta}) + \omega_{\beta\alpha}\theta - \sum_i{}' \frac{\beta_i^2 d_i^2}{2}\left[\coth\frac{\hbar\omega_i}{2kT}\right.\right.$$
$$\left.\left. - \operatorname{csch}\frac{\hbar\omega_i}{2kT}\cosh\left(\omega_i\theta - \frac{\hbar\omega_i}{2kT}\right)\right]\right\} \sum_{n_0=0}^{v_0} \frac{(v_0! v_0'!)^{1/2}}{n_0!(v_0-n_0)!(v_0'-n_0)!}$$
$$\times\; e^{(v_0-n_0)\omega_0\theta}\left[\frac{\beta_0 d_0}{\sqrt{2}}(1 - e^{-\theta\omega_0})\right]^{v_0+v_0'-2n_0} \tag{1.117}$$

In particular at low temperatures, Eq. (1.117) reduces to

$$I_{vv'}(\theta) = \exp\left[-\sum_i \frac{\beta_i^2 d_i^2}{2}(1 - e^{-\omega_i\theta}) + \omega_{\beta\alpha}\theta\right] \sum_{n_0=0}^{v_0} \frac{(v_0! v_0'!)^{1/2}}{n_0!(v_0-n_0)!(v_0'-n_0)!}$$
$$\times\; e^{(v_0-n_0)\omega_0\theta}\left[\frac{\beta_0 d_0}{\sqrt{2}}(1 - e^{-\omega_0\theta})\right]^{v_0+v_0'-2n_0} \tag{1.118}$$

4.   Discussion

In this section we show some numerical results of time-resolved emission spectra. For this purpose we consider the case in which only two lowest vibration levels of a particular mode are coherently pumped and the temperature effect is neglected. In this case we have

$$P(\omega, t)_n = \rho^{(s)}(\Delta t)_{\beta 0, \beta 0} F_{\beta 0, \beta 0}(\omega) + \rho^{(s)}(\Delta t)_{\beta 1, \beta 1} F_{\beta 1, \beta 1}(\omega) \quad (1.119)$$

and

$$P(\omega, t)_c = \rho^{(s)}(\Delta t)_{\beta 0, \beta 1} F_{\beta 0, \beta 1}(\omega) + \rho^{(s)}(\Delta t)_{\beta 1, \beta 0} F_{\beta 1, \beta 0}(\omega) \quad (1.120)$$

where for the displaced oscillator case,

$$
\begin{aligned}
F_{\beta 0, \beta 0}(\omega) &= \frac{2\omega^3}{3\pi\hbar c^3} |\vec{\mu}_{\alpha\beta}|^2 \int_{-\infty}^{\infty} \exp\left[it(\omega - \omega_{\beta\alpha}) - \sum_i S_i(1 - e^{it\omega_i}) \right] dt \\
&= \frac{2\omega^3}{3\pi\hbar c^3} |\vec{\mu}_{\alpha\beta}|^2 \sqrt{\frac{2\pi}{\sum_i S_i \omega_i^2}} \exp\left[-\frac{(\omega_{\beta\alpha} - \sum_i S_i \omega_i - \omega)^2}{2 \sum_i S_i \omega_i^2} \right]
\end{aligned}
\quad (1.121)
$$

and

$$
\begin{aligned}
F_{\beta 1, \beta 1}(\omega) &= \frac{2\omega^3}{3\pi\hbar c^3} |\vec{\mu}_{\alpha\beta}|^2 \int_{-\infty}^{\infty} [(1 - 2S_0) + S_0(e^{it\omega_0} + e^{-it\omega_0})] \\
&\quad \times \exp\left[it(\omega - \omega_{\beta\alpha}) - \sum_i S_i(1 - e^{it\omega_i}) \right] dt
\end{aligned}
\quad (1.122)
$$

Here $S_i = \frac{1}{2}\beta_i^2 d_i^2$, the Huaung–Rhys constant (or the coupling constant). $F_{\beta 1, \beta 1}(\omega)$ can be expressed as

$$
\begin{aligned}
F_{\beta 1, \beta 1}(\omega) &= (1 - 2S_0) F_{\beta 0, \beta 0}(\omega) + S_0 \left(\frac{\omega}{\omega + \omega_0} \right)^3 F_{\beta 0, \beta 0}(\omega + \omega_0) \\
&\quad + S_0 \left(\frac{\omega}{\omega - \omega_0} \right)^3 F_{\beta 0, \beta 0}(\omega - \omega_0)
\end{aligned}
\quad (1.123)
$$

Notice that from Appendix E we obtain

$$\rho^{(s)}(\Delta t)_{\beta 0, \beta 0} = (\hat{\rho}_0^{(s)})_{\beta 0, \beta 0} + (\hat{\rho}_0^{(s)})_{\beta 1, \beta 1}(1 - e^{-R_{\beta 1}\Delta t}) \quad (1.124)$$

and

$$\rho^{(s)}(\Delta t)_{\beta 1, \beta 1} = (\hat{\rho}_0^{(s)})_{\beta 1, \beta 1} e^{-R_{\beta 1}\Delta t} \quad (1.125)$$

where $R_{\beta 1}$ denotes the relaxation rate constant of the $\beta 1$ level. Here for simplicity we have assumed that $R_{\beta 0}$ is negligible in the time scale under consideration.

Next we consider the coherence contribution. Notice that in this case we obtain the real portion of $F_{\beta 0, \beta 1}(\omega)$ as

$F_{\beta 0, \beta 1}(\omega)$

$$
= F_{\beta 1, \beta 0}(\omega) = \left(\frac{\beta_0 d_0}{\sqrt{2}}\right) |\vec{\mu}_{\alpha\beta}|^2 \frac{2\omega^2(\omega - \omega_0)}{3\pi\hbar c^3} \sqrt{\frac{2\pi}{\sum_i S_i \omega_i^2}}
$$

$$
\times \left\{ \exp\left[-\frac{(\omega_{\beta\alpha} + \omega_0 - \sum_i S_i \omega_i - \omega)^2}{2\sum_i S_i \omega_i^2} \right] - \exp\left[-\frac{(\omega_{\beta\alpha} - \sum_i S_i \omega_i - \omega)^2}{2\sum_i S_i \omega_i^2} \right] \right\}
$$

$$
+ \frac{\beta_0 d_0}{\sqrt{2}} |\vec{\mu}_{\alpha\beta}|^2 \frac{2\omega^2(\omega + \omega_0)}{3\pi\hbar c^3} \sqrt{\frac{2\pi}{\sum_i S_i \omega_i^2}}
$$

$$
\times \left\{ \exp\left[-\frac{(\omega_{\beta\alpha} - \omega - \sum_i S_i \omega_i)^2}{2\sum_i S_i \omega_i^2} \right] - \exp\left[-\frac{(\omega_{\beta\alpha} - \omega_0 - \omega - \sum_i S_i \omega_i)^2}{2\sum_i S_i \omega_i^2} \right] \right\}
\tag{1.126}
$$

and $\qquad \rho^{(s)}(\Delta t)_{\beta 0, \beta 1} = (\hat{\rho}_0^{(s)})_{\beta 0, \beta 1} \exp\left[-\Delta t(i\omega_{\beta 0, \beta 1} + R_{\beta 0, \beta 1}) \right] \qquad (1.127)$

where $R_{\beta 0, \beta 1}$ denotes the dephasing constant and $\omega_{\beta 0, \beta 1} = -\omega_0$.

In previous papers[18, 34, 35] we have shown that if the pumping laser pulse has a constant amplitude with duration T, then

$$
(\hat{\rho}_0^{(s)})_{\beta v, \beta v'} = \frac{T^2}{\hbar^2} V'_{\beta v, \alpha 0} V'_{\alpha 0, \beta v'}
\tag{1.128}
$$

where $V'_{\beta v, \alpha 0}$ and $V'_{\alpha 0, \beta v'}$ denote the interaction matrix element involved in the pumping laser. Using the Condon approximation, we obtain

$$
\frac{(\hat{\rho}_0^{(s)})_{\beta 1, \beta 1}}{(\hat{\rho}_0^{(s)})_{\beta 0, \beta 0}} = \frac{|\langle X_{\beta 1_0} | X_{\alpha 0_0} \rangle|^2}{|\langle X_{\beta 0_0} | X_{\alpha 0_0} \rangle|^2}
\tag{1.129}
$$

and

$$
\frac{(\hat{\rho}_0^{(s)})_{\beta 0, \beta 1}}{(\hat{\rho}_0^{(s)})_{\beta 0, \beta 0}} = \frac{\langle X_{\beta 1_0} | X_{\alpha 0_0} \rangle}{\langle X_{\beta 0_0} | X_{\alpha 0_0} \rangle}
\tag{1.130}
$$

For the displaced oscillator we obtain

$$
\frac{(\hat{\rho}_0^{(s)})_{\beta 1, \beta 1}}{(\hat{\rho}_0^{(s)})_{\beta 0, \beta 0}} = \frac{\beta_0^2 d_0^2}{2} = S_0
\tag{1.131}
$$

and

$$
\frac{(\hat{\rho}_0^{(s)})_{\beta 0, \beta 1}}{(\hat{\rho}_0^{(s)})_{\beta 0, \beta 0}} = -\frac{\beta_0 d_0}{\sqrt{2}}
\tag{1.132}
$$

Equations (1.113)–(1.132) have been used for numerical calculations. For this purpose we introduce an average frequency in $\sum_i S_i \omega_i$ and $\sum_i S_i \omega_i^2$ so that

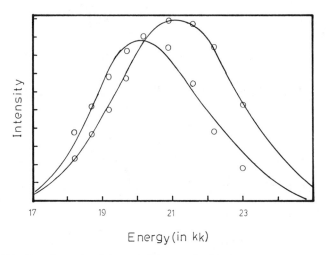

Figure 1.5. Experimental and theoretical time-resolved emission spectra. ———— Theory; ooo Experiment.

$\sum_i S_i \omega_i = S\bar{\omega}$ and $\sum_i S_i \omega_i^2 = S\bar{\omega}^2$, where $S = \sum_i S_i$. The numerical results are shown in Fig. 1.5 for the emission spectra as a function of time.

Recently solvation dynamics around the probe solute molecule, 7-(dimethylamino)coumarin-4-acetate ion, dissolved in water, have been measured on the femtosecond time scale by Barbara and coworkers.[24] Their reported time-resolved emission spectra have been fit by using Eq. (1.83) (see Fig. 1.6). For $F_{\beta 0, \beta 0}(\omega)$ and $F_{\beta 1, \beta 1}(\omega)$, for simplicity the Gaussian forms $F_{\beta 0, \beta 0} \propto \exp[-(\omega - \omega_0)^2/a_0]$ and $F_{\beta 1, \beta 1} \propto \exp[-(\omega - \omega_1)^2/a_1]$ have been used to obtain $\omega_0 = 19.7$ and $a_0 = 1.6$, and $\omega_1 = 21.2$ and $a_1 = 2.4$. The agreement between experiment and theory is reasonable. It is important to note that the bandwidth varies with time, as predicted by the theory.

For the purpose of demonstration, we carry out the single-mode case, using Eq. (1.83), and neglect all bath modes. Results are shown in Figs. 1.6 and 1.7. When comparing the two figures, the coherence effect is obvious in the early stage. Coherence is also shown in the oscillatory behavior of emission in Fig. 1.8. Note that 1 KK^{-1} in the time scale corresponds to 33 fs. Figures 1.9 and 1.10 show similar patterns of evolution for the upper state initially populated at $v = 1$ and $v = 2$. A large dephasing constant (Figs. 1.7 and 1.10) quickly damps the coherent part of emission intensity and therefore, allows us to see the evolution of the incoherent contribution.

The net dephasing effect, as caused by both relaxation and pure dephasing, can come from strong anharmonicity and intermode coupling, as well as strong vibronic coupling in the excited state. Nonlinear coupling among

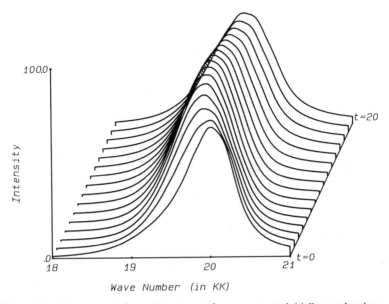

Figure 1.6. Time-resolved fluorescence spectra from upper state initially populated at $v = 0$, 1. $\omega_{\beta\alpha} = 30$ KK, $\omega_0 = R_{\beta 1} = 0.2$ KK, $R_{ph} = 0$, $S_0 = 2$. Time in units of KK^{-1}; intensity in arbitrary units.

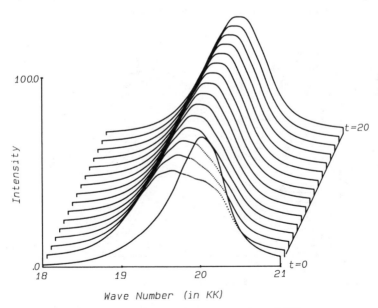

Figure 1.7. Same as Fig. 1.6, except $R_{ph} = 5$ KK.

163

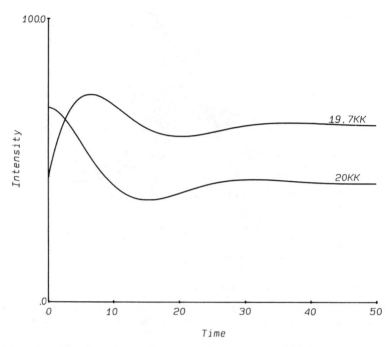

Figure 1.8. Time dependence of fluorescence from upper state initially populated at $v = 0$, 1 for two frequencies: 19.7 KK and 20 KK. For parameters, see Fig. 1.6.

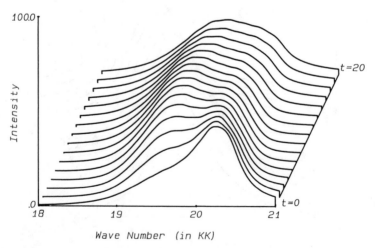

Figure 1.9. Time-resolved fluorescence spectra from upper state initially populated at $v = 1$, 2. $\omega_{\beta\alpha} = 20$ KK, $\omega_0 = R_{\beta1} = 0.2$ KK, $R_{ph} = 0$, $S_0 = 2$.

164

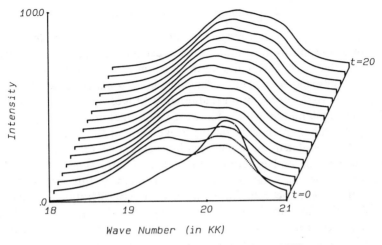

Figure 1.10. Same as Fig. 1.9, except $R_{ph} = 5 \, KK$.

different vibrational modes is especially strong when the molecule is pumped vertically from the lower state and is located in a highly anharmonic range of vibrational coordinates of the excited state. By this reasoning a planar chromophore having a nonplanar excited state would have a strong coupling, and hence fast relaxation dephasing takes place before or along with emission.

We have shown that coherence is significant at short time scales, especially when coupling between electronic motion and nuclear motion is large, as shown in our model calculation. With our calculation scheme steady-state emission spectra can also be treated. Furthermore, the present work can be applied to probe the solvation dynamics using time-resolved emission. In this case it is more convenient to separate $\sigma^{(s)}$ into two subsystems, one due to the solute molecule and the other due to the solvent cage molecules which affect the electronic energy levels of the solute molecule (i.e., the solvent effect). In this way we can see that the observed spectral shift as a function of time will describe the dynamics of the local solvent molecules.

In concluding this section, we have presented a general theoretical treatment for ultrashort time-resolved emission spectroscopy in which not only the emission maximum but also the whole spectra can be obtained as a function of time. To show the application of this theory, the time-resolved emission spectra for a system of displaced oscillators have been obtained for the low-temperature case. Furthermore, the theory can be used to study the polarization of emission as a function of time, and although the time-resolved emission spectroscopy is emphasized in this section, the theory can be used to study the steady-state emission spectroscopy.

C. Impulsive Raman Scattering

Temporal and spatial coherences (or modulation) can be created from a coherent scattering from a phonon mode excited via impulsive stimulated Raman scattering.[36] This temporal modulation of the transmitted intensity of a probe pulse has been demonstrated in a nonresonant interaction. Part of the spectrum of the probe beam is alternately Stokes shifted and anti-Stokes shifted by its interaction with the nuclear vibrational motion induced by the pump pulse. Thus the energy on the Stokes side of the peak of the probe spectrum is modulated out of phase with the energy on the anti-Stokes side.

In the femtosecond experiments creation of intra- and intermolecular vibrational coherences has been observed. The aim of this section is (1) to provide the general theoretical framework for the consideration of coherences created by the ultrashort pulses in various media and (2) to describe electromagnetic properties of the media in which coherences exist. There are properties described by generalized linear susceptibilities $\chi_{ab}(\vec{k}_1\omega_1; \vec{k}_2\omega_2)$, which are functions of two independent frequencies and wave vectors.

1. Density Matrix Jumps Induced by Ultrashort Pulses— Placzek Approximation

We assume that a system (molecules, liquids, and solids) subjected to the electromagnetic pulse can be represented by the energy scheme shown in Fig. 1.11. This scheme comprises of a group of levels E_{mv} with large energy differences,

$$\hbar\omega_{mv;nv'} = E_{mv} - E_{nv'}, \qquad m \neq n \qquad (1.133)$$

and relatively small differences,

$$\hbar\omega_{mv;mv'} = E_{mv} - E_{mv'} \qquad (1.134)$$

Usually m, n label various electronic states, while v, v' are indices for the states of vibrons, phonons, magnons, and so on. We assume that the laser pulse strength can be presented as a sum of quasi-monochromatic field components,

$$E_a(\vec{r}, t) = \sum_{\vec{k},\Omega} \varepsilon_a^{(\vec{k})}(\vec{r}, t) \exp[-i(\Omega t - \vec{k} \cdot \vec{r})] \qquad (1.135)$$

where $\varepsilon_a^{(\vec{k})}(\vec{r}, t)$ is a complex amplitude component of the pulse, which varies slowly in the optical period or wavelength. Here Ω is the optical frequency of the carrier,

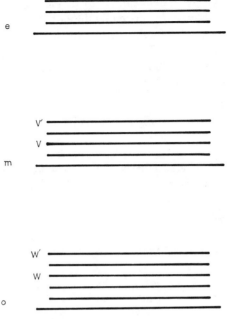

Figure 1.11. Energy-level scheme illustrating separation of fast and slow motions. Energy differences in the same electronic states $o, n, l, \ldots$ present slow motions.

$$\vec{k} = \vec{n}\frac{\Omega}{c}; \qquad |\vec{n}| = 1 \qquad (1.136)$$

is the wave vector, and c is the speed of light in the medium. We have in mind the situation where the frequency Ω is much larger than the frequency $|\omega_{mv;mv'}|$, which describes slow motions of the system, such as phonons and intra-molecular vibrations,

$$\Omega \gg |\omega_{mv;mv'}| \qquad (1.137)$$

At the same time Ω may be on the same order of magnitude as $|\omega_{mv;nv'}|$, $m \neq n$ (electronic motions), and not necessarily resonant with one of these eigen-frequencies. The Hamiltonian of the system interacting with the external electromagnetic field Eq. (1.135) has the form

$$H = H_0 + V(t) \qquad (1.138)$$

where $\qquad V(t) = -\sum_j \vec{d}_j \cdot \vec{E}(r_j, t) = \sum_j V_j(\vec{r}_j, t) \qquad (1.139)$

Thus we assume that the system can be divided into the cells represented by index j, each of which is characterized by the electric dipole moment $\vec{d}_j$. Such an assumption is general enough to cover the majority of interesting cases, including the crystal lattice.

In the interaction representation the von Neumann equation for the density matrix of the system has the form

$$i\hbar\frac{\partial \rho}{\partial t} = [V^I(t), \rho] \tag{1.140}$$

where the time dependence $V^I(t)$ is determined both by the explicit field dependence and by the unperturbed Hamiltonian H_0,

$$V^I(t) = e^{iH_0 t/\hbar} V(t) e^{-iH_0 t/\hbar} \tag{1.141}$$

Equation (1.140) can be rewritten in integral form,

$$\Delta\rho(t) = \frac{1}{i\hbar}\int_{t_i}^{t} [V^I(t_1), \rho(t_1)]\, dt_1 \tag{1.142}$$

where $\Delta\rho(t) = \rho(t) - \rho(t_i)$. Up to this point the consideration is quite general. Now we make certain assumptions. We assume that while the pulse duration T is short, it is still longer than the electronic motion period

$$\omega_{el} T \sim |\omega_{mv;nv'}| T \gg 1 \tag{1.143}$$

On the other hand we assume that, in a general case, it is shorter than any other time scales of the system,

$$T < \omega_{vib}^{-1}, \qquad T < \tau_c, \qquad T < T_{relax} \tag{1.144}$$

Here ω_{vib} is a characteristic vibrational frequency (or a frequency of some other slow motion) of the system, $\hbar\tau_c^{-1}$ is a characteristic bandwidth of some thermal bath (such as phonons or a dissociation continuum), and T_{relax} is a characteristic relaxation time.

Thus we assume that during the pulse duration only electronic motions proceed. Other kinds of motion (vibrational, rotational, relaxation, etc.) are "frozen." [It should be mentioned that it is not necessary that all conditions (1.144) be satisfied. Depending on the particular problem, only part of conditions may be necessary.]

A conventional consideration of the influence of the electromagnetic field on the system leads either to the introduction of field-dependent transition rates (see, e.g., Refs. 37) or to some other kind of continuous influence on the system dynamics (see, e.g., Refs. 38 and 39).

In the case of the ultrashort pulses satisfying conditions (1.143) and (1.144) the influence of the laser field cannot be accounted for as an additional transition rate. Its influence on the system can instead be described as the density matrix "jumps".[18] Having in mind conditions (1.143) and (1.144), we can calculate the density matrix jumps from Eq. (1.142) taking limits of integration $(-\infty, \infty)$. Then using the perturbation theory we get

$$\Delta\rho^{(1)} = \frac{1}{i\hbar} \int_{-\infty}^{\infty} [V^I(t_1)_1 \rho(-\infty)] \, dt_1 \tag{1.145}$$

$$\Delta\rho^{(2)} = \frac{1}{(i\hbar)^2} \int_{-\infty}^{\infty} \int_{-\infty}^{t_1} [V^I(t_1)[V^I(t_2), \rho(-\infty)]] \, dt_2 \, dt_1 \tag{1.146}$$

When we consider the interaction (1.139) linear in the field strength $\vec{E}$, the first-order density matrix jump (1.143) does not affect low-frequency motions due to condition (1.137). Nevertheless, we analyze Eq. (1.145), having in mind cases when $V(t)$ is proportional to the second order of the field strength (as illustrated further on).

We introduce the time Fourier expansion of some function by the relations

$$f(t) = \int_{-\infty}^{\infty} \hat{f}(\omega) e^{-i\omega t} \, d\omega \tag{1.147}$$

$$\hat{f}(\omega) = \frac{1}{2\pi} \int_{-\infty}^{\infty} f(t) e^{i\omega t} \, dt \tag{1.148}$$

Assuming also that the density matrix $\rho(-\infty)$, before the jump, was diagonal, with elements designated as ρ_{mm}^0, we get

$$\Delta\rho_{mn}^{(1)} = \frac{1}{i\hbar} (\rho_{nn}^0 - \rho_{mm}^0) \int_{-\infty}^{\infty} V_{mn}^I(t_1) \, dt_1$$

$$= \frac{2\pi}{i\hbar} \hat{V}_{mn}^I(\omega = 0)(\rho_{nn}^0 - \rho_{mm}^0) \tag{1.149}$$

In the Placzek approximation[40,41] (see also Genkin et al.[42]), a polarization induced by electric field has the form

$$P_a = \sum_b \alpha_{ab}(\vec{X}) E_b(\vec{r}, t) \tag{1.150}$$

where $\vec{X}$ designates the nuclear configuration of the system and $\alpha_{ab}(\vec{X})$ denotes the polarizability. Expanding $\alpha_{ab}(\vec{X})$ over vibrational modes of the system, one gets

$$\alpha_{ab}(\vec{X}) = \alpha_{ab}^0 + \sum_{\vec{k},l} \alpha_{ab}(\vec{k},l) Q(\vec{k},l) e^{-i\vec{k}\cdot\vec{r}} + \cdots \tag{1.151}$$

where $\cdots$ denotes the contribution of higher order Raman and Brillouin processes. The density of the interaction energy has the form[1,40] (see also Refs. 21, 41, and 42)

$$d\tilde{H} = -\sum_a P_a \, dE_a \tag{1.152}$$

Using this relation and Eq. (1.151), the interaction energy takes the form

$$
\begin{aligned}
V = &-\frac{1}{2} \sum_{\vec{f}_1, \vec{f}_2; a, b} U\Delta(\vec{f}_1 + \vec{f}_2)\alpha_{ab}^0 E_a(\vec{f}_1, t) E_b(\vec{f}_2, t) \\
&-\frac{1}{2} \sum_{\vec{k}, l; \vec{f}_1, \vec{f}_2; a, b} \alpha_{ab}(\vec{k}, l) Q(\vec{k}, l, t) E_a(\vec{f}_1, t) E_b(\vec{f}_2, t) U\Delta(-\vec{k} + \vec{f}_1 + \vec{f}_2) + \cdots
\end{aligned}
\tag{1.153}
$$

Here U is the volume of the crystal (assuming $U \to \infty$), and the nonvanishing value of $\Delta(\vec{f})$ is equal to unity. Normal mode operator Q is determined by

$$Q(\vec{k}, l) = \left(\frac{\hbar}{2\omega}\right)^{1/2} [a^+(-\vec{k}, l) + a(\vec{k}, l)] \tag{1.154}$$

$$[a(\vec{k}, l), a^+(\vec{k}', l')] = \delta^{\vec{k}\,\vec{k}'} \delta_{ll'} \tag{1.155}$$

while
$$\dot{a}(\vec{k}, l) = -i\omega(\vec{k}, l) a(\vec{k}, l) \tag{1.156}$$

Spatial Fourier components $E_a(\vec{f}, t)$ are defined as

$$\vec{E}(\vec{r}, t) = \sum_{\vec{f}} \vec{E}(\vec{f}, t) e^{i\vec{f}\cdot\vec{r}} \tag{1.157}$$

and
$$\vec{E}(\vec{f}, t) = \frac{1}{U} \int \vec{E}(\vec{r}, t) e^{-i\vec{f}\cdot\vec{r}} \, d^3\vec{r} \tag{1.158}$$

It should be mentioned that expansion over modes $e^{-i\vec{k}\cdot\vec{r}}$ can be performed for the infinite crystal, when boundary conditions do not play a role. In general, and especially for acoustic modes, we have to take into account the boundary conditions.

We will not perform here the full analysis of the excitation of low-frequency coherences in the Placzek approximation. We will briefly describe how the second sum in Eq. (1.153), using Eq. (1.154), generates phonon coherences.

Assuming that the system is in the ground electronic state, we can rewrite Eq. (1.149) as

$$\Delta\rho^{(1)}_{0v;0v'} \equiv \Delta\rho_{v;v'} = \frac{i}{\hbar}(\rho^0_{vv} - \rho^0_{v'v'})\int_{-\infty}^{\infty} V^I_{vv'}(t_1)\,dt_1 \tag{1.159}$$

Here v, v' designate the sets of phonon indices (phonon numbers) in various phonon modes. Since, according to Eq. (1.153), V is the sum of operators Q in various modes $(\vec{k}l)$, nonvanishing elements of $\Delta\rho_{v;v'}$ consist of coherences with phonon numbers v and v' different in one particular mode and diagonal in all other modes. Therefore Eq. (1.159) can be rewritten explicitly as

$$\Delta\rho^{(1)}_{v_1 v_2 \ldots v_{(\vec{k}l)}\ldots v_n\ldots;v_1 v_2\ldots(v_{\vec{k}l}\pm 1)\ldots v_n\ldots}$$
$$\equiv \Delta\rho^{(1)}_{vv\pm 1}$$
$$= \frac{-i}{2\hbar}\sum_{\vec{f}_1,\vec{f}_2;a,b}\alpha_{ab}(\vec{k}l)\int_{-\infty}^{\infty} Q_{vv\pm 1}(\vec{k},l,t)E_a(\vec{f}_1,t)E_b(\vec{f}_2,t)$$
$$\times U\Delta(-\vec{k}+\vec{f}_1+\vec{f}_2)\,dt(\rho^{(0)}_{vv}-\rho^{(0)}_{\pm1v\pm1}) \tag{1.160}$$

Expanding E_a and E_b into the Fourier integral and using Eq. (1.156), we get

$$\Delta\rho^{(1)}_{v;v+1} = \frac{-\pi i}{\hbar}\sum_{\vec{f}_1,\vec{f}_2;a,b}U\alpha_{ab}(\vec{k},\vec{l})\frac{\sqrt{\hbar(v+1)}}{2\omega_{\vec{k},l}}\int \hat{E}_a(\vec{f}_1\omega)\hat{E}_b(\vec{f}_2,-\omega_{kl}-\omega)$$
$$\times \Delta(-\vec{k}+\vec{f}_1+\vec{f}_2)\,d\omega(\rho^{(0)}_{vv}-\rho^{(0)}_{v+1v+1}) \tag{1.161}$$

where $v = v_{\vec{k},l}$.

On the other hand, we obtain for the specific component of Eq. (1.135)

$$\hat{E}_a(\vec{f},\omega) = \frac{1}{2\pi U}\sum_{\vec{K}}\sum_{\Omega}\int\int \varepsilon_a^{(\vec{K})}(\vec{r},t)e^{i(\omega-\Omega)t_e-i(\vec{f}-\vec{K})\vec{r}}\,dt\,d^3r$$
$$= \sum_{\vec{K}}\sum_{\Omega}\hat{\varepsilon}_a(\vec{f}-\vec{K},\omega-\Omega) \tag{1.162}$$

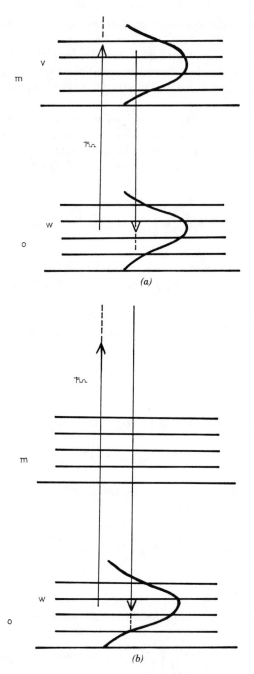

Figure 1.12. Excitation of coherences by electromagnetic field with carrier frequency Ω. Two arrows present second-order perturbation, proportional to field amplitude squared. (*a*) Resonant case. (*b*) Nonresonant case.

172

where the function $\sum_{\vec{k}\Omega} \hat{\varepsilon}_a(\omega)$ is substantially nonzero in the interval T^{-1}, T being the mean duration of the pulse. Therefore the coherence could be created for short pulses

$$\omega_{\text{vib}} T \lesssim 1 \tag{1.163}$$

The wave vectors of the phonon coherences are determined by the function $\Delta(-\vec{k} + \vec{f_1} + \vec{f_2})$. A certain range of phonon wave vectors are allowed due to the fact that $\hat{\varepsilon}_a(\vec{f}, \omega)$ is a continuous function of $\vec{f}$, substantially non-vanishing in the interval

$$\Delta f = \frac{1}{l} \tag{1.164}$$

where l is the spatial dimension of the pulse.

In the Placzek approximation, the dispersion of the susceptibility and the resonances with the electronic states is neglected. In this case, as we have seen, the laser pulse may create vibrational coherences in the ground electronic state of the system [see Eqs. (1.159)–(1.161) and Fig. 1.12b]. As we see in the next section, the laser pulse resonant to excited electronic vibrational (or excited electronic phonon) states will excite vibrational (phonon) coherences in both excited and ground states (Fig. 1.12a).

We assume that at a certain initial time $t = t_i$ the system is subjected to the ultrashort [Eq. (1.144)] pump pulse. Then at times $t > t_i$, when the pump pulse is over, the system evolves without being subjected to the laser field. At this stage it may radiate spontaneously from the upper state, exhibiting quantum beats[4,14] (Fig. 1.13b). At time t_p the system may be subjected to the probe pulse. This opens a number of possibilities depicted in Fig. 1.13. There may be Raman (or Brillouin) scattering on the ground state coherences[36a] (Fig. 1.13c). In principle there may also be Raman scattering on the upper electronic state vibrational (phonon) coherences (Fig. 1.13d). The probing pulse may also be absorbed from the excited coherent state to another upper electronic state,[4,35,36b] giving also rise to the spontaneous emission from this state (Fig. 1.13d, e). The absorption may also take place between the ground and upper level coherences at the frequency coinciding with that of the pump pulse[37] (Fig. 1.13f).

Considering action of the pump and probe pulses, we adopt the approximation

$$t_p - t_i \gg T \tag{1.165}$$

This condition enables us to consider action of the pump and probe pulses

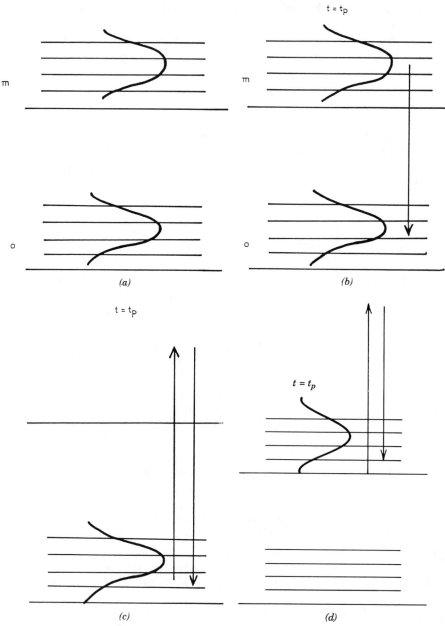

Figure 1.13. Classification of various possibilities after creation of coherences by pump pulse. Explanation is given in text.

I—at $t = t_i$.

I_a—spontaneous emission with quantum beats.

I_b—Raman (Brillouin) scattering on ground electronic state.

I_c—Roman (Brillouin) scattering on excited electronic state.

$I_{c,d}$—absorption to excited elctronic state l or absorption followed by spontaneous emission.

I_e—absorption of probe pulse at frequency of pump pulse.

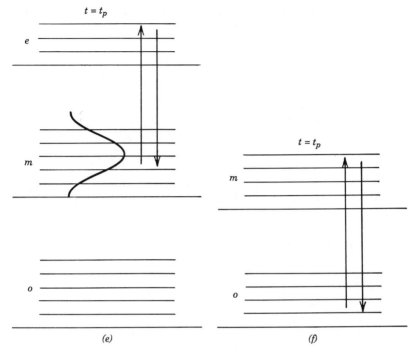

Figure 1.13. (*Continued*)

independently. Otherwise we would have to use the fourth-order perturbation theory in the region of overlap of the pump and probe pulses.

2. Density Matrix Jumps Induced by the Pump Pulse

In this section we consider the density matrix jumps without using the Placzek approximation. This enables us to consider the absorption processes accompanied by the creation of the coherences depicted in Fig. 1.12a. We start from the Hamiltonian (1.138), taking into account that $V(t)$ has a limited duration of magnitude on the order of T. Further we use a specific form [Eq. 1.139] of this interaction energy. However, we take into account that $V(t)$ is proportional to the first power of the field. In this case we need to use the second-order contribution to the coherences (1.146). We denote by $|0, v\rangle$ the ground electronic state 0 and the vibrational (or, generally, any low-frequency) state v. It is assumed that initially (before the pulse) the system was in the ground electronic state and its density matrix was diagonal,

$$\rho_{mv;m'v'}(-\infty) = \rho_{0v;0v}\delta_{mm'}\delta_{m0}\delta_{vv'} \qquad (1.166)$$

The second-order contribution to the density matrix (1.146) has the form

$$
\Delta\rho_{mv;mv'} = -\frac{1}{\hbar^2} \int_{-\infty}^{\infty} dt_1 \int_{0}^{\infty} d\tau \left[\sum_{lw} V^I(t_1)_{0v;lw} \right.
$$

$$
\times V^I(t_1 - \tau)_{lw;0v'} \rho_{0v';0v} \delta_{m0} + \sum_{lw} \rho_{0v;0v} \delta_{m0} V^I(t_1 - \tau)_{0w;lw}
$$

$$
\times V^I(t_1)_{lw;0v'} - \sum_{w} V^I(t_1)_{mv;0w} V^I(t_1 - \tau)_{0w;mv'} \rho_{0w;0w}
$$

$$
\left. - \sum_{w} V^I(t_1 - \tau)_{mv;0w} V^I(t_1)_{0w;mv'} \rho_{0w;0w} \right] \qquad (1.167)
$$

Using definitions (1.147) and (1.148), the convolution theorem

$$
\int_{-\infty}^{\infty} e^{i\omega\tau} d\tau \int_{-\infty}^{\infty} f_1(t_1) f_2(\tau - t_1) dt_1 = (2\pi)^2 \hat{f}_1(\omega) \hat{f}_2(\omega) \qquad (1.168)
$$

and the definition of the ζ function

$$
\zeta(\omega) = -i \int_{0}^{\infty} e^{i\omega\tau} d\tau = \lim_{\varepsilon \to +0} \frac{1}{\omega + i\varepsilon} \equiv \frac{P}{\omega} - i\pi\delta(\omega) \qquad (1.169)
$$

we get instead of Eq. (1.167),

$$
\Delta\rho_{mv;mv'}
$$

$$
= -\delta_{m0} \frac{2\pi i}{\hbar^2} \int_{-\infty}^{\infty} d\omega \, \zeta(\omega) \sum_{lw}
$$

$$
\times [\hat{V}^I(-\omega)_{0v;lw} \hat{V}^I(\omega)_{lw;0v'} \rho_{0v';0v} + \hat{V}^I(\omega)_{0v;lw} \hat{V}^I(-\omega)_{lw;0v'} \rho_{0v;0v}]
$$

$$
+ \frac{2\pi i}{\hbar^2} \int_{-\infty}^{\infty} d\omega \, \zeta(\omega) \sum_{w} [\hat{V}^I(-\omega)_{mv;0w} \hat{V}^I(\omega)_{0w;mv'}
$$

$$
+ \hat{V}^I(\omega)_{mv;0w} \hat{V}(-\omega)_{0w;mv'}] \rho_{0w;0w} \qquad (1.170)
$$

Taking into account that the sum in the second term is a symmetrical function of ω and relation (1.169), we get for the coherence created in the excited state $m \neq 0$ the following expression:

$$
\Delta\rho_{mv;mv'} = \frac{4\pi^2}{\hbar^2} \sum_{w} \hat{V}^I(0)_{mv;0w} \hat{V}^I(0)_{0w;mv'} \rho_{0w;0w}, \qquad (m \neq 0) \quad (1.171)
$$

while the coherence created at the ground state $m = 0$ takes the form

$$
\begin{aligned}
\Delta\rho_{0v;0v'} = \frac{2\pi i}{\hbar^2} \int_{-\infty}^{\infty} d\omega\zeta(\omega) &\Bigg\{ \sum_w [\hat{V}^I(-\omega)_{0v;0w} \\
&\times \hat{V}^I(\omega)_{0w;0v'}(\rho_{0w;0w} - \rho_{0v';0v'}) + \hat{V}^I(\omega)_{0v;0w} \\
&\times \hat{V}^I(-\omega)_{0w;0v'}(\rho_{0w;0w} - \rho_{0v;0v})] \\
&- \sum_{l \neq 0;w} [\hat{V}^I(-\omega)_{0v;lw} \hat{V}^I(\omega)_{lw;0v'}\rho_{0v';0v'} \\
&+ \hat{V}^I(\omega)_{0v;lw} \hat{V}^I(-\omega)_{lw;0v'}\rho_{0v;0v}] \Bigg\}
\end{aligned}
\tag{1.172}
$$

Having in mind applications to femtosecond experiments with ultrashort laser pulses, we can represent the interaction energy with the electromagnetic field in the form

$$
V = [V(\Omega)e^{-i\Omega t} + V(-\Omega)e^{i\Omega t}]L(t)
\tag{1.173}
$$

where the dimensionless function $L(t) \leq 1$ (its maximum value being unity) describes the pulse shape, while Ω is the laser frequency. Then the pulse duration is defined as

$$
\int_{-\infty}^{\infty} L(t)\, dt = T
\tag{1.174}
$$

Using the definitions (1.147) and (1.148) and relation (1.141), we find an expression for the Fourier transform of V:

$$
\hat{V}^I(\omega)_{mn} = V(\Omega)_{mn}\hat{L}(\omega + \omega_{mn} - \Omega) + V(-\Omega)_{mn}\hat{L}(\omega + \omega_{mn} + \Omega)
\tag{1.175}
$$

Now we are able to analyze the expressions for the density matrix jumps and the coherences, Eqs. (1.171) and (1.172), created by the electromagnetic pulse. There are two kinds of contributions to these expressions: resonant with $\omega = 0$ and nonresonant with ω substantially different from zero. As has been shown, the density matrix jump $\Delta\rho_{mv;mv'}$ ($m \neq 0$) has only the resonant part given by Eq. (1.171).

It follows from the definition of the ζ function (1.169) and Eq. (1.172) that the resonant contribution to the coherence created in the ground state has the form

$$\Delta\rho^{\text{res}}_{0v;0v'} = \frac{2\pi^2}{\hbar^2}\left\{\sum_w [\hat{V}^I(0)_{0v;0w}\hat{V}^I(0)_{0w;0v'}(\rho_{0w;0w} - \rho_{0v';0v'})\right.$$

$$+ \hat{V}^I(0)_{0v;0w}\hat{V}^I(0)_{0w;0v'}(\rho_{0w;0w} - \rho_{0v;0v})]$$

$$\left. + \sum_{l\neq 0; w} \hat{V}^I(0)_{0v;lw}\hat{V}^I(0)_{lw;0v'}(\rho_{0v';0v'} + \rho_{0v;0v})\right\} \qquad (1.176)$$

Now we adopt the rotating-wave approximation, assuming that frequencies $\omega_{nv;0w}$ are close to Ω, while Ω is large enough:

$$|\omega_{nv;0w} - \Omega| \lesssim T^{-1}, \qquad \Omega T \gg 1 \qquad (1.177)$$

At the same time we assume that the frequency $|\omega_{0v;0w}|$ of slow motions (vibrations, phonons, etc.) may be neglected in comparison with Ω,

$$\Omega \gg |\omega_{0v;0w}| \qquad (1.178)$$

In this approximation, taking into account Eq. (1.175), we get instead of Eqs. (1.171) and (1.176),

$$\Delta\rho_{nv;nv'} = \frac{4\pi^2}{\hbar^2}\sum_w \rho_{0w;0w} V(\Omega)_{nv;0w} V(-\Omega)_{0w;nv'}\hat{L}(\omega_{nv;0w} - \Omega)$$

$$\times \hat{L}(\Omega - \omega_{nv';0w}), \qquad n \neq 0 \qquad (1.179)$$

and $$\Delta\rho^{\text{res}}_{0v;0v'} = \frac{2\pi^2}{\hbar^2}\sum_w V(-\Omega)_{0v;nw} V(\Omega)_{nw;0v'}\hat{L}(\Omega - \omega_{nw;0v})$$

$$\times \hat{L}(\omega_{nw;0v'} - \Omega)(\rho_{0v;0v} + \rho_{0v';0v'}) \qquad (1.180)$$

Considering the nonresonant contribution to $\Delta\rho_{0v;0v'}$ we have to take into account that in the static case $\Omega = 0$, the first sum in Eq. (1.172) has the same order of magnitude as the second one. Hence for large Ω, as Eq. (1.178), the first sum becomes very small compared with the second one and may be neglected. (See similar considerations in Born and Huang.[41]) Therefore the nonresonant (n.r.) contribution to $\Delta\rho_{0v;0v'}$ has the form

$$\Delta\rho^{\text{n.r.}}_{0v;0v'} = -\frac{2\pi i}{\hbar^2}\int_{-\infty}^{\infty}\frac{d\omega}{\omega}\sum_{l\neq 0; w}\hat{V}^I(\omega)_{0v;lw}$$

$$\times \hat{V}^I(-\omega)_{lw;0v'}(\rho_{0v';0v'} - \rho_{0v;0v}) \qquad (1.181)$$

where $\oint$ means the principal value of the integral. In the rotating-wave approximation we get, similar to Eq. (1.180), the following expression:

$$
\Delta \rho_{0v;0v'}^{n.r.} = -\frac{2\pi i}{\hbar^2} \int_{-\infty}^{\infty} \frac{d\omega}{\omega} \sum_{l \neq 0, w} [V(\Omega)_{0v;lw} V(-\Omega)_{lw;0v'}
$$
$$
\times \hat{L}(\omega - \omega_{lw;0v} - \Omega)\hat{L}(-\omega + \omega_{lw;0v'} + \Omega) + V(-\Omega)_{0v;lw} V(\Omega)_{lw;0v'}
$$
$$
\times \hat{L}(\omega - \omega_{lw;0v} + \Omega)\hat{L}(-\omega + \omega_{lw;0v'} - \Omega)](\rho_{0v';0v'} - \rho_{0v;0v})
$$

$$(1.182)$$

The nonresonant contribution to the coherences in the ground state, Eq. (1.181) or Eq. (1.182), may be calculated using the Placzek approximation.[40] In this case one may replace $\omega_{lw;0v}$ in Eq. (1.181) or (1.182) by ω_{l0}. Then we get from Eq. (1.182)

$$
\Delta \rho_{0v;0v'}^{n.r.} = -\frac{i}{\hbar^2} \left\{ \left\langle v \left| \sum_n \left[\frac{V(\Omega)_{0n} V(-\Omega)_{n0}}{\Omega + \omega_{n0}} + \frac{V(-\Omega)_{0n} V(\Omega)_{n0}}{\omega_{n0} - \Omega} \right] \right| v' \right\rangle \right.
$$
$$
\left. \times (\rho_{0v';0v'} - \rho_{0v;0v}) \int_{-\infty}^{\infty} L^2(t)\, dt \right\}
$$
$$
= \frac{i}{\hbar^2} (\rho_{0v;0v} - \rho_{0v';0v'}) \int_{-\infty}^{\infty} \sum_n \langle v | V_{0n}^2(\Omega) | v' \rangle \frac{2\omega_{n0}}{\omega_{n0}^2 - \Omega^2} L^2(t)\, dt
$$

$$(1.183)$$

Here it has been assumed that

$$
|\Omega \pm \omega_{n0}| T \gg 1 \tag{1.184}
$$

Indeed, this expression is similar to Eq. (1.159), while the role of the interaction energy matrix elements (in the second-order perturbation theory) is given by the following expression:

$$
V_{0v;0v'}^{(2)} = \frac{1}{\hbar} \sum_n \frac{2\omega_{n0}}{\omega_{n0}^2 - \Omega^2} \langle v | V_{0n}^2(\Omega) | v' \rangle L^2(t) \tag{1.185}
$$

In the case where we can neglect the dispersion

$$
\Omega \ll |\omega_{n0}| \tag{1.186}
$$

we get

$$
V_{0v;0v'}^{(2)} = 2 \sum_n \frac{\langle v | V_{0n}^2(\Omega) | v' \rangle}{E_n - E_0} L^2(t) \tag{1.187}
$$

3. Creation of Space Coherences

The pump pulse creates coherent motions in the system. The time development of the system during the time after the pump pulse (and before the probe
pulse) is characterized by the density matrix $\Delta\rho_{nv;nv'}$ (n could label both ground
and excited electronic states). It means that the time dependence of the mean
value of some operator A would exhibit coherent quantum beats,

$$\langle A(t) \rangle = \sum_{vv'} \Delta\rho_{mv;mv'} A_{mv';mv} e^{i\omega_{mv';mv}t} \qquad (1.188)$$

where $\omega_{mv';mv}$ is the frequency of slow motions, such as vibrational or phonon.
At the same time the ultrashort pump pulse can create, a coherent space
structure.

To consider space coherences created by ultrashort electromagnetic pulses
we start from the interaction energy presented by Eq. (1.139). Assuming that
the system (crystal) is divided into quasi-independent cells j, we use the
approximate relation

$$\langle m|V_j|0\rangle\langle 0|V_{j'}|m\rangle = \langle m|V_j|0\rangle\langle 0|V_j|m\rangle\delta_{jj'} \qquad (1.189)$$

Here electronic states are characterized by the indices in various cells,

$$|m\rangle = |m_1, m_2, \dots, m_j, \dots\rangle \qquad (1.190)$$

and V_j acts on the variables at cell j. Apparently this "local" property of matrix
elements refers to the interband transitions if the description by the Bloch
functions is used.

We start from expression (1.171). Substituting it into Eq. (1.139) and using
property (1.189), we get

$$\Delta\rho_{mv;mv'} = \frac{4\pi}{\hbar^2} \sum_{wj} \langle mv|\hat{V}_j^I(0)|0w\rangle\langle 0w|\hat{V}_j^I(0)|mv'\rangle\rho_{0w;0w},$$

$$m \neq 0 \qquad (1.191)$$

Each V_j depends on the electromagnetic field at point X_j of the nuclear
configuration. Similarly, Eq. (1.172) becomes

$$\Delta\rho_{0v;0v'} = -\frac{2\pi i}{\hbar^2} \int_{-\infty}^{\infty} \zeta(\omega) \sum_{nwj} [\langle 0v|\hat{V}_j^I(-\omega)|nw\rangle\langle nw|\hat{V}_j^I(\omega)|0v'\rangle$$

$$\times \rho_{0v';0v'} + \rho_{0v;0v}\langle 0v|\hat{V}_j^I(\omega)|nw\rangle\langle nw|\hat{V}_j^I(-\omega)|0v'\rangle \, d\omega \qquad (1.192)$$

In this expression we used approximations leading to expressions (1.180) and (1.181). In the case of dipole interactions in cell j, we have

$$V_j = -\vec{d}_j(X_j) \cdot \vec{E}(\vec{r}_j, t) \tag{1.193}$$

where X_j denotes a nuclear configuration. Introducing the coordinate of the displacement from the equilibrium configuration X_j^0 we get

$$\vec{u}_j = \vec{X}_j - \vec{X}_j^0 \tag{1.194}$$

Then the dipole moment can be presented as

$$\vec{d}_j(X_j) = \vec{d}_j + \vec{d}_j' : \vec{u}_j + \cdots \tag{1.195}$$

where : means the tensor product of vectors $\vec{d}_j'$ and $\vec{u}_j$.

$$d_{ab}' = \frac{\partial d_a}{\partial u_b}, \qquad (\vec{d}' : \vec{u})_a = \sum_b \frac{\partial d_a}{\partial u_b} u_b \tag{1.196}$$

Employing, as usual, an expansion of $\vec{u}$ over normal coordinates,

$$Q(\vec{k}, l)e^{-i\vec{k}\cdot\vec{r}} \tag{1.197}$$

where $\vec{k}$ is a (quasi) wave vector and l denotes a certain mode branch. We can present the dipole moment as

$$\vec{d}_j(X_j) = \vec{d}_j + \sum_{lk} \vec{M}(\vec{k}, l)Q(\vec{k}l)e^{-i\vec{k}\cdot\vec{r}_j}$$

$$+ \sum_{l_1 l_2 \vec{k}_1 \vec{k}_2} \vec{M}(\vec{k}_1 l_1, \vec{k}_2 l_2)Q(\vec{k}_1 l_1)Q(\vec{k}_2 l_2)e^{-i(\vec{k}_1 + \vec{k}_2)\cdot\vec{r}_j} + \cdots \tag{1.198}$$

where $\vec{d}_j$ and $\vec{M}$ are electronic operators. At the same time the spatial Fourier transformation of the field is given by Eq. (1.157). Therefore the interaction energy (1.193) can be presented as

$$V_j = \sum_{\vec{f}} V(\vec{f})e^{i\vec{f}\cdot\vec{r}_j} + \sum_{\vec{f}\,kl} V(\vec{f}; \vec{k}l)e^{i\vec{f}\cdot\vec{r}_j - i\vec{k}\cdot\vec{r}_j}$$

$$+ \sum_{\vec{f}\,\vec{k}_1 \vec{k}_2 l_1 l_2} V(\vec{f}; \vec{k}_1 l_1; \vec{k}_2 l_2)e^{-i(\vec{k}_1 + \vec{k}_2 - \vec{f})\cdot\vec{r}_j} + \cdots \tag{1.199}$$

where

$$V(\vec{f}) = -\vec{d}_j \cdot \vec{E}(\vec{f},t); \qquad V(\vec{f};\vec{k},l) = -[\vec{M}(\vec{k},l)\cdot\vec{E}(\vec{f},t)]Q(\vec{k},l)$$

$$V(\vec{f};\vec{k}_1 l_1;\vec{k}_2\, l_2) = -[\vec{M}(\vec{k}_1 l_1;\vec{k}_2 l_2)\cdot\vec{E}(\vec{f},t)]Q(\vec{k}_1,l_1)Q(\vec{k}_2,l_2) \qquad (1.200)$$

The summation over j in expressions (1.191) and (1.192) may be transformed to the integration

$$\sum_j \to n \int_U d^3\vec{r} \qquad (1.201)$$

where n is the (uniform) density of the cells and U is the volume of the system. Using Eqs. (1.199)–(1.201), we get from Eqs. (1.191) and (1.192)

$$
\begin{aligned}
\Delta\rho_{mv;mv'} = \frac{4\pi}{\hbar^2} N \sum_w \rho_{0w;0w} \Bigg\{ & \sum_{\vec{f}_1\vec{f}_2} \Delta(\vec{f}_1+\vec{f}_2)\langle m|\hat{V}^I(\omega=0,\vec{f}_1)|0\rangle \\
& \times \langle 0|\hat{V}^I(\omega=0,\vec{f}_2)|m\rangle\langle v|w\rangle\langle w|v'\rangle \\
& + \sum_{\vec{f}_1\vec{f}_2\vec{k}l} \Delta(\vec{f}_1+\vec{f}_2-\vec{k})[\langle m|\hat{V}^I(\omega=0,\vec{f}_1)|0\rangle \\
& \times \langle 0w|\hat{V}^I(\omega=0,\vec{f}_2,\vec{k}l)|mv'\rangle\langle v|w\rangle \\
& + \langle mv|\hat{V}^I(\omega=0,\vec{f}_1,\vec{k}l)|0w\rangle\langle 0|\hat{V}^I(\omega=0,\vec{f}_2)|m\rangle\langle w|v'\rangle]\Bigg\} + \cdots
\end{aligned}
$$

$$(1.202)$$

$$
\begin{aligned}
\Delta\rho_{0v;0v'} = \frac{2\pi i}{\hbar^2} N \int_{-\infty}^{\infty} d\omega\,\zeta(\omega) \Bigg\{ & \sum_{n\vec{f}_1\vec{f}_2 w} \Delta(\vec{f}_1+\vec{f}_2) \\
& \times [\rho_{0v';0v'}\langle 0|(\hat{V}^I(-\omega,\vec{f}_1)|n\rangle\langle n|\hat{V}^I(\omega,\vec{f}_2)|0\rangle\langle v|w\rangle\langle w|v'\rangle \\
& + \rho_{0v;0v}\langle 0|\hat{V}^I(\omega,\vec{f}_1)|n\rangle\langle n|\hat{V}^I(-\omega,\vec{f}_2)|n\rangle\langle v|w\rangle\langle w|v'\rangle] \\
& + \sum_{nw\vec{k}l\vec{f}_1\vec{f}_2} \Delta(\vec{f}_1+\vec{f}_2-\vec{k})[\langle 0|\hat{V}^I(-\omega,\vec{f}_1)|n\rangle \\
& \times \langle nw|\hat{V}^I(\omega,\vec{f}_2,\vec{k}l)|0v'\rangle\langle v|w\rangle\rho_{0v';0v'} \\
& + \langle 0v|\hat{V}^I(-\omega,\vec{f}_1,\vec{k}l)|nw\rangle\langle n|\hat{V}^I(\omega,\vec{f}_2)|0\rangle\langle w|v'\rangle\rho_{0v';0v'} \\
& + \langle 0|\hat{V}^I(\omega,\vec{f}_1)|n\rangle\langle nw|\hat{V}^I(-\omega,\vec{f}_2,\vec{k}l|0v'\rangle\langle v|w\rangle\rho_{0v;0v} \\
& + \langle 0v|\hat{V}^I(\omega,\vec{f},\vec{k}l)|nw\rangle\langle n|\hat{V}^I(-\omega,\vec{f}_2)|0\rangle\langle w|v'\rangle\rho_{0v;0v}]\Bigg\} \quad (1.203)
\end{aligned}
$$

Here $N = Un$ is the total number of cells, and if we neglect Umklapp processes, the Δ function operates as a Kronecker delta,

$$\Delta(\vec{f}) = \delta_{\vec{f},0} \qquad (1.204)$$

Brackets $\langle v|w \rangle$ denote integrals over vibrational functions of excited and ground electronic states (Frank–Condon factors). In the case of weak electron coupling $\langle v|w \rangle = \delta_{vw}$. We illustrate the application of these formulas in the case of resonant excitation of the electronic state $|m\rangle$ by the short laser pulse. It follows from Eq. (1.202) that time–space coherence characterized by the wave vector

$$\vec{k} = \vec{f}_1 + \vec{f}_2 \qquad (1.205)$$

of branch l has the form (in the weak coupling case)

$$\langle m, v_{\vec{k}} | \Delta \rho | m, v_{\vec{k}}' \rangle = \beta_{m0} Q_{vv'}(\vec{k}, l) \qquad (1.206)$$

where

$$\begin{aligned}
\beta_{m0} = \frac{4\pi N}{\hbar^2} \sum_{\vec{f}_1 \vec{f}_2(\vec{f}_1 + \vec{f}_2 = \vec{k})} & \{ [\vec{d}_{m0} \cdot \hat{\vec{E}}(\vec{f}_1, \omega_{m0})] \\
& \times [\vec{M}_{0m}(\vec{k}, l) \cdot \hat{\vec{E}}(f_2, \omega_{0v;mv'})] \rho_{0v;0v} \\
& + [\vec{M}_{m0}(\vec{k}, l) \cdot \hat{\vec{E}}(\vec{f}_1, \omega_{mv;0v'})] [\vec{d}_{0m} \cdot \hat{\vec{E}}(f_2, \omega_{0m})] \rho_{0v';0v'} \}
\end{aligned} \qquad (1.207)$$

Thus, the space–time coherence (1.206) or (1.207) is created, provided there is an overlap of the Fourier components of the type

$$\hat{\vec{E}}(\vec{f}, \omega_{mv';0v'}) \qquad \text{and} \qquad \hat{\vec{E}}(\vec{k} - \vec{f}, \omega_{0m}) \qquad (1.208)$$

which is possible for pulses resonant to ω_{n0} and of duration T, satisfying the condition

$$\omega_{\text{vib}} T \lesssim 1 \qquad (1.209)$$

where ω_{vib} is a characteristic vibrational (or any other slow-motion) frequency. The mean value of displacement $\vec{u}$ given by Eq. (1.194) at point $\vec{r}$ taken with the aid of the density matrix (1.206) will exhibit spatial periodic behavior $e^{i\vec{k} \cdot \vec{r}}$ apart from the coherent time behavior.

4. Probe Pulse—Generalized Susceptibilities of Nonstationary and Nonhomogeneous Media

As a result of the ultrashort femtosecond pump laser pulses, the space–time coherences are created in the medium. The probe pulse, or continuum electromagnetic field, will be affected by the coherences created by the pump pulse. In the linear approximation (for weak enough probe pulses) electromagnetic properties of the medium activated by the pump pulse may be described by the generalized susceptibilities. These susceptibilities connect polarization and electric field (or magnetization and magnetic field). These connections are called material equations of the medium. General properties of the linear connection between the polarization (magnetization) and the electric (magnetic) field have been considered by Kubo et al.[43] and others.[1,21,44,45]

In general the linear connection between the polarization and the electric field has the form[8]

$$P_a(\vec{r},t) = \sum_b \int \int_{-\infty}^{t} \psi_{ab}(\vec{r},t;\vec{r}_1,t_1)E_b(\vec{r}_1,t_1)\,dt_1\,d^3\vec{r}_1 \qquad (1.210)$$

Here, as before, the field and the polarization are assumed to be classical quantities. This means that we neglect quantum fluctuations of these quantities. Functions ψ_{ab} are characteristics of the medium, determining its linear electrodynamic properties. Another representation of the connection between polarization and electric field, equivalent to Eq. (1.210), can be obtained using the Fourier representation of the electric field,

$$E_a(\vec{r},t) = \int \int \hat{E}_a(\vec{k},\omega)e^{-i(\omega t - \vec{k}\cdot\vec{r})}\,d\omega\,d^3\vec{k} \qquad (1.211)$$

This representation includes as a specific case a presentation in the form of a sum over various spacially monochromatic waves (1.157),

$$E_a(\vec{r},t) = \sum_{\vec{f}} \int E_a(\vec{f},\omega)e^{-i(\omega t - \vec{f}\cdot\vec{r})}\,d\omega \qquad (1.212)$$

where

$$\hat{E}_a(\vec{k},\omega) = \sum_{\vec{f}} \delta(\vec{k}-\vec{f})E_a(\vec{f},\omega) \qquad (1.213)$$

Substituting this relation into Eq. (1.211), we get Eq. (1.212). Developing the polarization in the Fourier integral (and sum) and substituting Eqs. (1.211) and (1.212) into Eq. (1.210), we get

$$P_a(\vec{r},t) = \sum_b \int d^3\vec{k} \int d\omega \int d^3\vec{k}_1 \int d\omega_1 \tilde{\chi}_{ab}(\vec{k}\omega,\vec{k}_1\omega_1)\hat{E}_b(\vec{k}_1,\omega_1)e^{-i(\omega t - \vec{k}\cdot\vec{r})}$$

$$(1.214)$$

or

$$P_a(\vec{r}, t) = \sum_b \sum_{\vec{f}\,\vec{f}_1} \int d\omega \int d\omega_1 \chi_{ab}(\vec{f}\omega, \vec{f}_1\omega_1) E_b(\vec{f}_1\omega_1) e^{-i(\omega t - \vec{f}\cdot\vec{r})} \quad (1.215)$$

where generalized linear susceptibilities are expressed through the function ψ_{ab} in the form

$$\tilde{\chi}_{ab}(\vec{k}\omega, \vec{k}_1\omega_1) = \frac{1}{(2\pi)^4} \int d^3\vec{r} \int_{-\infty}^{\infty} dt \int d^3\vec{r}_1 \int_{-\infty}^{t} dt_1\, e^{i(\omega t - \vec{k}\cdot\vec{r})}$$
$$\times\, e^{-i(\omega_1 t_1 - \vec{k}_1\cdot\vec{r}_1)} \psi_{ab}(\vec{r}, t; \vec{r}_1, t_1) \quad (1.216)$$

$$\chi_{ab}(\vec{f}\omega, \vec{f}_1\omega_1) = \frac{1}{2\pi U} \int d^3\vec{r} \int_{-\infty}^{\infty} dt \int d^3\vec{r}_1 \int_{-\infty}^{t} dt_1\, e^{i(\omega t - \vec{f}\cdot\vec{r})}$$
$$\times\, e^{-i(\omega_1 t_1 - \vec{f}_1\cdot\vec{r}_1)} \psi_{ab}(\vec{r}, t; \vec{r}_1, t_1) \quad (1.217)$$

These generalized susceptibilities can describe electrodynamic properties in any linear medium, including nonstationary and inhomogeneous ones, that is, the medium for which the properties depend on time t and coordinates $\vec{r}$ (even when the external field is not applied). The bulk of physical literature deals with stationary and uniform media. In this case functions ψ_{ab} characterizing such a medium (which is not perturbed by the external field) will depend, as is well known, only on the differences of their arguments,

$$\psi_{ab}(\vec{r}, t; \vec{r}_1, t_1) = \psi_{ab}(\vec{\rho}, \tau) \quad (1.218)$$

where $\vec{\rho} = \vec{r} - \vec{r}_1, \tau = t - t_1$. Substituting Eq. (1.218) into Eq. (1.216) we find that, for stationary and uniform media,

$$\tilde{\chi}_{ab}(\vec{k}\omega, \vec{k}_1\omega_1) = \delta(\vec{k} - \vec{k}_1)\delta(\omega - \omega_1)\bar{\chi}_{ab}(\vec{k}, \omega) \quad (1.219)$$

where
$$\tilde{\chi}_{ab}(\vec{k}, \omega) = \int d^3\vec{\rho} \int_0^{\infty} d\tau\, \psi_{ab}(\vec{\rho}, \tau) e^{i(\omega\tau - \vec{k}\cdot\vec{\rho})} \quad (1.220)$$

and relation (1.214) takes the form

$$P_a(\vec{r}, t) = \sum_b \int\int \tilde{\chi}_{ab}(\vec{k}, \omega)\hat{E}_b(\vec{k}, \omega) e^{-i(\omega t - \vec{k}\cdot\vec{r})}\, d\omega\, d^3\vec{k} \quad (1.221)$$

or
$$P_a(\vec{r}, t) = \sum_{\vec{f}\, b} \int \bar{\chi}_{ab}(\vec{f}, \omega) E_b(\vec{f}, \omega) e^{-i(\omega t - \vec{f}\cdot\vec{r})}\, d\omega \quad (1.222)$$

Therefore we see that description of the electromagnetic properties of the media by susceptibility χ_{ab} as a function of $\vec{k}$ and ω only (and in the majority of cases by ω only) is valid for stationary and uniform media only.

As has been mentioned, the medium subjected to ultrashort pulses is neither stationary nor uniform. Properties of these media have to be described by the generalized susceptibilities.

To derive these susceptibilities, we again will start from the interaction energy (1.139) or (1.193),

$$V = -\sum_j \vec{d}_j(\vec{x}_j, \vec{X}_j) \cdot \vec{E}(\vec{r}_j, t) \tag{1.223}$$

Here X_j describes a specific nuclear configuration and x_j, the electronic coordinates. As in the previous section, we expand the dipole moment in the powers of displacements (1.194) from the equilibrium configuration X_j^0. To find the polarization of the media, we first find the mean value of $\vec{d}_j$ at cell j and configuration X_j. Using Eqs. (1.142) and (1.223), we get in the usual way

$$\langle d_{ja} \rangle = \frac{i}{\hbar} \sum_{j'b} \int_{-\infty}^{t} \langle [d_{ja}(\vec{x}_j, \vec{X}_j, t), d_{j'b}(\vec{x}_{j'}, \vec{X}_{j'}, t_1)] \rangle E_b(\vec{r}_{j'}, t_1) dt_1 \tag{1.224}$$

where the mean value in the right-hand side is taken over the unperturbed density matrix (without electric field E_b). Notice that for the diagonal unperturbed density matrix (i.e., without coherences present) and for uniform media, the term in brackets in Eq. (1.224) depends only on the difference of their arguments. From now on we use the approximation (1.189), which means that we neglect a spatial dispersion connected with the electronic motion. Thus the mean value of the dipole moment of point j induced by the electric field takes the form

$$\langle d_{ja} \rangle = \frac{i}{\hbar} \sum_{b} \int_{-\infty}^{t} \langle [d_{ja}(\vec{x}_j, \vec{X}_j, t), d_{jb}(\vec{x}_j, \vec{X}_j, t_1)] \rangle E_b(\vec{r}_j, t_1) dt_1 \tag{1.225}$$

To find the polarization P_a, we have to multiply $\langle d_{ja} \rangle$ by the density of cells n, which we assume to be constant,

$$P_a(\vec{r}, t) = n \langle d_{ja} \rangle \tag{1.226}$$

Now we can determine the function $\psi_{ab}(\vec{r}, t; \vec{r}_1 t_1)$ [see Eq. (1.210)] in our model,

$$\psi_{ab}(\vec{r},t;\vec{r_1}t_1) = \frac{ni}{\hbar}\delta(\vec{r}-\vec{r_1})\langle[d_a(\vec{X},t),d_b(\vec{X},t_1)]\rangle \qquad (1.227)$$

Here $d_a(\vec{X},t)$ are electronic operators that depend on $\vec{X}$ as parameters; the latter are functions of the coordinates $\vec{r}$. We skipped index j, assuming that dipole moment operators of all cells are identical. The operator $\vec{d}$ can be expressed similarly to Eq. (1.198),

$$\vec{d}(\vec{X},t) = \vec{d} + \sum_{lk} \vec{M}(\vec{k},l)Q(\vec{k}l)e^{-i\vec{k}\cdot\vec{r}}$$

$$+ \sum_{l_1\vec{k}_1;l_2\vec{k}_2} \vec{M}(\vec{k_1}\,\vec{l_1};\vec{k_2}l_2)Q(\vec{k_1}\,l_1)Q(\vec{k_2}l_2)e^{-i(\vec{k_1}+\vec{k_2})\cdot\vec{r}} + \cdots \qquad (1.228)$$

For function ψ_{ab} presented by Eq. (1.227), susceptibilities (1.216) and (1.217) have the forms

$$\tilde{\chi}_{ab}(\vec{k}\omega,\vec{k_1}\omega_1) = \frac{n}{(2\pi)^4}\frac{i}{\hbar}\int d^3\vec{r}\,e^{-i(\vec{k}-\vec{k_1})\cdot\vec{r}}\int_{-\infty}^{\infty} dt \int_{0}^{\infty} d\tau\, e^{l(\omega-\omega_1)t}e^{i\omega_1\tau}$$

$$\times \langle[d_a(\vec{X},t),d_b(\vec{X},t-\tau)]\rangle \qquad (1.229)$$

and

$$\chi_{ab}(\vec{f}\omega,\vec{f_1}\omega_1) = \frac{n}{2\pi U}\frac{i}{\hbar}\int d^3\vec{r}\,e^{-i(\vec{f}-\vec{f_1})\cdot\vec{r}}\int_{-\infty}^{\infty} dt \int_{0}^{\infty} d\tau\, e^{l(\omega-\omega_1)t}e^{i\omega_1\tau}$$

$$\times \langle[d_a(\vec{X},t),d_b(\vec{X},t-\tau)]\rangle \qquad (1.230)$$

As we see in our approximation (1.189), which is equivalent to the neglect of the electronic part of spatial dispersion, the susceptibilities (1.229) and (1.230) depend on the differences of their wave vectors.

To analyze susceptibilities (1.229) and (1.230), we have to consider functions

$$\tilde{\psi}_{ab} = \langle[d_a(\vec{X},t),d_b(\vec{X},t-\tau)]\rangle \qquad (1.231)$$

while the response function (1.227) is equal to

$$\tilde{\psi}_{ab}(\vec{r},t;\vec{r_1},t_1) = \frac{ni}{\hbar}\delta(\vec{r}-\vec{r_1})\tilde{\psi}_{ab} \qquad (1.232)$$

In the adiabatic approximation, and using expansion (1.228), we get (in the weak coupling case)

$$\tilde{\psi}_{ab} = \sum_{v'v} \langle mv|d_a(\vec{X},t), d_b(\vec{X},t-\tau)]|mv'\rangle\rho_{mv';mv}$$

$$= \sum_v \langle mv|[d_a(\tau), d_b(0)]|mv\rangle\rho_{mv;mv}$$

$$+ \sum_{v'vk l} \langle mv|[d_a(\tau), M_b(\vec{k},l,0)]|mv\rangle\langle v|Q(\vec{k},l,t-\tau)|v'\rangle e^{-i\vec{k}\cdot\vec{r}}\rho_{mv';mv}$$

$$+ \sum_{vv'kl} \langle mv|[M_a(\vec{k},l,\tau), d_b(0)]|mv\rangle\langle v|Q(\vec{k},l,t)|v'\rangle e^{-i\vec{k}\cdot\vec{r}}\rho_{mv';mv} + \cdots$$

$$(1.233)$$

where the omitted terms are proportional to products of Q and $e^{-i(\vec{k}_1+\vec{k}_2)\cdot\vec{r}}$. It is easy to see that the first sum in Eq. (1.233) leads to the usual (electronic) susceptibility when substituted into Eq. (1.229),

$$\tilde{\chi}_{ab}(\vec{k},\omega;\vec{k}_1,\omega_1) = \delta(\vec{k}-\vec{k}_1)\delta(\omega-\omega_1)\sum_v \frac{ni}{\hbar}\int_0^\infty e^{i\omega\tau}$$

$$\times \langle mv|[d_a(\tau), d_b(0)]|mv\rangle\rho_{mv;mv}$$

$$= \delta(\vec{k}-\vec{k}_1)\delta(\omega-\omega_1)\chi_{ab}(\omega) \qquad (1.234)$$

We now concentrate our attention on the second and third sums in Eq. (1.233). Substituting them into Eq. (1.230), we get

$$\chi_{ab}(\vec{f}\omega;\vec{f}_1\omega_1)$$

$$= \sum_{v'v,l} \frac{ni}{2\pi\hbar}\rho_{mv';mv}\int_{-\infty}^\infty dt \int_0^\infty d\tau\{\langle mv|[d_a(\tau), M_b(\vec{f}_1-\vec{f},l,0)]|mv\rangle$$

$$\times \langle v|Q(\vec{f}_1-\vec{f},l,t-\tau)|v'\rangle$$

$$+ \langle mv|[M_a(\vec{f}_1-\vec{f},l,\tau), d_b(0)]|mv\rangle\langle v|Q(\vec{f}-\vec{f}_1,l,t)|v'\rangle\}e^{i(\omega-\omega_1)t}e^{i\omega_1\tau}$$

$$(1.235)$$

It is easy to see that in the Placzek approximation one can neglect τ in the matrix element of Q (first term). In this case we get

$$\chi_{ab}(\vec{f},\omega;\vec{f}_1,\omega_1) = \sum_l \langle \hat{Q}(\vec{f}_1-\vec{f},l;\omega-\omega_1)\rangle\chi_{ab}(\vec{f}_1-\vec{f},l,\omega_1) \quad (1.236)$$

Here $\chi_{ab}(\vec{f}_1-\vec{f},l,\omega_1)$ is the susceptibility, defined by

$$\chi_{ab}(\vec{f}-\vec{f}_1,l,\omega_1) = \frac{ni}{\hbar}\int_0^\infty e^{i\omega_1\tau}\langle m|[d_a(\tau), M_b(\vec{f}_1-\vec{f},l,0)]$$

$$+ [M_a(\vec{f}_1-\vec{f},l,\tau), d_b(0)]|m\rangle\, d\tau \qquad (1.237)$$

and

$$\langle \hat{Q}(\vec{f}_1 - \vec{f}, l, \omega - \omega_1) \rangle = \frac{1}{2\pi} \int_{-\infty}^{\infty} \sum_{v,v'} \rho_{mv';mv} \langle v | Q(\vec{f}_1 - \vec{f}, l, t) | v' \rangle e^{i(\omega - \omega_1)t}$$

(1.238)

In general (when the Placzek approximation is not necessarily valid), the generalized susceptibility (1.235) is essentially nonzero, when there exists a coherence on the mode l with

$$\vec{k} = \vec{f}_1 - \vec{f}$$

(1.239)

and

$$\omega_1 - \omega = \omega(\vec{k}, l)$$

(1.240)

where $\omega(\vec{k}, l)$ is the vibrational frequency of the lth branch and with wave vector $\vec{k}$.

Now we are able to give the expression for the polarization as function of the electric field in our model. Having in mind that the mean value of the pure electronic dipole moment $\langle \vec{d} \rangle$ vanishes, we get from Eqs. (1.228), (1.214), (1.215), (1.239), (1.235), and (1.236)

$$P_a(\vec{r}, t) = n \sum_{lk} M_a(\vec{k}, l) \langle Q(\vec{k}l, t) \rangle e^{-i\vec{k} \cdot \vec{r}}$$

$$+ n \sum_{l_1 \vec{k}_1 l_2 \vec{k}_2} M_a(\vec{k}_1 l_1; \vec{k}_2 l_2) \langle Q(\vec{k}_1 l_1, t) \rangle \langle Q(\vec{k}_2 l_2, t) \rangle e^{-i(\vec{k}_1 + \vec{k}_2) \cdot \vec{r}} + \cdots$$

$$+ \sum_{b\vec{k}\vec{f}_1} \int \int \chi_{ab}(\vec{f}_1 - \vec{k}, \omega; \vec{f}_1 \omega_1) E_b(\vec{f}_1 \omega_1) e^{-i(\omega t - \vec{f}_1 \cdot \vec{r})} e^{-i\vec{k} \cdot \vec{r}} \, d\omega_1 \, d\omega$$

(1.241)

According to the previously mentioned property of χ_{ab} (when electronic spatial dispersion is neglected), it does not depend on $\vec{f}_1$. This means that, the electric-field-dependent part of the polarization can be written as

$$\tilde{P}_a(\vec{r}, t) = \sum_{bk} \int \int \chi_{ab}(\vec{k}\omega; \omega_1) E_b(\vec{r}, \omega_1) e^{-i\vec{k} \cdot \vec{r}} e^{-i\omega t} \, d\omega_1 \, d\omega$$

(1.242)

This expression can be further transformed, performing summation over $\vec{k}$,

$$\sum_{k} \chi_{ab}(\vec{k}, \omega; \omega_1) e^{-i\vec{k} \cdot \vec{r}} = \chi_{ab}(\vec{r}, \omega; \omega_1)$$

(1.243)

and using the relation[35] connecting frequency–frequency susceptibility

$\chi_{ab}(\omega, \omega_1)$ with frequency–time susceptibility,

$$\chi_{ab}(\omega, \omega_1) = \frac{1}{2\pi} \int_{-\infty}^{\infty} e^{i(\omega - \omega_1)t} \, dt \, \chi_{ab}(\omega_1, t) \qquad (1.244)$$

As a result we obtain

$$\tilde{P}_a(\vec{r}, t) = \sum_b \int \chi_{ab}(\vec{r}, \omega_1, t) E_b(\vec{r}, \omega_1) e^{-i\omega_1 t} \, d\omega_1 \qquad (1.245)$$

This expression is quite general (but neglecting electronic spatial dispersion). In the Placzek approximation the time and coordinate dependence of $\chi_{ab}(\vec{r}, \omega, t)$ is proportional to the displacement $\vec{u}(\vec{r}, t)$, which in turn was created by the pump pulse. The latter case has been considered by Nelson[36] and corresponds to the case illustrated in Figs. 1.12 and 1.13.

It should be mentioned that the polarization contains not only the part (1.245) induced by the probe pulse, but also another part connected with time- and coordinate-dependent displacements [first sums in Eq. (1.244)], which was created by the pump pulse.

Analysis of the propagation, absorption, and scattering of the probe pulse, with material Eqs. (1.241) and (1.245), is beyond the scope of this chapter. It should however be mentioned that the absorption rate determined by the generalized susceptibilities has been considered in the literature.[35,34] The spontaneous emission rate from the excited coherent state has been analyzed in Lin et al. The scattering of the coherence has been considered in Refs. 36 (in the Placzek approximation).

5. Discussion

To summarize the results, we will again look at Figs. 1.11–1.13. Using the adiabatic separation we can distinguish between fast and slow motion of certain systems: molecules, solid, and liquid. We denote electronic (or other fast motion) levels by $o, m, l, \ldots$; energy levels describing slow motions are denoted by $w, v, v', \ldots$. These may be levels of intramolecular vibrations or rotations. They can also be levels of collective motions, such as phonons or magnons. The analysis performed in this section is quite general and does not depend on the specific model of the molecule or solid.

We consider the action of two laser pulses induced on the system: pump and probe pulses. At least the pump pulse is considered to be short enough, which means that the pump pulse duration T satisfies some or all of conditions (1.144). In this case a change in the density matrix of the system has the character of a jump. Density matrix jumps are analyzed in previous sections. In second-order perturbation theory a change in the density matrix is proportional to the square of the laser field. The spectrum of the field quadrature contains frequencies of the slow motions: it is "on speaking terms" with intramolecular vibrations, phonons, and so on. Two possible cases are depicted on Fig. 1.12. In Fig. 1.12a the carrier frequency is resonant with electronic frequencies ω_{m0}. In this case coherent slow motions are created in both the ground and the excited states. In Fig. 1.12b the carrier frequency is out of resonance with all levels of the system. In this case coherences are created only in the ground state.

General formulas have been obtained for the density matrix jumps describing coherences in both ground and excited states (Section I.C.3). It has been shown that a pump pulse creates not only coherent temporal quantum beats, but a coherent space structure as well.

The subsequent development of the system is depicted in Fig. 1.13. There may be spontaneous emission from the upper state.[24,35] The probe pulse may be scattered on the coherence in the ground state[3a] (Fig. 1.13c) or on the coherence in the excited state[36b] (Fig. 1.13d). The probe pulse may be absorbed, exciting another electric state[4,36b] (Fig. 1.13a), subsequently giving rise to the spontaneous emission (Fig. 1.13e). Finally, the probe pulse may be absorbed at the frequency of the pump pulse[37] (Fig. 1.13f).

Certainly this classification does not cover all the possible cases. Recently Dantus et al.[46] observed experimentally the time evolution of alignment and angular momenta of fragments in dissociation reactions of HgI_2 and ICN, which were induced by the ultrashort laser pulses. The evolution of the dipole alignment has been observed via fluorescence signals. The fluorescence has been observed from the state reached by the probe pulse transition. Therefore in general terms this process can be classified as illustrated in Fig. 1.13e. Chesnoy and Mokhtary[47] analyzed both experimentally and theoretically induced processes of absorption and refraction of the probing pulse at the frequency of the pump pulse (Fig. 1.13f). The following processes have been analyzed: induced dichroism, induced birefringence, frequency modulation, and amplitude modulation.

Hayashi et al.[48] developed a theory of time-resolved coherent anti-Stokes Raman scattering from molecules in liquids. Effects of the coherence transfer induced by the molecule–heat-bath interactions on the ultrashort time-resolved coherent anti-Stokes Raman scattering (CARS) from molecules in liquids are studied theoretically. Based on the perturbative density matrix formalism, an expression for the CARS intensity is derived taking into account the coherence transfer between the Raman active vibrational transitions of two molecules in liquids. The coherence transfer constants and dephasing constants are properly incorporated with the aid of Liouville space Feynman diagrams. The structure of the coherence transfer matrix element, which expresses the time evolution of the coherence between the relevant transitions, is clarified by solving the master equation with the coherence transfer and dephasing constants in the Markov approximation. Frequency shifts of the quantum beats appear in the time-resolved CARS as a result of the coherence transfer. A multispherical layer model is adopted in evaluating the coherence transfer effects in liquids in femtosecond time domains. Model calculations of time-resolved CARS spectra have been carried out to demonstrate the coherence transfer effects in both short- and long-range coherence transfer cases. It is predicted that the quantum beats are amplified in the time-resolved CARS spectra of molecules in liquids in a long-range coherence transfer case when there exist differences in the coherence transfer constants between spherical layers.

Subsequently effects of vibronic coherence transfer induced by the heat bath on ultrafast time-resolved resonant light scattering (RLS) spectra were theoretically investigated by Fujimura's group using the master equation approach.[49] The vibronic coherence initially created by a coherent optical excitation transfers to other vibronic coherent states due to inelastic interactions between the vibronic system concerned (the relevant system) and the heat bath. The vibronic coherence transfer results in the quantum beats in the time-resolved RLS spectra. The bath-induced vibronic transition operator is derived in the double space representation of the density matrix theory. Model calculations of the femtosecond time-resolved RLS spectra were performed to demonstrate the effects of the bath-induced vibronic coherence transfer.

Recently Kono and Fujimura[50] have treated the resonance secondary emission (RSE) in femtosecond laser excitation in reference to the motion of the created wave packet moving on the excited-state potential surface. The density matrix of emitted light for the multi-intermediate-level system is obtained, from which the emission correlation function is derived. The correlation function is put into the theoretical expression of the time-dependent "physical spectrum" for the Fabry–Perot interferometer (which is used in order to consider temporal and energetic resolution inherent in detection). The compact and practical expressions obtained connect the time- and frequency-resolved spectrum with the time evolution of the wave packet. Numerical results for a displaced harmonic oscillator model indicate that the time- and frequency-resolved spectrum can reveal how the wave packet created by a femtosecond laser pulse travels on the excited potential surface if the response time $1/\Gamma_d$ of the photodetector satisfies the relation that $\Omega <$ $\Gamma_d <$ca. the Stokes shift (where Ω is the vibrational frequency). They have shown that the excited-state wave function can be split into two terms, the one that adiabatically follows the temporal change in incident light (the adiabatic term) and the one that represents the effect of spectral broadening of light (the Fourier broadening term). It is only the Fourier broadening term that survives after the termination of incident light and reflects the motion of the created wave packet on the excited potential surface. In off-resonance excitation, the adiabatic term produces Raman-like emission and the Fourier broadening term produces fluorescencelike emission. In resonance excitation, these two terms are indistinguishable from each other with respect to emission frequency: for the duration of incident light, the adiabatic term offsets the Fourier broadening term, leading to a slow buildup of intensity in the time- and frequency-resolved spectrum (which is slower than the initial rise of the incident pulse profile).

II. TRANSITION-STATE SPECTROSCOPY

The possibility of direct spectroscopic observation of transient species which are intermediate between reactants and products, usually referred to as the transition state or reaction complex, has received considerable attention in the past few years.

Chemical reaction events usually occur on a subpicosecond time scale, making observation during the event difficult. Therefore until recently all experimental knowledge of chemical reactions had been gleaned from observations on the asymptotic states of the reactants and products. Unfortunately reconstruction of the reactive event from such asymptotic observations, no matter how detailed, is difficult and probably not unique. A deeper understanding of the reaction process seems to require a more direct probe of the reaction event.

Several groups[51-55] have recently become interested in developing methods for the direct spectroscopic study of systems in the process of chemical reaction. The methods explored to date include photoexcitation during reaction with detection either by fluorescence[51,52] or by photoionization,[53] spontaneous emission during chemical reaction,[54] and resonance Raman spectroscopy through molecules in the process of dissociating.[55] Each approach has sought to elucidate the nature of the reaction event itself by obtaining a spectrum of the system in the midst of reaction. Spectroscopic interrogation of these transition states requires the absorption or emission of a photon during collision. This phenomenon is comparatively well known (usually as line broadening) during nonreactive collisions, and can be studied relatively easily at high pressures of the colliding pairs. For simple atomic systems, the line profile can be used to determine the difference between ground and excited electronic potential energy curves. (See, for example, references contained in Refs. 1b, 4c, and 5b, and Ref. 56.) If the ground-state potential is known, then both potentials can be mapped out and the collision completely characterized.

Very recently the measurement of transition-state spectra in real time has become possible due to the rapid development in picosecond and femtosecond spectroscopy. Due to our limited expertise in this area only some systems will be reviewed.

A. NaI and NaBr

While not the first of the femtosecond experiments, the NaI and NaBr studies have proven to be some of the most interesting. This is due to their unique electronic structure. Most alkali halides have an exited continuum absorption spectrum. However, NaI in particular has a dense discrete line spectrum in the energy range given by the wavelengths of 300–400 nm. This is due to the fact that the ionic ground state crosses over the covalent excited state (Fig. 2.1). In the process of the excited molecule decaying into products, the transition-state molecules can experience two different situations. The state can be trapped on the adiabatic surface formed by the ground and excited states, or it can pass through the crossing point of these two potential energy surfaces (PES) and remain on the dissociative covalent surface. In this latter case it moves directly to the product, while in the former situation the molecule remains in the adiabatic well oscillating.

The earlier studies of NaI by Berry[20] and Schaefer et al.[19] laid the ground work for interpretation of the femtosecond work. Schaefer et al. gave much

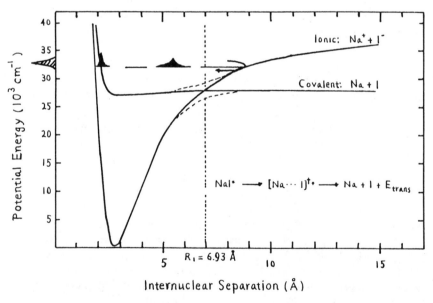

Figure 2.1. PESs of NaI and wave-packet propagation. Hatched curve at left shows typical pump energy. Dispersion of wave packet as it moves from its initial state on the excited surface is shown schematically. Level crossing between ionic and covalent surfaces occurs at $R_1 = 6.93$ Å.

of the necessary information about the excited PES in the 29,250–33,250-cm^{-1} range.

The first results of the femtosecond transition-state spectroscopy, (FTS) of NaI and NaBr came in two papers by Zewail's group.[4b,4c] With equipment similar to their earlier ICN work, they studied the reaction

$$NaI(^1\Sigma_{10}^+) + \hbar\omega \rightarrow [NaI(O^+)] \rightarrow Na(^2S) + I(^2P). \tag{2.1}$$

Using a femtosecond pump probe FTS technique. (A similar reaction can be written for NaBr.) In this process a very short laser pulse was used to initiate photodissociation of the gaseous NaI or NaBr. The evolution of the excited system was monitored with a delayed probe of a different wavelength. In the case of the NaI the pump pulse was used with wavelengths in the range of 200–328 nm, and the system was probed with light with wavelengths in the range of 500–612 nm. The probe pulse wavelength is centered around the D lines of free sodium (589 nm).

Briefly what occurs after the pump pulse is that the excited species undergoes dissociation. However, the NaI molecules become trapped in the adiabatic well formed by the ground- and excited-state energy surfaces and they oscillate until they eventually decay out of the well onto the outer Na(^2S) + I(^2P) surface. On the other hand, the NaBr system decays very rapidly out of the adiabatic well.

This behavior shows up in the spectra developed by Zewail's group. One can see these oscillations in the NaI system by observing the laser-induced fluorescence (LIF) generated by the probe laser pulse. The probe pulse is introduced at various time delays and will induce fluorescence depending on the condition of the transition state. Basically when the transition state is in a configuration such that it is in resonance with the probe, it is excited to a higher energy state and the system fluoresces. In the case of earlier work on ICN there was a direct relation between the fluorescence and the internuclear separation of the dissociating fragments.

Figures 2.2 and 2.3 show the fluorescence signals for NaI at a number of different probe and pump wavelengths obtained by Zewail's group. They are generally similar in nature except for Fig. 2.2b, which shows a steplike increasing signal. This is a result of the fact that the probe pulse wavelength in Fig. 2.2b is the on-resonance wavelength for free sodium. In essence Fig. 2.2b is a measure of the buildup of free sodium which has dissociated out of the adiabatic well. The steps show that the leakage out of the well is in synchronization with the oscillations of the NaI trapped in the well. A corresponding behavior is not seen for NaBr. It appears that while NaI is adiabatically trapped in the well, NaBr is "diabatically" free and leaks rapidly from the well. This behavior is indicated in Fig. 2.4. In this figure there exists only

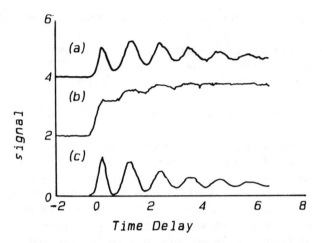

Figure 2.2. FTS results from NaI. (*a*) $\lambda_{probe} = 580$ nm. (*b*) $\lambda_{probe} = 590$ nm (on-resonance). (*c*) $\lambda_{probe} = 612$ nm. In each case $\lambda_{pump} = 310$ nm. Time delay, in picoseconds, has been set to zero arbitrarily in each case at the first oscillation peak.

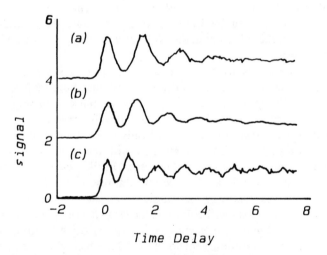

Figure 2.3. FTS results from NaI. (*a*) $\lambda_{pump} = 300$ nm. (*b*) $\lambda_{pump} = 310$ nm. (*c*) $\lambda_{pump} = 328$ nm. In each case $\lambda_{probe} = 580$ nm. Time delay, in picoseconds, has been set to zero arbitrarily in each case at the first oscillation peak. Oscillation period corresponds to energy spcings of 24, 28, and 34 cm^{-1}, respectively.

196

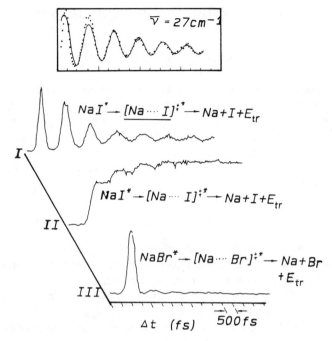

Figure 2.4. Experimental results for two reactions of NaI and NaBr. For NaI reaction, both on-resonance and off-resonance Na atom detection (LIF) is given, indicated by underlining of relevant species. Modulaton depth depends on probe wavelength. Signal is (essentially) linearly dependent on probe and pump intensities.

one main peak with virtually no additional oscillation. Thus the excited state appears to leak out during one period of motion. However, very few results have been placed into the literature for NaBr. In fact, only the one off-resonance result for NaBr has ever been given, and even that was given without stating the probe or pump frequencies. These frequencies would be helpful in understanding why the NaBr results show no asymptotic contribution from free sodium.

While there is some dependence on pump and probe frequencies, in general, the oscillations in the NaI FTS persist for 10 or more periods, but the NaBr signal is damped with two or three periods of motion.[4c] From the oscillation periods of the signals at different pump frequencies the energy level spacings of the adiabatic well can be determined. For pump frequencies from 300 to 328 nm the energy levels spacings went from 24 to 34 cm^{-1}, respectively.[4c] These compare well with the spacing calculated from the work of Schaefer et al.[19] (25–35 cm^{-1}).

Zewail's group also made some estimates of the PES based on an anharmonic Morse-type potential. They took the vibrational energy levels to be given by

$$E_n = h v_0 (n + \tfrac{1}{2}) - \chi h v_0 (n + \tfrac{1}{2})^2 \qquad (2.2)$$

where n is the quantum number, v_0 is the oscillator harmonic frequency, and χ is the anharmonicity parameter. They found the energy level spacing to decrease with higher energy, which is consistent with this type of well. This is opposite in sense to the parameters found by Schaefer et al. who used an RKR calculation to determine the adiabatic PES.[19]

In the classical limit, the relationship between energy and oscillation frequency becomes

$$v^2 = v_0^2 - 4\chi v_0 (v_{\text{pump}} - v_{\text{diss}}) \qquad (2.3)$$

where v_{diss} is the frequency corresponding to the dissociation energy. Using this relation, Zewail's group reported a value of the anharmonicity on the order of 0.1 cm^{-1}. This value was consistent in magnitude with that found by Schaefer et al. The final element of information extracted from these results was an estimate of the Landau–Zener probability P_{LZ}, which should give an indication of the rate of crossing from the adiabatic curve into products.

The Landau–Zener probability is given by

$$P_{\text{LZ}} = \exp \left[-\frac{2\pi V_{12}^2}{\hbar v(E, J) \Delta F} \right] \qquad (2.4)$$

where V_{12} is the coupling matrix element between the two surfaces, $v(E, J)$ is the recoil velocity at the crossing point given as a function of energy E and rotational quantum number J, and ΔF is the difference in the slope for the two PESs. As the covalent surface is basically flat, the couple element V_{12} can be estimated for the ground-state curve at the crossing point. The values found for P_{LZ} of approximately 0.1 lead to a value of $V_{12} = 350 \text{ cm}^{-1}$ for NaI, which is somewhat lower than the value found earlier by Schaefer et al. ($V_{12} = 434 \text{ cm}^{-1}$).

Shortly after the experimental FTS work was completed, a fair amount of activity was initiated to determine whether theory could explain the processes being seen in these experiments. Early theories to explain the FTS results for ICN had been mainly classical in nature.[6] The first article to present an analysis of the NaI result was a first-order quantum calculation using a wave-packet treatment.[4d] In this analysis the lower electronic state is denoted by $|2\rangle$ and the upper covalent state by $|1\rangle$. The coupling between the diabatic states is given by

$$V_{12}(R) = A \exp\left[-\beta(R - R_x)^2\right] \tag{2.5}$$

with $A = 0.055$ eV, $\beta = 0.69$ Å^{-1}, and $R_x = 6.93$ Å. The state of the system caused by the pump pulse is given to first order by

$$|\Psi(t)\rangle = \frac{i}{\hbar} \int_0^t dt_1 \, U(t - t_1)\vec{\mu} \cdot \vec{E}(t_1)|\Phi_i\rangle e^{-iE_1 t_1/\hbar} \tag{2.6}$$

where $|\Phi_i\rangle$ is the initial state, E_i is its energy, and $\vec{\mu}$ is the transition dipole to the covalent state. $U(t)$ is the propagator describing the motion on the two coupled electronic states. The electric field vector is given by

$$\vec{E} = \vec{E}_0 \exp(-i\omega_0 t) \exp\left[-\alpha(t - t_0)^2\right] \tag{2.7}$$

where $\vec{E}_0$ is a constant vector, $t_0 = 80$ fs, ω_0 is the pulse frequency, and $\alpha = 1.1 \times 10^{-3} fs^{-2}$. Here α was chosen so that the full-width half maximum of the pulse was 50 fs.

The evaluation of the propagator was done with a fast Fourier transform technique developed earlier for curve-crossing problems.[57] The state of the system is represented by a linear combination of two electronic states with time-dependent coefficients

$$|\Psi(t)\rangle = \chi_1(R; t)|1\rangle + \chi_2(R; t)|2\rangle \tag{2.8}$$

The basic assumption imposed to calculate the final result is that the LIF is proportional to the populations created by the pump pulse. It is further assumed that only those populations that are in the covalent state are capable of contributing to the LIF signal. The ionic states are considered to be energetically unavailable under the conditions of the experiment. Three populations were subsequently defined:

$$P_b(t) = \int_0^{R_x} |\chi_1(R; t)|^2 \, dR \tag{2.9}$$

$$P_f(t) = \int_{R_x}^{\infty} |\chi_1(R; t)|^2 \, dR \tag{2.10}$$

$$P_i(t) = \int_0^{\infty} |\chi_2(R; t)|^2 \, dR \tag{2.11}$$

The last population is for the ionic part of the total system and does not contribute to the LIF. $P_b(t)$ is the population of the bound species with

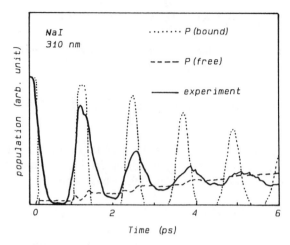

Figure 2.5. Bound and free neutral populations and observed LIF signal. A Gaussian pump pulse with a wavelength of 310 nm and half-width at half maximum of 50 fs was used. Initial state $v = 0$.

covalent character. $P_f(t)$ is the population of the covalent states with R greater than the crossing point, and therefore represents the free sodium. In summary then $P_i(t)$ gives no contribution to the LIF. $P_f(t)$ gives a contribution to the on-resonance probe signal. Finally $P_b(t)$ is related to the off-resonance signal.

The oscillations in the off-resonance signal are then given by the populations in the adiabatic well oscillating in and out of the spectroscopically bright (covalent) state and the spectroscopically dark (ionic) state (Fig. 2.5). The on-resonance signal is given simply by $P_f(t)$, which is the populations of covalent states that have achieved an internuclear separation greater than the crossing point.

Engel et al.[4d] also consider the dependence of the vibration states on the pump wavelength and found it only important for the 310-nm pump pulses. They also found that while the shape of the pulse was not extremely important, *the temporal width of the pump pulse had an extremely important effect on the populations $P_b(t)$ and $P_f(t)$.* They also reported that the main difference between the signals from NaI and NaBr had to do with the coupling, V_{12}.

Engel et al. also performed a semiclassical analysis of the motion of the wave packet based on a method presented separately by Marcus.[58] They found the time of oscillation for the wave packet to be 1.3 ps, which is within 5% of the oscillation time calculated by their quantum analysis.

In a follow-up paper Engel and Metiu gave a more extensive report of the previous analysis.[59] They detail their quantum solution and show more completely the time evolution of the diabatic wave functions. In addition, they compare their results with the Landau–Zener approximation. Finally they give a more detailed discussion of the dependence of the LIF signal on temperature, pulse length, and pulse shape.

In their analysis Engel and Metiu considered only the electronic state with the $\Omega = 0^+$ angular momentum absorption. They believe that this was responsible for not accurately reproducing the early population changes in the free sodium atom population. The experimental curves always show a sudden burst of sodium atoms within 10–20 fs, which cannot be explained by the slow decay from the adiabatic curve. Therefore this must be due to the $\Omega = 1$ projection of the neutral state, which is not coupled to the ionic state and subsequently produces the initial sudden on-resonance signal. Their results that scan the $\Omega = 1$ projections are reproduced in Fig. 2.6 for various pump frequencies. It should be remembered that they calculate the population of the various components of the pumped state and assume that this corresponds directly to the LIF signal.

In this paper Engel and Metiu go into much greater detail on the time development of the wave trains and the preparation of wave packets by the

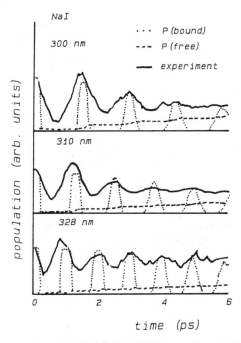

Figure 2.6. Computed bound $[P_v(t)]$ and free $[P_f(t)]$ neutral populations and experimental off-resonance LIF signals as a function of delay time. Curves in each panel are normalized so that heights of first maxima coincide.

pump laser pulse. This is necessary as they wish to discuss in detail the dependence of the spatial width of the wave trains on the pulse length and the nuclear dynamics of the excited state. Their discussion is quite detailed and the interested reader should consult their paper. Basically they show how the wave train is broadened by time development of the wave packet and the temporal length of the laser pulse. Then a description of the wave-packet development on the adiabatic surface is given, which shows the interaction with the crossing point with the ionic energy surface and the eventual loss of probability to the free sodium region of the covalent PES. They give a number of intricate figures of the development of systems up to 4.2 ps.

Engel and Metiu also test further the Landau–Zener prediction with their exact quantum calculations. They found that the Landau–Zener approximation gives good results in the pump frequency of 328 nm. It appears to be good

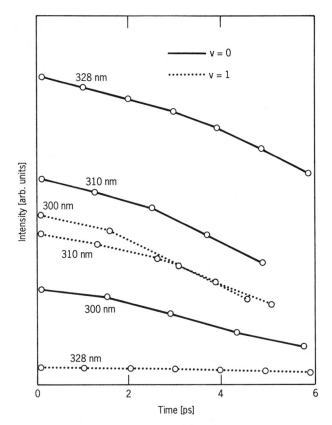

Figure 2.7. Relative intensities of population $P_v(t)$ for 328, 310, and 300 nm for initial states of $v = 0, 1$. Points on graphs give heights of maxima in $P_v(t)$ curves. Lines serve only to connect points belonging to same $P_v(t)$ curve.

also for a pump frequency of 300 nm, but the result deteriorates with the number of oscillations through the crossing point.

Engel and Metiu examined the temperature dependence of populations $P_b(t)$ and $P_f(t)$. They used only a Boltzmann weighting of the various vibrational levels for 600 °C, the experimental temperature. They basically calculated the temperature-dependent populations using

$$\bar{P}_\alpha(t; T) = \sum_v P_\alpha(v, t) e^{-\beta(\varepsilon_v - \varepsilon_0)} \tag{2.12}$$

for $\alpha = f$ or b, where $\beta = 1/kT$, ε_v are the vibrational energies, ε_0 is the ground-state energy, and $P_\alpha(v, t)$ is calculated for the initial vibration state. They found that all the curves keep relatively the same shape and the widths of the peaks are increased only slightly with increasing vibrational quantum number. It appears that the wave train is constructed in such a way that the details of the ground vibration state are not important. However, in contrast to this, the magnitudes of the excited population are strongly dependent on both the laser wavelength and the details of the initial state. These effects are demonstrated in Figs. 2.7 and 2.8 taken from their work.[59] Figure 2.7 shows the rather strong dependence of the $P_b(t)$ population on initial vibrational

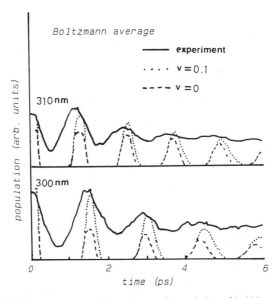

Figure 2.8. Boltzmann average of bound neutral populations for 300- and 310-nm pump pulses at a temperature of 873 K. These are compared with curves obtained with $v = 0$ state only and experimental results. Averaged function is normalized so that height of its first peak coincides with that of first peak in LIF signal.

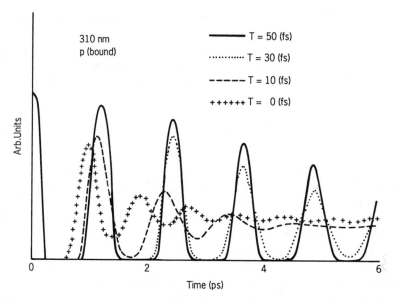

Figure 2.9. Bound neutral populations calculated for Gaussian laser pulses with different widths and a wavelength of 310 nm. All curves are normalized so that heights of first peaks coincide.

states for the different pumping frequencies, while Fig. 2.8 shows that the positioning and shape are not affected greatly by the quantum number, although the amplitude effect is clearly shown.

Engel and Metiu also investigated the dependence of $P_b(t)$ on the pump pulse length. Their results show very important effects on the total character of neutral bound populations. These effects are serious enough to warrant further investigation; see Fig. 2.9.

The paper by Engel and Metiu certainly highlights the fact that there are still many differences between theory and experiments, which need to be answered.

Shortly after the initial presentation of the theory in Engel et al.,[4d] Choi and Light[60] presented a theory with a similar nature. However, the work of Engel and Metiu differs in a number of details from that of Choi and Light. First, Engel and Metiu used Gaussian laser pulses, while Choi and Light considered a δ-function pulse. In addition, Choi and Light used a quantum version of Heller's[61] general approach to time-dependent wave-packet propagation while, as mentioned earlier, Engel and Metiu used a fast Fourier transform method developed by Alvarellos and Metiu.[57] Finally, Engel and Metiu calculated populations of states, while Choi and Light calculated the norms of the wave packets. In spite of the differences the overall character of the results was the same. Even the difference between Gaussian and δ-function laser pulses seems to have little effect on the qualitative character of the results.

The actual technique used by Choi and Light, the Gauss–Chebyshev discrete variable representation (DVR), should be very useful. Contrary to the normal basis function approach, the DVR approach expresses the solution to the time-dependent Schrödinger equation as amplitudes at a well-defined set of coordinate points. The evolution of the system is carried out on the diabatic curves rather than on the adiabatic curve. Choi and Light present an extensive analysis of these curves where RKR-calculated curves are fitted to the analytic expressions developed earlier by Faist and Levine.[62] In the Franck–Condon region the adiabatic and diabatic curves are essentially the same, so there should be little effect from the choice of the curves of the calculated excitation due to the pumping pulse.

The propagation of the wave packet was carried out using the split time evolution operator. The amplitudes of the evolving wave packet at a set of points $\{R_a\}$ of configuration space was recorded at the set of discrete-time steps $\{t_n\}$. A set of values for the overlap integral $\langle \phi_0 | \phi(t) \rangle$ for $\{t_n\}$ was Fourier transformed by discrete fast Fourier transform (DFFT). As compared to the time-dependent picture, which represents dynamics of the wave packet under the influence of the diabatically coupled Hamiltonian, the DFFT of the time-dependent overlap integral reveals details of the PES that governs the dynamics of the wave packet. This required Choi and Light to make the detailed analysis of the PESs, as mentioned. For details of the split time evolution operator scheme one should consult Choi and Light's paper and the references therein to the work of Feit and Fleck. The time step used was 1–5 fs, and the propagation was carried out for 100–1000 time steps, that is, up to approximately 5 ps, giving about five oscillations of the wave packets in the excited-state well.

Choi and Light also varied the strength of the coupling parameter V_{12} between the two surfaces over a range of 250–450 cm^{-1}. They found that a coupling strength of $V_{12} = 350$ cm^{-1} resulted in a diabatic crossing probability of $\sim 9.4\%$ for a wave packet propagating on the surface proposed by Faist and Levine.[62] This value is very close to the $\sim 10\%$ value estimated by Rosker, Rose, and Zewail from their FTS results.[4] It is interesting to note that they only obtained a $\sim 1.7\%$ probability for diabatic crossing of the wave packet on an RKR calculated excited adiabatic surface.

The actual dynamics of the norms of the wave packets on the covalent and ionic surfaces are given in Figs. 2.10 and 2.11 for various values of the coupling parameter V_{12}. These norms of the wave function are related to the probabilities investigated by Engel and Metiu. In addition one should also compare the total time-integrated flux of the dissociating products on covalent and

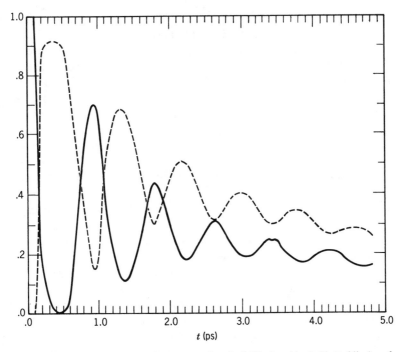

Figure 2.10.   Norms of wave packets on covalent (solid line) and ionic (dotted line) surfaces. $V_{12} = 350 \, \text{cm}^{-1}$.

ionic states as a function of time and for the various values of the coupling parameter V_{12}. Again these figures are similar to the free sodium results of Engel and Metiu. Finally it should be noted that the Fourier transform of the overlap integrals determined in the analysis of Choi and Light gives the vibrational-state-resolved absorption spectra. The vibrational energy spacings in these spectra range between 31 and 42 cm^{-1}, with the energy levels given by $31{,}250 \pm 2000 \, \text{cm}^{-1}$. This compares well with values determined by other researchers.

At the time Engel and Metiu presented their quantum description of the FTS of NaI, Lee et al.[63] produced a paper on a classical theory for the femtosecond photodissociation of NaI*. They assume that LIF is proportional to the absorption cross section and the particles, each contributing a Lorentzian to the absorption, and the sum of the Lorentzian giving the desired result. The treatment also assumes that the pump and probe processes are instantaneous. The resulting absorption cross section is then given by a convolution of the envelopes of the pump and probe fields. The absorption

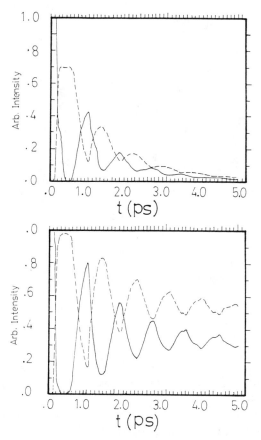

Figure 2.11.   Norms of wave packets on covalent (solid line) and ionic (dotted line) surfaces. (a) $V_{12} = 250 \text{ cm}^{-1}$. (b) $V_{12} = 450 \text{ cm}^{-1}$.

cross section can be simplified further by assuming that the pump and probe fields have Gaussian envelopes.

Lee et al. solved for the FTS spectra and obtained reasonable results, which were qualitatively equivalent to those determined by the quantum theories. The important point made by these authors is that they have taken into account both the pump and the probe excitations, whereas the quantum theories have considered only the development of the pumped states. Lee et al. contend that the classical model should work well because the wave packet prepared by the pump excitation is relatively narrow compared to the potential well, the mean de Broglie wavelength is short, and the curve crossing

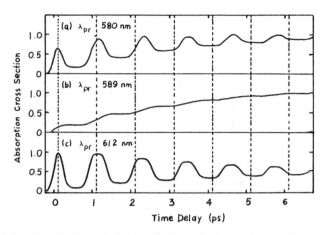

Figure 2.12. Calculated transient absorption cross sections. (*a*) λ_{probe} = 580 nm. (*b*) λ_{probe} = 589 nm (on-resonance). (*c*) λ_{probe} = 612 nm. In each case λ_{pump} = 310 nm. Time delay, in picoseconds, is set to zero at maximum of pump pulse. Dotted guide lines correspond to arbitrary settings of time delay zeros at first oscillation maxima, because the zero of time is undetermined in the experiment.

region is narrow. An example of their calculated absorption cross sections is shown in Fig. 2.12.

In an effort to improve the comparison between theory and experiment, Engel and Metiu extended their previous work by considering a two-photon excitation of NaI.[64] In all earlier FTS work it had been assumed that the LIF signal was caused solely by the absorption of the probe pulse and was, therefore, proportional to the population of NaI in the excited covalent state. While the straightforward single-photon theory produces reasonable results, Engel and Metiu recognized that the temporal domains of the pump and probe pulses could overlap. Thus the LIF measurements with ultrashort pulses do not necessarily give a signal that is proportional to the population of the pumped covalent state.

In a two-photon process the molecule starts in the ground electronic state $|2\rangle$, which is the ionic $^1\Sigma^+$ state. The first pulse excites the molecule to the covalent electronic state $|1\rangle$, which has a quantum number $\Omega = 0^+$ and is asymptotically correlated with $\text{Na}(^2S) + \text{I}(^2P_{3/2})$. The second pulse then excites the molecule from state $|1\rangle$ to state $|3\rangle$, which correlates with $\text{Na}^*(^2P) + \text{I}(^2P_{3/2})$. This third state then undergoes fluorescence of the form $\text{Na}^*(^2P) \rightarrow \text{Na}(^2S)$, which gives radiation in the sodium D line of 589 nm.

To calculate the wave function $|3\rangle$ created on the third level by the two-photon absorption, the previous technique of Engel and Metiu was extended to a second-order perturbation solution given by

$$|\psi(t)\rangle = -\frac{1}{\hbar^2} \int_0^t \int_0^{t_2} U_3(t - t_2) H_{31}(t_2)$$

$$\times \; U_{12}(t_2 - t_1) H_{12}(t_1) \exp(-i\omega_v t_1) |\phi_v\rangle \, dt_1 \, dt_2 \qquad (2.13)$$

Here $\exp(i\omega_v t)|\phi_v\rangle$ is the eigenfunction and phase factor of the electronic ground state. The perturbation caused by the laser pulse is represented by

$$H_{\alpha\beta}(t) = |\alpha\rangle \vec{\mu}_{\alpha\beta} \cdot \vec{E}(t) \langle \beta| \qquad (2.14)$$

which is the coupling of the dipole moment with the electric field vector. The expression for $|\psi(t)\rangle$ represents a transition due to the first pulse at t_1 and then its time development to t_2 when the second pulse occurs followed by a final time development to t_3.

Besides extending the wave-function solution to second order, the PES of the third electronic state had to be specified. They chose the surface proposed by Bower et al.,[65] which is given functionally as

$$V_3(R) = D\{1 - \exp[-\alpha(R - R_e)]\}^2 - D + \Delta E \qquad (2.15)$$

For the details of the function, Engle and Metiu[64] should be consulted. The important point to note is that at large R the difference between V_1 and V_2 becomes a constant. This means that at longer R the difference between [Na*I] and [NaI]$^+$ is very close to that of the free sodium emission energy. The similarity between the excited states and the lack of information on these states will remain a difficulty in the analysis of the FTS results, at least for NaI.

In addition to the constant differences in the energies of the excited states, there is the problem that ultrashort radiation is not monochromatic, so there exist no pure off- or on-resonance experiments. As an example, the off-resonance probe wavelengths of 580 and 612 nm differ from the resonance wavelength of 589 nm by 9 and 23 nm, respectively. However, the Fourier transform of a 50-fs Gaussian pulse has a linewidth at half-height of 20 nm. Thus some of the photons are in the off-resonance with the free Na.

Noting these problems Engel and Metiu carried out an analysis[64] similar to their single-photon excitation study.[59] They obtained for their newly calculated populations some rather pronounced doublets in some of the peaks (Fig. 2.13). The calculated results still deviate from those obtained experimentally. Even Engel and Metiu recognized the discrepancies between the calculated peak widths and those observed experimentally. They state that inadequate knowledge about the excited-state and the transition dipole moments makes it difficult to determine how to improve the calculated results. It may also be doubtful whether the efforts required in the second-order

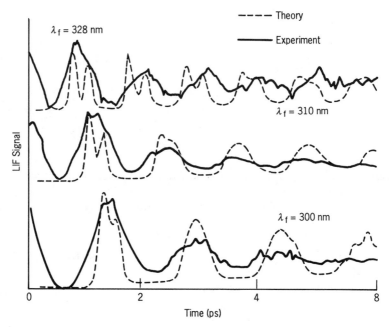

Figure 2.13. Experimental LIF signal and calculated population. Curves correspond to pump laser wavelengths of 300, 310, and 328 nm and a probe laser wavelength of 580 nm. Calculations used a width of 50 fs (FWHM) for both pulses. Calculated curves are Boltzmann averages over initial vibrational states.

calculations are justified by the improvement in the results over those obtained by the single-photon excitation calculations.

An additional contribution by Engel and Metiu was concerned with a proposed femtosecond experiment.[66] Here they present an analysis to show that the FTS results for the NaI system could be extended if the probe pulse were tuned so as to ionize the pumped state. In this way more than just the covalent part of the adiabatic curve could be accessed, that is, the ionic part of the predissociating state could be attacked with $[Na^+ \ldots I^-] \rightarrow Na^+ + I^- + e$ process. As Engel and Metiu present a rather extensive analysis of the proposed process, all that remains is to see how soon the experimental community carries it out.

B. ICN

The ICN molecule was the first to be exposed to the FTS technique.[4a] This was probably a result of the large amount of activity with ICN in the picosecond range.[67,68] The reaction studied in the femtosecond transition-state spectroscopy of ICN is the photofragmentation reaction given by

$$I - CN \rightarrow [I \ldots CN]^{\ddagger} \rightarrow I + CN \qquad (2.16)$$

In these experiments a pump pulse excites the ICN molecule to the A continuum, which can lead to dissociation via two channels denoted by the following reactions:

$$ICN + h\nu \rightarrow I(^2P_{3/2}) + CN(X^2\Sigma^+)$$
$$\rightarrow I^*(^2P_{1/2}) + CN(X^2\Sigma^+) \qquad (2.17)$$

The ultimate outcome of the reactions would be to produce iodine in the ground spin-orbit state or in the excited state I*. In either case the CN fragment is found in the $(X^2\Sigma^+)$ state.[67] According to Dantus et al.[4a] the excited iodine channel is not produced by using either the 285-nm or the 306-nm pump laser pulse. Thus the process involves only the I + CN channel. The pump pulse also determines the initial internuclear separation R_0, on the excited PES as the absorption of the pump photon will only be appreciable at a separation that corresponds to an energy difference between the ground state and the excited surface equal to hc/λ. In the FTS experiment the system is allowed to evolve for a while, and then it is subjected to a second probe laser pulse. At this point if the absorption is appreciable, the internuclear separation corresponds to the energy difference between this excited state and a higher excited state. The absorption of the probe is not usually measured, but instead the LIF is measured and is assumed to be in proportion to the population of transition-state species excited to the upper level. The frequency of the probe pulse will determine the internuclear distance at which the absorption is effective. These configurations are often referred to as those in the optically coupled region (OCR). In addition the choice of frequency determines which of two different types of experiments are carried out: clocking experiments or detuning experiments.

In the clocking experiments the probe laser is set to on-resonance with the final product. For ICN the probe wavelength is set to 388.5 nm, the band head of the P branch of the free CN. The clocking experiments do not measure the actual dynamics of the transition state, but instead give the average time to form the products or the "time or flight" to the product state. The probe pumps the CN to the upper PES at internuclear separation indicative of the products, and the LIF produced from this state is a measure of the amount of product present at probe time. Typical results are presented in Fig. 2.14. There are actually two curves in the experimental results. The curve to the left is the result of measuring the multiphoton ionization transients of N, N-diethylaniline (DEA), which serves as the zero of time in this experiment. The second curve is from the measured CN transients. The accuracy of the whole

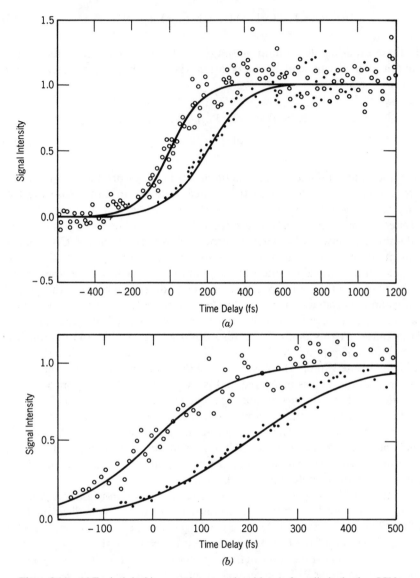

Figure 2.14. (a) Typical clocking experiment results with transform-limited pulses. ICN transient (solid squares) in this case shows an observed delay of $\tau_{1/2} = 205 \pm 30$ fs from DEA transient (open circles). (b) Shows expanded scale near $t = 0$. Lines are fits to data.

process is dependent on the response time of the DEA. Zewail's group found the half-life of the ICN to be $\tau_{1/2} = 205 \pm 30$ fs when excited by a 306-nm pump pulse.[69,70] For a pump pulse of 285 nm the transition-state half-life is reduced to $\tau_{1/2} = 160 \pm 30$ fs. From the analysis of clocking experiments with different-wavelength pump pulses, information can be extracted on recoil velocities (more correctly recoil forces) and available energy of the free fragments. However, these analyses are very involved and complicated by the fact that at some of the wavelengths the excited spin-orbit state I* may become accessible. However, from a simple analysis of the recoil velocities and the half-lives of different pumped states, Dantus et al. have made an approximate determination that the ICN bond is broken at 4 Å, which is less than previously proposed.[70]

In detuning from the resonance wavelength of the free CN the shifting can be done toward the red or the blue side. As the energy difference between the upper excited states of the transition complexes are directly proportional to the internuclear distance, tuning to the low-energy (red) side of the free CN resonance means setting the OCR to smaller values of internuclear separation. As the wavelength of the probe is increased, smaller separations and shorter event times are probed. Typical off-resonance ICN transients are shown in Fig. 2.15. Here as the probe wavelength is increased, the transients peak at shorter delay times. In these situations the actual transition-state dynamics are being probed. As in the on-resonance experiments the internuclear distances being probed are a function of the wavelength of both the pump and the probe. In detuning to the blue side of the resonant wavelength the free CN fragment is being probed. However, at the shorter wavelengths a spectrum of rotational states is being observed. As far as transition-state spectroscopy is concerned, these experiments are of value for determining the apportionment of rotation energy in the free CN fragment. At this point the meaning of the various data obtained for these experiments is not clear. For example the "time of flight" for probe wavelengths of 387.4 nm is reduced to nearly 0 fs.[70]

Both classical and quantum theories have been developed to explain the results found in these experiments.[6,71–73] The classical approach of Bersohn and Zewail provides a rather accurate description of the FTS results of ICN. The value of the quantum mechanical treatment is the insight that is gained on the behavior of the wave packets on the energy surface of the excited molecule.

In general there are three energy surfaces involved in the FTS results for ICN: the ground state V_0, the pumped dissociating state V_1, and the upper excited state, V_2 (Fig. 2.16). The energy of the pump laser pulse being equal to the energy difference between V_0 and V_1 at a certain internuclear separation R dictates the starting configuration of the excited molecule on the dissociat-

214   S. H. LIN, B. FAIN, AND N. HAMER

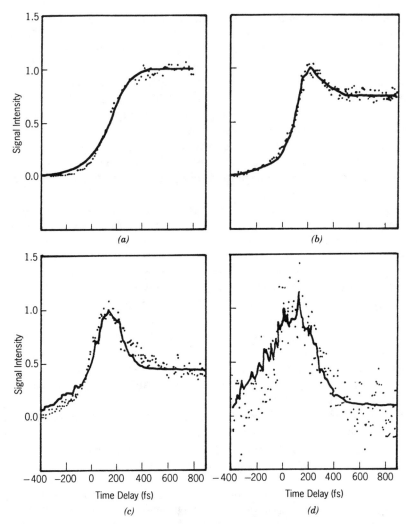

Figure 2.15. Typical ICN transients as a function of probe wavelength. (a) $\lambda_2 = 388.9$ nm. (b) $\lambda_2 = 389.8$ nm. (c) $\lambda_2 = 390.4$ nm. (d) $\lambda_2 = 391.4$ nm. Solid lines were generated by fitting classical model convolved with measured experimental response for each transient.

ing energy surface given by V_1. As the molecule begins to fall apart, the internuclear separation increases. With this increase in R follows an increase in the energy difference between V_1 and V_2. If the probe pulse has an energy equal to the difference between V_1 and V_2 at the specified R, then the molecule will be excited to the upper level and will produce an LIF signal. Thus the

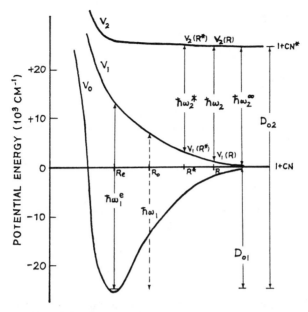

Figure 2.16. Principal distances and frequencies in ICN molecule, typical of many other molecules. R_e—equilibrium distance between center of mass of CN and I in ICN; R_0—point on upper surface at which molecule is found immediately after dissociation; R^*—point at which molecule would be in resonance with a probing photon of frequency ω_2^*. Actual probing at time t is being carried out at point $R(t)$ by a photon of frequency ω_2.

dynamics of the transition state can be followed by studying the LIF signals as a function of probe time and energy.

To interpret the observed results Bersohn and Zewail carried through the following classical theory. First they estimated the two excited states as being of exponential nature with the form

$$V_1(R) = V_1 e^{-R/L_1} \tag{2.18}$$

$$V_2(R) = V_2 e^{-R/L_2} + \hbar\omega_2^\infty \tag{2.19}$$

where $\hbar\omega_2^\infty$ is the excitation energy of a particular free fragment. The probe pulse whose absorption is maximum when the fragments are at a distance R^* will have a distribution in energy that we take to be Lorentzian with a half-width γ. The absorption at the distance R is

$$A(R) = C\{\gamma^2 + [V_2(R) - V_1(R) - V_2(R^*) + V_1(R^*)]^2\}^{-1} \tag{2.20}$$

where C is a constant.

Substituting Eqs. (2.18) and (2.19) into Eq. (2.20), one obtains

$$A(R) = C[\gamma^2 + (V_2 e^{-R/L_2} - V_1 e^{-R/L_1} + \hbar\omega_2^\infty - \hbar\omega_2^*)^2]^{-1} \quad (2.21)$$

where

$$\hbar\omega_2^* = \hbar\omega_2^\infty + V_2 e^{-R^*/L_2} - V_1 e^{-R^*/L_1} \quad (2.22)$$

The behavior of the system between pump and probe pulses is determined by solving the Hamiltonian equation of motion on the V_1 surface,

$$\frac{1}{2}\mu\left(\frac{dR'}{dt}\right)^2 + V(R') = E \quad (2.23)$$

where E is the terminal kinetic energy. Upon integrating the Hamiltonian expression, one obtains

$$e^{-R-R_0/L_1} = \text{sech}^2\left(\frac{vt}{2L_1}\right) \quad (2.24)$$

where R_0 is the initial pumped configuration at time t. This expression relates R to the time. We substitute Eq. (2.24) into Eq. (2.21) to obtain

$$A(t) = C[\gamma^2 + W^2(t)]^{-1} \quad (2.25)$$

and

$$W(t) = V_2\left(\frac{E}{V_1}\right)^{L_1/L_2}\left[\text{sech}^{2L_1/L_2}\left(\frac{vt}{2L_1}\right) - \text{sech}^{2L_1/L_2}\left(\frac{vt^*}{2L_1}\right)\right]$$

$$- E\left[\text{sech}^2\left(\frac{vt}{2L_1}\right) - \text{sech}^2\left(\frac{vt^*}{2L_1}\right)\right] \quad (2.26)$$

where $R = R^*$ at $t = t^*$.

This expression takes on a much simpler form if we assume that the upper excited-state potential is essentially flat, that is, L_2 is thus infinite. Equation (2.25) reduces to

$$A(t) = C\left(\gamma^2 + \left\{E\left[\text{sech}^2\left(\frac{vt}{2L_1}\right) - \text{sech}^2\left(\frac{vt^*}{2L_1}\right)\right]\right\}^2\right)^{-1} \quad (2.27)$$

Bersohn and Zewail applied this to the ICN experiments of Dantus et al. with E being around 7000 cm^{-1} and γ around 300 cm^{-1}. The results are given in Fig. 2.17, where a set of absorption intensities are plotted versus the dimensionless quantity $vt^*/2L_1$. The curves are calculated for different values of $\text{sech}^2 vt^*/2L_1$, which is the fraction of the potential energy remaining at

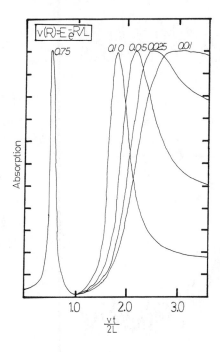

Figure 2.17. Plots of
$C\{\gamma^2 + E^2[\mathrm{sech}^2(vt/2L_1) - \mathrm{sech}^2(vt^*/2L_t)]^2\}^{-1}$
for various values of the fraction of potential
energy remaining, $\mathrm{sech}^2(vt^*/2L_1)$ ranging from
0.75 to 0.01.

separation R^* at the time of the probe pulse. The graphs calculated for values
of 0.75–0.10 are all strongly peaked. The physical meaning ascribed to this is
that the absorption coefficient is changing so rapidly as a function of the
distance that the system is only briefly able to absorb. When the probing
wavelength corresponds to a value of R^* where the potential varies more
slowly, then the system stays in resonance for a much longer time. Note that
the middle curve in Fig. 2.18 from Dantus et al.[4a] is very similar to the 0.05
curve in Fig. 2.17. Therefore, the Bersohn and Zewail theory handles the
general features of the FTS results reasonably well.

Bersohn and Zewail noted the correspondence between the absorption and
the PES. They showed how to extract the PES from the experimental results.
However, this process was presented with more rigor and detail in Bernstein
and Zewail.[71] While the Bernstein and Zewail method was based on the
concepts presented in the Bersohn and Zewail work, several new ideas were
introduced that made this approach more rigorous. Bersohn and Zewail
assumed a Lorentzian form for the laser line wavelength profiles. In addition,
they assumed functional forms for the excited-state PESs. Therefore the signal function that was matched with experiment had the assumed potential
parameters and Lorentzian half-width built in. The Bernstein and Zewail
procedure was to produce a direct inversion of the experimental data from

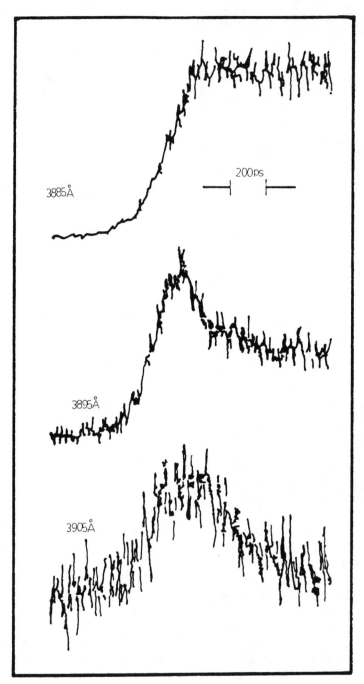

Figure 2.18. Schematic for femtosecond probing of transition state using a pump probe (t, λ tuning) method. PESs are drawn to inidcate different probe wavelengths at free and perturbed CN transitions.

the detuned transients to obtain quantitative information on the excited-state PESs without making assumptions with regard to their functional form and without assuming anything about the wavelength profile.

By analyzing the asymptotic signals at different degrees of detuning Bernstain and Zewail were able to show that the probe laser had a near-Lorentzian spectral profile. This was possible because the detuned pulse excites the free CN in proportion to the signal in the wings of the profile. For the experimental data of Dantus et al.[4a] the laser pulse was found to be essentially Lorentzian with a half-width of 38 cm^{-1}.

The determination of the excited-state energy surface is straightforward but somewhat tedious. The absorption signal is related to a line shape function which in turn is a function of the deviation from the normal probe frequency. The deviance from the normal probe frequency is expressed in terms of the PES, which is given as a function of time as the spatial coordinate can be related to time through the velocity of the dissociating fragments. This demonstrates that the potential function can be related to the absorption signal. However, the complete analysis is rather involved and the interested reader is referred to Bernstein and Zewail who not only consider the ICN data of Dantus et al., but also a series of computer-simulated experiments in model systems.

In two rather sizable papers, Williams and Imre[72,73] extended their previous work in continuous-wave excitations to the pulsed laser processes in femtosecond transition-state spectroscopy. This can be seen by looking at the main equation for the time development of the wave function $|\psi_{ex}\rangle$ on the excited-state PES,

$$|\psi_{ex}(t)\rangle = -\frac{i}{\pi} \int_{-\infty}^{t} e^{-iH_{ex}(t-t')}\mu E(t')|\psi_{gr}(t')\rangle \, dt' \qquad (2.28)$$

where H_{ex} is the Hamiltonian on the excited surface, μ is the transition dipole moment, and $E(t')$ is the electric field strength. However, Williams and Imre solved the first-order Schrödinger equation by the modified fast Fourier transform method of Kosloff and Kosloff.[74,75]

In their first paper[72] Williams and Imre present a general study of the evolution in time of populations and emission spectra of excited states due to two different laser pulses. In their second paper[73] they apply this analysis to the specific case of the ICN molecule. They calculated the effect of pulse duration for pulse widths of 40 and 125 fs. They also performed calculations for a number of probe frequencies. (They used a constant pump frequency of 306 nm.) They presented a number of graphs of the wave-packet evolution on the excited energy surfaces and also gave the populations as a function of time on the PESs. These latter quantities are related to the experimentally obtained FTS signals. Their results compare well with the work of Dantus et al.[4a]

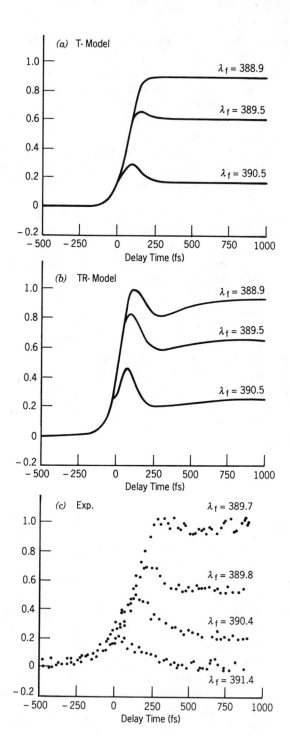

220

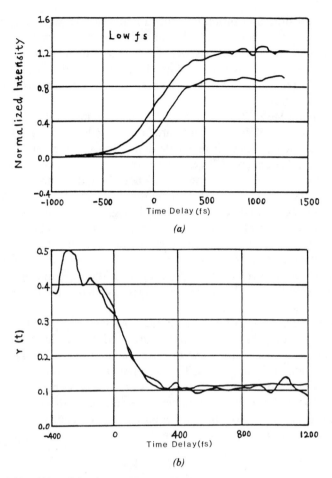

(a)

(b)

Figure 2.20. (*a*) Parallel and perpendicular ICN FTS transients obtained for $\lambda_1 = 305.3$ nm, $\lambda_2 = 387.9$ nm. Data are shown after normalization of asymptotes to 1.2 and 0.9 for parallel and perpendicular transients, respectively. (*b*) $r(t)$ obtained from top transients and calculated $r(t)$ using a Gaussian for rotational distribution centered at $j = 2$ with $\Delta j = 5$ (as estimated by comparing spectrum of probe laser and absorption spectrum of CN).

◁ **Figure 2.19.** Total CN fluorescence versus probe delay. (*a*) T model (includes only translational motion). (*b*) TR model (includes both translational and rotational motion). (*c*) Experimental results of Rosker et al. Pump wavelength is $\lambda_1 = 306$n nm; both pump and probe pulse widths are 125 fs (FWHM). Uppermost curve in each plot corresponds to most on-resonance probe frequency.

Williams and Imre do caution that they found the FTS signals were strong functions of both excited states. Modifications to the second excited state would change the fluorescence signal as a function of delay time. They also noted that they could have obtained much better agreement with experiment if they had been willing to adjust their excited-state potentials from those found in the literature.

Heather and Metiu[76] have carried out a quantum mechanical study of ICN extending the work of Williams and Imre to investigate rotational effects. They represented calculations which indicate that at longer delay times the dependence of the LIF signal on delay time also contains information regarding the rotational motion of the CN fragment. They show that this is the case by calculating the LIF signal using a model that is identical to that of Williams and Imre (T model) and next calculating the signal for a model in which the I–CN bond is allowed to bend and also the CN fragment is allowed to rotate (TR model). The difference in the signals calculated by these two models should highlight the importance of rotational effects. The results of these two calculations are shown in Fig. 2.19 along with the experimental results of Rosker et al.[4b,c] It is not clear from a comparison with the standard FTS signal that rotational effects are present.

However, a new type of FTS experiment has been reported by Dantus et al.[4a] In this work polarized laser pulses are used. Using this technique, the alignment during dissociation can be measured and then related to the final rotational state of the fragments. The experimental process is very similar to the normal FTS experiment. After dissociation has been initiated with a polarized pump laser pulse, the fragments are detected with a polarized probe pulse. The probe pulse is oriented either parallel or perpendicular to the pump pulse. From these two probe orientations the signal polarization anisotropy $r(t)$

$$r(t) = \frac{(I_{11} - I_L)}{(I_{11} + 2I_L)}$$

$$(2.29)$$

is determined. The value of the anisotropy can be calculated as a function of time. Some of the results of this type of experiment are given for ICN in Fig. 2.20. These results do show that rotational effects are present. However, the rotational effects are much more pronounced for HgI_2.

C. $M(CO)_n$

Recently Joly and Nelson have used the femtosecond time-resolved absorption spectroscopy to study the photochemistry of $Cr(CO)_6$ and $Mn_2(CO)_{10}$ in methanol.[36b,c] CO ligand dissociation and metal–metal bond cleavage in these compounds serve as prototypes for much of organometallic photochemistry. First, it is hoped to elucidate qualitative or semiquantitative reactive PESs. Although the reaction dynamics are heavily influenced by the solvent, one may expect to understand at least the initial stages of CO ligand loss in terms of a weakly perturbed PES and nearly collision-free wave-packet propagation. Second, it is hoped to examine solvent effects on reaction dynamics. These effects are expected to be particularly apparent in the metal–metal bond cleavage of $(CO)_5Mn–Mn(CO)_5$ since separation of the large photofragments must lead to substantial solvent displacement. Finally, knowledge of the reacton dynamics is of value in its own right to both physical and inorganic chemists. In the compounds studied, basic questions persist about whether reactions proceed from the initially excited states or from other electronic levels reached through relaxation or intersystem crossing. The factors that influence the relative quantum yields for the competing ligand loss and metal–metal bond cleavage reactions in $Mn_2(CO)_{10}$ are also poorly understood.

Femtosecond pulses at 308 nm excite predominantly the $^1A_{1g} \rightarrow {}^1T_{1g}$ ligand field transition[77] in $Cr(CO)_6$. This transition, which is formally symmetry forbidden but occurs through vibronic coupling, reduces π backbonding and places density in a Cr–CO σ^* orbital. This labilizes the Cr–CO bonds. In methanol solution, a single CO dissociates and a solvent molecule coordinates to yield $Cr(CO)_5(MeOH)$ within 3 ps.[78] If the initial pulse is short compared to the time required for CO loss, then a coherent wave packet is formed upon the excited-state surface. Wave-packet propagation can be monitored by excited-state absorption of variably delayed probe pulses.

In $Mn_2(CO)_{10}$ in solution, both CO loss and homolytic metal–metal bond cleavage are known to occur within 25 ps.[79] Femtosecond pulses at 308 nm excite predominantly the intense $(\varepsilon - 3 \times 10^4)$ Mn–Mn $\sigma \rightarrow \sigma^*$ transition.[80] This spectral region may also include a weaker $\sigma \rightarrow \pi^*$ transition. Recent measurements[81] indicate wavelength-dependent quantum yields, with higher excitation energies favoring CO loss. This suggests that the two photoprocesses begin from different electronic states. Joly and Nelson note that while the $Mn(CO)_5$ radical is stable for microseconds,[82] $Mn_2(CO)_9$ undergoes further rearrangement in hydrocarbon solvents to form a CO bridged species.[83] In methanol, solvent complexation may occur to yield $Mn_2(CO)_9MeOH$ instead of the bridged compound.[79]

Experimentally a continuous-wave mode-locked Nd:YAG laser synchronously pumps a femtosecond dye laser with an antiresonant ring.[36a] The dye laser output consists of 65-fs pulses centered at 615 nm with a repetition rate of 82 MHz. These pulses are amplified in a three-stage dye amplifier pumped with the frequency-doubled output of a Nd:YA regenerative amplifier. 70% of the amplified output is frequency-doubled to yield 308-nm excitation pulses. These are focused to a 100-μm-diameter spot inside a 2-mm quartz flow cell containing the sample. The remaining 30% of the red light is focused into a 2-mm cell of D_2O liquid to generate a white-light continuum. 10-nm interference filters are used to select the wavelength of the probe pulses, which are variably delayed along a 1-μm stepping-motor delay line and overlapped with the pump pulses inside the sample. The polarizations of the beams are adjusted to be 54.7° apart to eliminate effects of molecular rotation on the signal. The probe-light intensities before and after the sample (I_0 and I, respectively) were measured, and $(I - I_0)/I_0$ was determined. Cross-correlation involving the ultraviolet and probe light indicates an instrument response time of approximatley 95 fs, which is consistent with the fastest rise times observed in the transient absorption measurements. The samples were saturated ($\sim 10^{-3}M$) solutions of $Cr(CO)_6$ and $Mn_2(CO)_{10}$ in high-purity methanol. Typical values of $(I - I_0)/I_0$ were between 2 and 6%.

1. $Cr(CO)_6$

Transient absorption measurements for $Cr(CO)_6$ in methanol were recorded with 10 different probe wavelengths between 400 and 800 nm. Figure 2.21

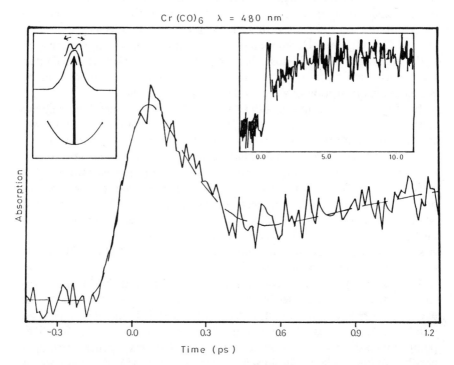

Figure 2.21. Transient absorption data at $\lambda = 480$ nm for $Cr(CO)_6$. Solid line—data showing spectral evolution as CO dissociates; dotted line—fit to functional form for potential, $V(Q) = \text{sech}^2(aQ)$, shown in left inset. Right inset shows subsequent exponential signal increase associated with solvent complexation which yields $Cr(CO)_5(MeOH)$.

shows 480-nm data illustrating the important short-time features. The data show a pulse-duration-limited rise in absorption. A rapid nonexponential fall in excited-state absorption during the first 500 fs reflects the process of CO dissociation. The slower exponential rise is interpreted (consistent with earlier results[78]) in terms of solvent complexation to form $Cr(CO)_5(MeOH)$.

The rapid and nonexponential short-time decay is taken as a signature of coherent wave-packet propagation in the initially excited PES, which is un-stable with respect to CO loss. Due to the nature of the (g-symmetry) excited state and the molecular symmetry, there cannot be any a priori preference for which CO dissociates. This means that the initially formed wave packet must be centered at a local potential maximum with respect to any asymmetric CO stretching motions involved in ligand loss. Since the transition is allowed only through vibronic coupling, ground-state molecules, which are somewhat dis-torted along asymmetric stretches, have higher transition probabilities than undistorted molecules, and the initially excited wave packet should have separated regions of high density, as illustrated qualitatively in the left inset in Fig. 2.21. The wave packet then "bifurcates" down the two sides of the potential energy surface, that is, most molecules with one CO bond "stretched" will continue to distort following photoexcitation until the same CO is lost.

Unfortunately, little is known about the reactive PES, the likely "reaction coordinate," or the PES of the high-lying state into which excited-state ab-sorption occurs. Reliable quantitative analysis of the transient absorption dynamics is therefore impossible. For purposes of illustration only, in Fig. 2.21 a fit to the data based on the kinematics of wave-packet propagation down an unstable potential (i.e., a ball rolling down a hill) is included. For mathematical convenience, a functional form of $V(Q) = \text{sech}^2(aQ)$ is assumed, where Q is the one-dimensional reaction coordinate and a is a constant. The classical equation of motion for this potential can be solved exactly. The fit was generated assuming a simple, one-dimensional reaction coordinate, which is identical for all molecules (i.e., no inhomogeneity), and that the signal strength varies linearly with progress along Q. The actual Q-dependent varia-tion in excited-state absorption, which is not necessarily monotonic, must depend on the details of the reactive and high-lying PESs, the probe central wavelength and spectral bandwidth, the width of the excited-state wave packet, and other factors. The fit is included merely to illustrate the manner in which a wave-packet dynamics analysis could proceed and the fact that such an analysis can explain the nonexponential time-dependent signal.

Figure 2.22 shows short-time transient data at 500, 440, and 410 nm. Each sweep shows a pulse-duration-limited rise followed by rapid, nonexponential decay at times shorter than 0.5 ps. It appears from these data that the excited-state absorption spectrum blue-shifts as ligand loss occurs.

The data in Fig. 2.22 show continued change between 0.5 and 5 ps, but the change is smaller in magnitude and in the opposite direction than that at

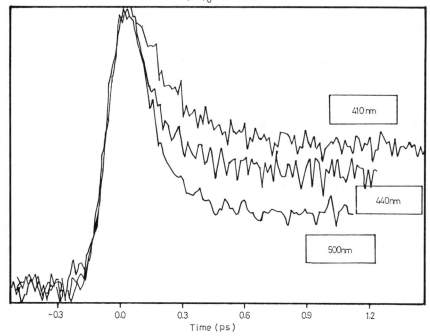

Figure 2.22. Short-time data for $Cr(CO)_6$ taken at 410, 440, and 500 nm. Signal differences are interpreted as blue-shifting of excited-state absorption spectrum as a CO dissociates.

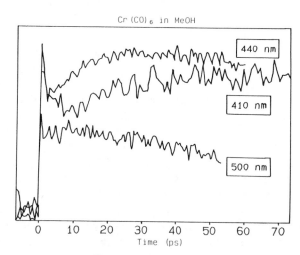

Figure 2.23. Long-time data for $Cr(CO)_6$ taken at 410, 440, and 550 nm. Continued changes in signal levels are attributed to vibrational cooling of $Cr(CO)_5 MeOH$ species.

480 nm. Presumably the absorption of the bare species is slightly stronger at these wavelengths (and weaker at 480 nm) than that of $Cr(CO)_5(MeOH)$.

Figure 2.23 shows long-time transient measurements at 500, 440, and 410 nm. At 500 nm a monotonic decrease in absorption is observed between 10 and 60 ps. At 440 nm absorption first increases and then declines. At 410 nm monotonically increasing absorption is observed. The data indicate a gradual blue-shifting of the absorption spectrum. Joly and Nelson[36c] interpret this in terms of vibrational cooling of the initially "hot" $Cr(CO)_5(MeOH)$. Similar observations associated with vibrational cooling in I_2 in solution following photoexcitation have been reported.[84] The rather complex effects of vibrational cooling on electronic absorption spectra have been treated in detail recently.[85] The vibrational energy relaxation time appears to be roughly 50 ps.

2. $Mn_2(CO)_{10}$

Irradiation of $Mn_2(CO)_{10}$ with 308-nm light may result in cleavage of the metal–metal bond, yielding long-lived $Mn(CO)_5$ radical fragments, or CO loss, yielding $Mn_2(CO)_9$, which either assumes a bridged configuration[9] or complexes with a solvent molecule. Fortunately the different products absorb light at very different wavelengths. Figure 2.24 shows transient absorption data from $Mn_2(CO)_{10}$ at 500 nm, the absorption maximum of $Mn_2(CO)_9$. The data show a fast rise followed by a 530-fs exponential decay to a level

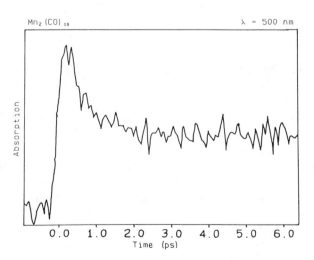

Figure 2.24.   Transient absorption data for $Mn_2(CO)_{10}$ at $\lambda = 500$ nm, the absorption maximum of the CO bridged species $Mn_2(CO)_g$. Decay time of 530 fs reflects CO bridge formation or solvent complexation which follows CO dissociation.

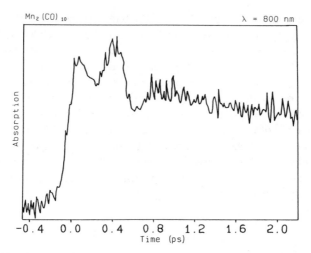

Figure 2.25. Transient absorption data for $Mn_2(CO)_{10}$ at 800 nm, the absorption maximum of the $Mn(CO)_5$ radical. Oscillatory features are interpreted tentatively as metal–metal bond oscillations in excited state.

that remains constant for at least 50 ps. The signal-to-noise ratio is not high enough to determine whether the rise time is pulsewidth limited. However, it is probable that the decay time of 530 fs reflects the time required for the formation of the bridged or solvent-complexed species. The exponential time dependence suggests phase-incoherent reaction kinetics, which is plausible since coherence is not expected to persist after the initial CO loss event. The overall time scale is sufficiently slow that relaxation of the initially excited state may occur prior to reaction. However, direct optical excitation of electronic states from which CO loss occurs cannot be ruled out.

Figure 2.25 shows data taken at 800 nm, the absorption maximum of the $Mn(CO)_5$ radical. The signal shows a pulse-width-limited rise followed by two oscillations and a gradual decay to a nonzero signal level. It is likely that the gradual decay represents vibrational relaxation of the initially "hot" $Mn(CO)_5$ radical photofragments. It is believed that there are three possible sources for the oscillations. First, the metal–metal bond may have some probability of vibrating before dissociation, which would produce oscillatory changes in the absorption spectrum. Second, other vibrational modes (e.g., low-frequency CO bends) may be excited through absorption or impulsive stimulated Raman scattering, and their oscillations could influence the absorption spectrum in the same way. Third, direct nonoscillatory metal–metal bond breakage could give rise to coherent "umbrella" type vibrations in the $Mn(CO)_5$ fragments, similar to those that occur in CH_3 fragments following photodissociation of

CH_3I. Joly and Nelson tentatively postulate that the oscillations in absorption are due to metal–metal vibrations. If the initially excited state is a σ^* state, then it should be repulsive with respect to metal–metal cleavage. However, the solvent could produce a shallow minimum in the potential corresponding to a restoring force which opposes separation of the large fragments. Inhomogeneity in the liquid would lead to variation in the solvent "barrier" height and in the amount of metal–metal "vibrational" energy dissipation suffered on approach to the barrier. Thus some metal–metal bonds could break on the first stretch, while others are turned back. Some molecules undergoing a second stretch could also dissociate since the barrier could be lowered substantially due to solvent disruption (essentially, pushing away of neighboring solvent molecules) during the first stretch. As in $Cr(CO)_6$, little is known about the details of the reactive PES or excited-state absorption.

In addition to further time-domain experiments, complementary frequency-domain spectroscopy of these compounds is called for. In particular, resonance Raman spectra would reveal much about the nature of the reaction coordinates and about what reactions occurred from the initially excited states.

D. $H + H_2$

Recently Mayne, Poirier, and Polanyi[86] have computed the transition-state absorption spectra for the $H + H_2 \rightarrow H_3^* \rightarrow H_2 + H$ reaction. They restricted themselves to the collinear configurations of the H_3 system. Calculations were carried out using one-dimensional classical trajectories on the accurate, ab initio Siegbahn–Liu–Truhlar–Horowitz (SLTH) H_3 ground-state surface.[87–89] The absorption intensity $I(v)$ was assumed to be proportional to the time the trajectory spent at the geometry where the vertical energy difference between the lower and upper potentials corresponded to the transition frequency v. A number of parameters that could affect the spectra, such as the relative translational energy or the reagent vibrational excitation, were varied. Several peaks present in the spectra were interpreted in terms of various features of the ground-state surface and the collision dynamics on that surface.

The emphasis in the study by Mayne et al.[86] was on the effect of the variations in PES features on the absorption spectrum, as well as the dependence on reagent mass and energy. For this reason the $H_3^*(D_{\infty h}^2 \Pi_u)$ surfaces used are model surfaces, though based on the data obtained from self-consistent-field (SCF) calculation. Four different model PES for H_3^* have been proposed; they are the three rotated-Morse PES, G1, G2, and SG (Gausiann 1, Gaussian 2, and shifted Gaussian) together with CG (coupled Gaussian). The transition spectra using these four surfaces have been reported by Mayne et al. Collinear classical trajectory calculations were carried out on the lower (SLTH) potential surface. The transition moment was in general assumed to be a constant.

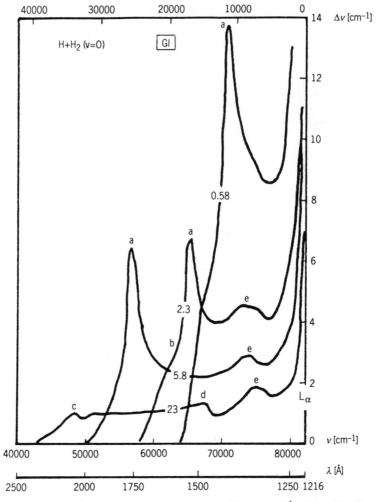

Figure 2.26. Computed absorption spectra for transition states $H_3^{\ddagger} \rightarrow H_3^*$ at $H + H_2$ collision energies $E_T = 0.58$, 2.3, 5.8, and 23 kcal/mol for the Gl H_3^* surface. Ordinate is in arbitrary
units. Transition at 1216 Å is the Lyman-$\alpha(L_\alpha)$ line. Scale across figure top gives displacement of
frequency (Δv) from L_α.

Figure 2.26 we shows the absorption spectra for collinear $H + H_2(v = 0)$
using the G1 surface for the excited state H_3^* at four collision energies. It can
be seen that increasing the collision energy tends both to decrease the total
signal and to shift the spectrum to the red. The former effect, decreased
intensity in the $H_3^{\dagger}$ absorption spectrum, is due simply to the fact that the total

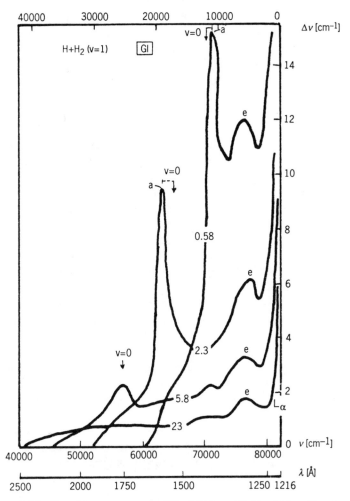

Figure 2.27.   Transition-state spectra of Fig. 2.26, but with $H + H_2(v = 1)$ partners.

signal reflects the duration of the collision, and that this decreases with increasing translational energy.

The transition-state spectra for $H + H_2(v = 1)$ on the G1 surface are given in Fig. 2.27. Indicated on the figure are the locations of the spectral peaks for $H_2(v = 0)$ (Fig. 2.26). It can be seen that, at low E_T, the peaks in the spectrum occur at almost the same wavelength for $v = 0$ and for $v = 1$. For $H + H_2(v = 0)$ the principal peak in the transition-state spectrum was found to be due to the ridge of density at the translational turning point. This is also the case for $H + H_2(v = 1)$. The difference in the dynamics is that the extra reagent vibrational energy broadens the translational turning point ridge. However, for low collision energy it does not significantly affect the distance to which the trajectories penetrate. For higher values of the collision energy the zero-point energy in $H + H_2(v = 0)$ did make a significant contribution. Consequently, increase in reagent vibration, at high E_T, shifts the main spectral peak a to the red. Moreover, vibrational excitation lowers the reactive threshold, so that for $v = 1$, $E_T = 5.8$ kcal/mol, there is a considerable fraction of reactive trajectories. As a consequence, the spectrum for $E_T = 5.8$ kcal/mol, $H + H_2(v = 1)$, exhibits the tail out to the long-wavelength characteristic of ("bobsledding") reactive trajectories. The spectra for nonreactive collisions at $E_T = 0.58$ and 2.3 kcal/mol are also noticeably broader than for $v = 0$. For reactive encounters this is due to a modification in the path of the bobsled.

Stimulated by the work of Mayne et al.,[86] Engel et al.[90] have calculated the absorption spectra quantum mechanically using the same PESs, so that direct comparison could be made with the corresponding classical trajectory spectra of Mayne et al. (MPP). The main aim of this study is to test the applicability of classical mechanics to the calculation of absorption spectra because fully quantal calculations for more complicated systems, especially in three dimensions, are probably not feasible at present. The effects of varying the relative translational energy and the H_2 vibrational excitation are examined. Their spectra, while exhibiting the same basic trends evident in the MPP spectra, show additional features which are due to the quantum nature of the $H + H_2$ system. In this work, laser fields are assumed to be weak so that the first-order perturbation theory can be applied to calculate the absorption cross section.

The continuum-bound absorption cross section for collinear collisions, which follows from the more general expressions for three-dimensional situations is given by (apart from some constants unimportant for the present purpose)

$$\sigma(v) \sim v |\langle \psi_n^{ex} | \mu_\varepsilon | \psi_E^{gr} \rangle|^2 P(E)$$

$$= vI(n, E)P(E) \tag{2.30}$$

where ψ_n^{ex} is the wave function describing the nth bound state of the upper, electronically excited H_3 surface, with the corresponding eigenenergy E_n, and ψ_E^{gr} is the continuum wave function with the relative translational (collision) energy E on the ground H_3 surface. The transition moment is denoted by μ_ε. Mayne et al. assumed that μ_ε is constant. Therefore for the purpose of comparison with the work by Mayne et al., Engle et al. also treat μ_ε as constant. The matrix element in Eq. (2.30) thereby reduces to a bound-continuum overlap integral. The $P(E)$ function in Eq. (2.30) is the energy probability function which, for thermal distribution of collision energies, is given by the Maxwellian expression appropriate for collinear collisions,

$$P(E) = E^{-1/2}e^{-E/kT} \tag{2.31}$$

Actually, since only the relative magnitudes of $\sigma(v)$ are considered, the spectra at any particular collision energy are calculated without taking $P(E)$ into account.

Both ψ_n^{ex} and ψ_E^{gr} are expanded in an orthonormal set of vibrational wave functions $\phi_i^{ex}(r)$ and $\phi_j^{gr}(r)$, respectively, as

$$\psi_n^{ex}(R,r) = \sum_i \phi_i^{ex}(r)\chi_{n,i}^{ex}(R) \tag{2.32a}$$

$$\psi_E^{gr}(R,r) = \sum_j \phi_j^{gr}(r)\chi_{E,j}^{gr}(R) \tag{2.32b}$$

The r and R coordinates refer to the H_2 intramolecular bond length and the H–H_2 center-of-mass distance, respectively. The expansion functions are defined by

$$\left[\frac{1}{2\mu}\frac{d^2}{dr^2} - v_{ex}(r) + \varepsilon_{ex,i}\right]\phi_i^{ex}(r) = 0 \tag{2.33a}$$

$$\left[\frac{1}{2\mu}\frac{d^2}{dr^2} - v_{gr}(r) + \varepsilon_{gr,j}\right]\phi_j^{gr}(r) = 0 \tag{2.33b}$$

where μ is the H_2 reduced mass. $v_{gr}(r)$ is taken to be identical to the potential of the isolated H_2 molecule. The $v_{ex}(r)$ can be chosen at will, so as to provide a compact set of expansion functions $\phi_i^{ex}(r)$. For the purpose of comparing classical and quantal results, Engel et al. consider only nonreactive $H + H_2$ collisions. The relative translational energies for which the spectra are calculated lie in the 25–251-meV (0.58–5.8-kcal/mol) range. Thus those regions of the ground-state PES, which are traditionally referred to as transition state, are not probed in this study. The absorption solely takes place in the entrance channel. These energies are well below or just slightly above the reaction threshold for collinear $H + H_2(v = 0)$ on the SLTH surface, about 0.22 eV. Consequently the reaction probability is very small, and the restriction to nonreactive collisions cannot measurably influence comparison with the results of Mayne et al. Neglecting reactive collisions should not qualitatively affect the spectra for $H + H_2(v = 1)$ either, particularly since in this case the chosen collision energies do not exceed 100 meV (2.3 kcal/mol).

The absorption spectra for $H + H_2(v = 0)$ for three collision energies are shown in Fig. 2.28. They terminate abruptly at 75–80,000 cm^{-1}, depending

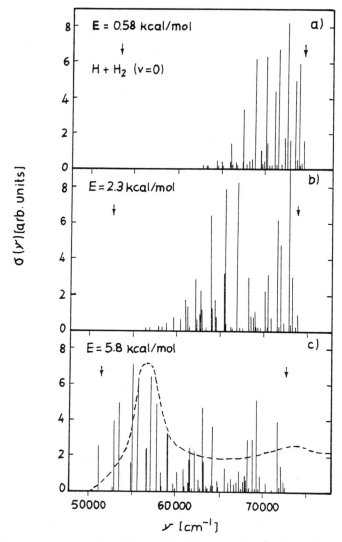

Figure 2.28. Calculated transition-state spectra for H + H₂($v = 0$) at three collision energies. Arrows at low and high frequency ends of spectra are transition to ground and highest bound levels calculated, respectively. Dashed line represents classical spectrum as read from Fig. 2.26. Both spectra are arbitrarily normalized at main maximum.

on the collision energy, at frequencies of the continuum-bound transitions to the highest bound level calculated (indicated by the arrow at the high-frequency end of the spectrum). Nevertheless the frequency range over which the spectra extend is sufficiently broad to permit comparison with the classical counterparts shown in Fig. 2.26. The most prominent feature of the MPP spectra is a rather sharp peak labeled a, which shifts to the red for higher collision energies. According to Mayne et al., it arises from a large buildup of probability density at the translational turning point, which moves toward smaller values of the difference potential ΔV and, consequently, lower transition frequencies as the collision energy is increased. Not surprisingly, the same basic trend is clearly present in our quantum mechanically calculated spectra in Fig. 2.28. For all three collision energies the major low-frequency transition amplitude, which is the equivalent of the MPP peak a, is centered at a frequency that agrees very well with the peak a position in the corresponding MPP spectra ($\sim 71,000$, $65,000$, and $57,000$ cm^{-1} at collision energies of 0.58, 2.3, and 5.8 kcal/mol, respectively, as estimated from Fig. 2.26). However, this amplitude, in comparison with the classical peak, is much broader, encompassing many individual continuum-bound transitions. It is part of a broad oscillatory pattern particularly evident for $E = 5.8$ kcal/mol, which is superimposed over the whole spectrum, thus determining its overall appearance. A direct comparison between the classical and the quantum spectra is shown in Fig. 2.28c for $E = 5.8$ kcal/mol.

Although this is not evident from Fig. 2.28, the total intensity of the absorption spectra increases significantly with decreasing collision energy, which is also observed in the MPP spectra. In the classical trajectory calculations, according to Mayne et al., this simply reflects the duration of the collision, which decreases as the collision energy is increased. In the quantum calculations it is due to the $(2\mu E)^{-1/2}$ normalization constant, which appears in the asymptotic expression for the continum wave function at a collision energy E. To get the actual relative intensities, the spectra in Fig. 2.28a and b have to be multiplied by the scale factors $(E_c/E_a)^{1/2} = 3.16$ and $(E_c/E_b)^{1/2} = 1.58$, respectively, where $E_a = 0.58$, $E_b = 2.3$, and $E_c = 5.8$ kcal/mol.

The features of the quantal spectra in Fig. 2.28 are most easily understood by consulting Fig. 2.29, where the continuum wave functions, actually their open $v' = 0$ channel component $[\chi_{E,j}^{gr}(R)$ in Eq. (2.32b) for $j = 0]$, are displayed for two collision energies 0.58 and 5.8 kcal/mol. Shown in the same figure are the regions of R values in which the wave functions of the ground level and the highest bound level calculated have appreciable amplitude. One sees that the continuum wave function for $E = 0.58$ kcal/mol can have no significant overlap with the low-lying levels of the G1 potential, which accounts for the absence of any transitions below $\sim 61,000$ cm^{-1}. In contrast to this, the higher collision energy (5.8 kcal/mol) continuum wave function penetrates much

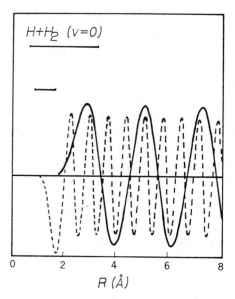

Figure 2.29.   The $v' = 0$ component of radial continuum wave functions for $H + H_2(v = 0)$ at $E = 0.58$ kcal/mol (solid line) and $E = 5.8$ kcal/mol (dashed line). Scales of wave functions are arbitrary. Regions of R values where wave functions of ground and highest calculated bound levels have a significant amplitude are denoted by horizontal bars.

further. Its large amplitude at smaller R values allows it to overlap significantly with essentially all calculated bound levels, hence the shift of the spectrum toward red.

The amplitude pattern of the continuum wave functions is responsible for the broad oscillation present in the absorption spectra. It is evident from Fig. 2.29 that within the region of R values spanned by the wave function of the highest calculated bound level (where continuum-bound overlap is possible), the low collision energy (0.58 kcal/mol) wave function completes less than half an oscillation, whereas the higher energy wave function has three to four maxima. The corresponding spectra clearly reflect this difference. Naturally, since this feature has purely quantum origin, it is not present in the MPP classical trajectory spectra.

The absorption spectra for $H + H_2(v = 1)$ are shown in Fig. 2.30. They have the same high-frequency cutoff ($\sim 70,000$ cm^{-1}) caused, as in the $H_2(v = 0)$ case, by the finite number (i.e., 74) of calculated bound levels of the G1 surface. With the increase of collision energy, the spectra exhibit the same shift to the red as well. The explanation for this shift is identical to that given for $H + H_2(v = 0)$. It is interesting to compare the spectra for $H + H_2(v = 1)$ in Figs. 2.30a and 2.30b with their $H + H_2(v = 0)$ counterparts shown in Figs.

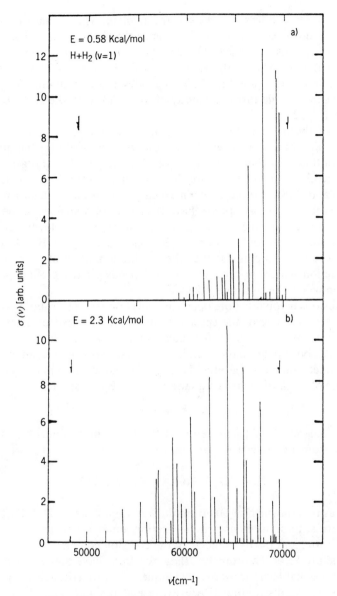

Figure 2.30. Transition-state spectra for $H + H_2(v = 1)$.

2.28*a* and *b*. One notices that the spectra involving vibrationally excited H_2 extend to frequencies that are much closer to the lowest possible continuum-bound ($n = 1$) transition (marked by the arrow at the low-frequency end of the spectrum), than the $H_2(v = 0)$ spectra. In particular, for $E = 2.3$ kcal/mol, the $H + H_2(v = 1)$ spectrum (Fig. 2.30*b*) includes the continuum–G1 ground level transition, while this is evidently not the case with the $H + H_2(v = 0)$ spectrum (Fig. 2.28*b*).

Some of the more distinct features and trends evident in the classical trajectory spectra of Mayne et al. manifest themselves also in the quantum spectra. In this sense there is a very satisfactory degree of quantitative agreement between the two sets of spectra. Yet in the overall appearance the quantum and classical spectra differ markedly. Very prominent, narrow peaks of the classical trajectory spectra have their quantum analogs in the rather extended maxima (at the appropriate frequencies) of the slowly oscillating transition intensity profile. Its broad features, especially for $H + H_2(v = 0)$, appear to be very simply related to the amplitude pattern of the continuum wave function which, in turn, of course depends on the ground potential surface on which the collisions take place.

In collisions involving more open channels (for example, rotational states) the oscillatory profile of the spectrum would be averaged over amplitude patterns of many components of the continuum wave function, as well as over the internal coordinates of the colliding partners. This would result in the loss of the distinctly quantum features of the spectra and would very likely increase the degree of similarity between quantum and classical spectra.

E. H + CO₂

Recently Scherer et al.[91] here reported the direct measurement of product formaton from a bimolecular reaction,

$$H + CO_2 \rightarrow HOCO \rightarrow HO + CO \qquad (2.34)$$

They measured the birth of OH from the hot atom reaction H + OCO.

In contrast to unimolecular fragmentation, bimolecular reactions involve a full collision, and the dynamics[92] of the collision complex is crucial to the fate of the products. Traditionally, these reactions have been studied by the crossed molecular beam scattering technique.[93,94] From the anisotropy of the products' angular distribution (PAD) in the center-of-mass coordinate system it is possible to infer whether the reaction proceeds by the direct mode (subpicosecond ineraction time) or via a "long-lived" (few rotational periods) collision complex.[95,96]

Sherer et al have employed the pump-probe method. In their pump-probe experiments, a picosecond laser pulse initiates the reaction via photodissociation of the van der Waals (vdW) molecule IH ... OCO, prepared in a seeded supersonic beam expansion. The "clocking" of the reaction is accomplished by the use of a picosecond probe pulse (for OH LIF monitoring), delayed in time with respect to the initiation pulse. Ultraviolet photolysis of such a vdW precursor molecule ensures that the initial H velocity (~ 20 km/s) is preferentially directed along the hydrogen bond to the OCO, limiting the range of impact parameters and "orienting" the reactants. The essential element in the experiment is that the ultraviolet photolysis picosecond (pump) pulse acts as a precise trigger setting off the H atom "projectile," thus establishing the zero of time for the bimolecular collision.

The seeded molecular beam was obtained by expanding mixtures of 5% HI and 8% CO_2 in He ($P \simeq 1800$ torr). The beam was characterized by electron impact (15–30 eV) time of flight (TOF) mass spectrometry. The mass spectra displayed peaks for HI, CO_2, $HI \cdot CO_2$, $(CO_2)_n$, and very little of $HI \cdot (CO_2)_2$. The complex $HI \cdot CO_2$ concentration was typically 3–5% relative to HI.

The picosecond pump pulse (239 nm) was generated by mixing the second harmonic of the 616-nm light with the 1.06 μm of a YAG laser. The probe was the second harmonic of the 616-nm pulse. The cross correlation between 616 and 308 nm was obtained using a $LiIO_3$ crystal. The probe was tuned to the $Q_1(1)$ transition of OH using a cell of H_2O_2. On-resonance enhancement for OH LIF was evident from probe wavelength tuning experiments.

The experimental results are displayed in Fig. 2.31. The rise time of the OH signal after deconvolution is found to be ~ 5–15 ps. This first direct observation of the time evolution of the nascent product of a bimolecular reaction represents the transient decay time of the collision complex, formed with the given available energy. For the photolysis wavelength used the initial relative translational energy E_{tr} ($\simeq E_{av}$) of the H + OCO reagents is calculated to be 200 kJ/mol assuming collinear incidence. The slower H atoms from the I* branch, in contrast, yields $E_{tr} = 112$ kJ/mol, only slightly above threshold.[97]

For the subject reaction there is considerable knowledge of the PES based on theory and experiment.[97-103] The minimum energy path has an entrance barrier corresponding to a threshold energy of ~ 105 kJ/mol, followed by a potential well supporting a complex HOCO (itself studied by matrix isolation spectroscopy),[104] followed by a small exit barrier to OH + CO. The initial H atom velocity is ~ 200 Å/ps, and the available energy of collision with CO_2 is above the transition-state barrier. Thus the initial step is assumed to be complete within the pulse, but the rapid H-atom attack slows down depending on the vibrational motion in HOCO, which leads to a slower rate for the formation of HO + CO. The measured OH rise time can now be used to model the product state distribution (PSD) and possible deviations from statistical behavior. Experiments at different photolysis wavelengths will yield the energy dependence of the formation and decay of the complex.

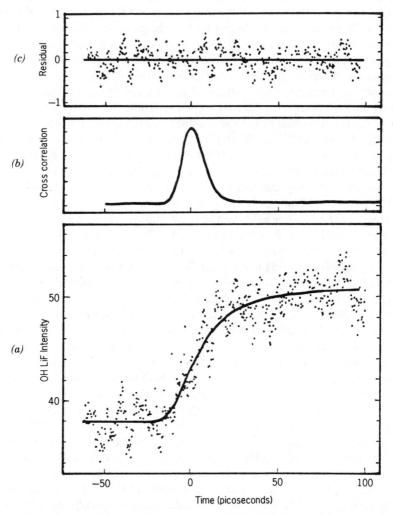

Figure 2.31. (a) Typical "transient" showing the OH LIF signal versus delay time τ (experimental results. (b) Experimental data points have been fitted (solid curve) assuming an exponential rise time (15 ps) and convoluting with measured cross-correlation function. (c) Weighted residuals versus τ_e.

240

F. HgI_2

Femtochemistry of the reaction

$$IHgI^* \rightarrow [IHg \ldots I]^{\ddagger *} \rightarrow HgI + I \qquad (2.35)$$

has been reported by Bowman, Dontus, and Zewail.[105] The experiments were performed using femtosecond pulses of two colors, the first pulse to initiate the reaction and the second tunable pulse to probe the products or the transition-state region. They used pulses as short as 50-fs duration in order to resolve dynamics on the 100–200-fs time scale. The pump pulse was typically centered at 310 nm, while the probe pulse was used at different wavelengths (390–620 nm). The experiments were performed at $\approx$ 130 °C. The LIF spectra are characteristic of the HgI, $B \rightarrow X$, photoproduct. For both Na–I and I–C≡N, one (or effectively one) vibrational coordinate describes the dissociation dynamics. If instead the reaction dynamics involve more than one coordinate, an interesting question arises. Suppose there is a real transition state at the saddle point, can one observe the motion (to final products) along these coordinates in real time on the global PES? The HgI_2 system is suitable for this purpose.

Figure 2.32 shows typical transients obtained for a pump wavelength (λ_1) of 310 nm. The probe (λ_2^*) was at 390 nm. The signal builds up and decays to an asymptotic value at longer times, as observed before in FTS experiments. The asymptote can be used to invert the data to obtain the potential, as recently discussed by Bernstein and Zewail.[71]

In addition to the decays, Bowman et al.[105] observe an oscillatory modulation in FTS. Keeping λ_1 the same, but changing the probe λ_2 to be far away $(\lambda_2^* = 620$ nm) from the absorption of the free fragment, they observe a fast modulated decay (period = 220 fs), not to an asymptote, but to zero intensity. These observations can be related to the dynamics of the reaction in real time.

In the first excited state the dissociation to HgI + I is considered direct, and the yield of the I and I* channels varies depending on λ_1.[106] The HgI is produced rotationally and vibrationally hot. On the femtosecond time scale the reaction may not be direct in that any vibrational structure or delay of trajectories can be observed.

The physics can be described as follows. Basically the potential surface is two-dimensional (for a given bend configuration) and repulsive along the antisymmetric (Q_2) coordinate. A cut along the symmetric (Q_1) coordinate at the saddle point shows the characteristics of a quasi-bound state. The initial ground-state wave function can overlap the two coordinates. Following the pump excitation in femtoseconds one expects vibrational motion along Q_1 and translational motion along Q_2. If the time for going down the hill is

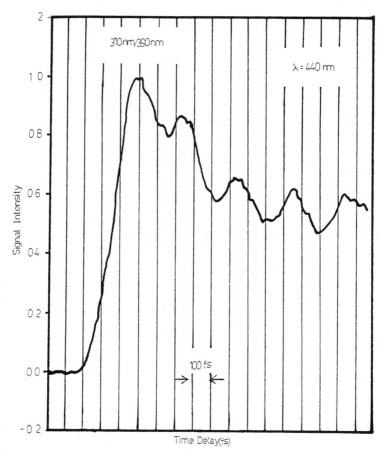

Figure 2.32. FTS transients at $\lambda_1 = 310$ nm and $\lambda_2 = 390$ nm. Detection wavelength was centered at 440 nm.

somewhat longer than the Q_1 vibrational time, then the FTS should show both a decay and an oscillatory behavior if probing is done in the zone of Q_1, Q_2 reached at early times. At longer times, when the trajectories reach the free fragment zone, then the vibrational frequency is that of HgI. Hence a change in the period is possible. Probing different vibrational–rotational states of HgI will monitor different trajectories on the PES. Bowman et al. also probed the dynamics by detecting different fluorescence (λ_{det}) of the LIF of the probe. As mentioned, the HgI is produced vibrationally excited and the fluorescence extends over a wide wavelength range; it reflects the different vibrational–rotational states produced in dissociation.

The dissociation trajectories of the reaction produce, on the femtosecond time scale, the fragments in a coherent state. The success of FTS probing relies on the fact that the two surfaces involved in λ_2^* absorption are much different and "selective" points can be proved, just as in the case of quantum beat detection in large molecules.[107] The HgI potentials in both the X and B states have been deduced.[108] In the X state the fundamental vibrational frequency is 125 cm^{-1} (corresponding to 267 fs), but at higher energies, near dissociation of HgI (3120 cm^{-1}),[109] the frequency becomes 34 cm^{-1} (corresponding to 0.98 ps). This is consistent with the observations and with the assignment of the fluorescence.

G.   K + NaCl

Jiang and Hutchinson have recently calculated the transition-state absorption spectrum and the dynamics of the excited state for a one-dimensional reactive system with the basic feature of the reaction[110]

$$K + NaCl \rightarrow KCl + Na \qquad (2.36)$$

Their system for the reaction Eq. (2.36) consists of two one-dimensional electronic potential energy curves coupled by a radiation field. The potential energy curves for both ground and excited levels used in the calculations are the well known Eckert potentials[111] and are chosen to have the basic features of the reaction coordinates on the PESs for the (K, NaCl) reaction.

Reaction surfaces for the (K, NaCl) system have been calculated by Child and Roach.[112] Jiang and Hutchison have chosen the Eckert potentials for the following reasons. First, the calculation is only one-dimensional, which is a greater deviation from reality than the simplicity of the potential. Second, the Eckart potentials give satisfactory fits to the Child–Roach potentials in terms of the energy differences, well depths, and widths, and the relative position of the two curves. In addition, the associated eigenvalue equations are exactly soluble.

For the same reasons, the reduced mass of the system along the reaction coordinate need not be defined exactly. The mass of the system is taken to be the reduced mass of the (K, NaCl) three-particle system defined as

$$\frac{1}{m} = \frac{1}{m_1} + \frac{1}{m_2} + \frac{1}{m_3} \qquad (2.37)$$

where m_1, m_2, and m_3 are the masses for K, Cl, and Na atoms, respectively. The mass has a value of 10.28 amu. In this work the upper potential curve will be an Eckart well, which also has an exact solution. The potential is defined by

$$V(X) = \frac{-AZ}{1 - Z} - \frac{BZ}{(1 - Z)^2} + V_E \qquad (2.38)$$

where

$$Z = Z(X) = -\exp\left[\frac{2\pi(X + X_E)}{l}\right] \qquad (2.39)$$

and X denotes the reaction coordinate. Among the five parameters appearing in the expression, A gives the asymptotic energy difference of the curve, B governs the actual shape of the potential in the interactive region, l is the half-width of this region, and V_E and X_E are introduced to shift the curve to fit the real system. The shape of the potential can be adjusted easily by varying these parameters. Jiang and Hutchinson performed the calculations for several sytems. In all of these, the excited potentials remained the same while the ground potentials were changed in terms of well depth, width, and relative position to the upper minimum. The ground potential curve I, shown in Fig. 2.33, is the best fit to the collinear configuration, while curve II is the best fit to the configuration for a KNaCl angle of 90°. The parameters used to obtain these curves are listed in Table 2.1.

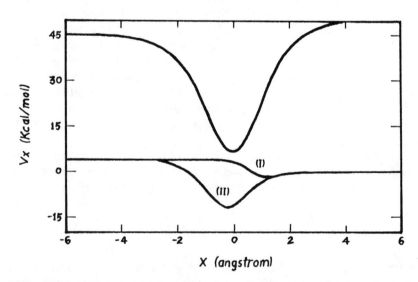

Figure 2.33. Eckart potential functions used in calculation. Lower two curves—ground electronic state $|\alpha\rangle$; upper curve—excited state $|\beta\rangle$. Curve I—best fit to collinear configuration of (K, NaCl) reaction; curve II—best fit to configuration of KNaCl of 90°.

TABLE 2.1
Parameters for Ground and Excited Eckart Potential Curves of Two Systems

	A (kcal/mol)	B (kcal/mol)	V_E (kcal/mol)	l (Å)	X_E (Å)
Ground level					
System I	−4	−15	4	2	−0.9
System II	−4	−55	4	3.3	0.3
Excited level	5	−165	45	4.5	0

For the ground level the wave functions of interest are those with high energies in the continuum spectrum. Although exact solutions are available for these states in terms of the hypergeometric series, the direct summation is again very difficult. Successful application of the WKB approximation requires that the wavelength of the particle change only slightly over one wavelength. Intuitively, one expects that this criterion will be satisfied since high energies are considered and because the ground KNaCl potential curve varies very slowly. The WKB wave functions on the α electronic state is well known and takes the form

$$|k_\alpha\rangle = \frac{N_{k\alpha}}{\sqrt{P(X)}} \exp\left[\frac{i}{\hbar} \int_{x_r}^{x} P(X)\, dX \right] \qquad (2.40)$$

where X_r is far away from the interactive region and serves as the reference point for the integration, $P(X)$ is the momentum corresponding to energy $E_{k\alpha}$, and $N_{k\alpha}$ is the normalization constant. A few selected WKB states have been checked by comparison to "exact" numerical integration of the eigenstates, and have found good agreement.

The initial wave packet $\Psi(X, 0)$ is taken to be a minimum uncertainty Gaussian packet defined by

$$\Psi(X, 0) = \left(\frac{\delta}{\pi} \right)^{1/4} \exp\left[-\frac{\delta}{2}(X - X_0)^2 - \frac{i}{\hbar} P_0 X \right] \qquad (2.41)$$

where P_0 and X_0 are initial momentum and position of the wave packet and δ governs the initial width. The velocity corresponding to P_0 (group velocity V_0) is chosen to be in a range that is comparable to the mean velocity of a particle at the experimental temperature. For example, a typical velocity of 1350 m/s used in the calculation corresponds to the mean velocity of a particle with mass 10.28 amu at the temperature of about 1100 K. The initial width is assigned arbitrarily to 1.0 Å.

First Jiang and Hutchison demonstrate that absorption of radiation by the transition state is expected and that, even for this simple system in one degree of freedom, a spectrum with partially resolvable maxima and minima is observed. The absorption spectra were calculated for several systems which differ only in the ground potential curves. Figure 2.34 compares the transition probability spectra for system I (cf. Fig. 2.33) with that for system II at three successive time steps. The initial group velocity V_0 of the wave packet is

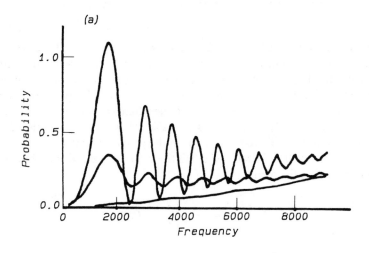

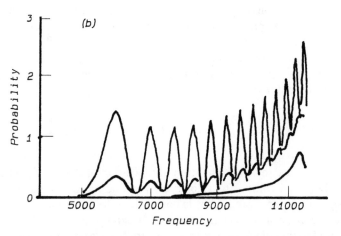

Figure 2.34. Transition probability spectra for systems I and II at three successive time steps. (*a*) System I with wave packet of $V_0 = 2500$ m/s, at 202, 242, and 592 fs (asymptotic). (*b*) System II with wave packet of $V_0 = 1350$ m/s, at 407, 472, and 1286 fs (asymptotic).

2500 m/s for system I and 1350 m/s for system II. Both spectra show an interesting oscillating behavior over a large frequency range when the wave packet has totally passed through the interactive region. After comparing these calculations for several variations of this system, these spectra are believed to be expected as the basic feature of unbound–bound excitation of transition states. The calculations also indicate that the spectra are not very sensitive to the changes on the potential curves; it is believed that a similar absorption pattern should be obtained even with the exact reaction path potential curves. It should be noted that these prominent spectral features differ significantly from the experimental results for the transition state of the reaction (K, NaCl).[51c, 113] However, the frequency region investigated in the experiment was very different: the absorption in that region is due mainly to free–free transitions instead of the free–bound transitions Jiang and Hutchinson are observing.

The transition spectra of the reaction K + NaCl were obtained by Maguire et al.[51c, 113] The basic experiment consists of crossing molecular beams of K and NaCl in the cavity of a continuous-wave dye laser and observing any fluorescence of Na* through a filter. Angular distribution studies of this class of reactions have shown that complexes are formed with lifetimes of a few picoseconds, and it is believed that these transient reaction complexes are excited by the laser, opening a new channel to Na*, which is then detected in fluorescence:

$$K + NaCl \rightarrow [KNaCl] \qquad (2.42)$$

$$[KNaCl] + h\nu \rightarrow [KNaCl]^* \qquad (2.43)$$

$$[KNaCl]^* \rightarrow KCl + Na^* \qquad (2.44)$$

$$Na^* \rightarrow Na + h\nu' \qquad (2.45)$$

In this reaction scheme [KNaCl] denotes the set of transient species intermediate between reagents and products, and [KNaCl]* denotes a similar set of excited species. Because the dark reaction (K + NaCl → Na + KCl) is exoergic, the laser photon can be chosen to be to the red of Na D lines ($v < v'$), which is a considerable aid in eliminating possible artifact processes.

The desired photoexcitation signal is the fluorescence at the Na D lines which requires the simultaneous presence of all three beams. Maguire et al. have observed these three-beam signals when the laser is tuned more than 1000 Å to the red of the D lines, and they established that the signals arise from processes that are first order with respect to each of the reagents. Various other experimental tests have ruled out many artifacts, and it is now believed that reactions (2.42)–(2.45) represent the most plausible explanation of their experiments.

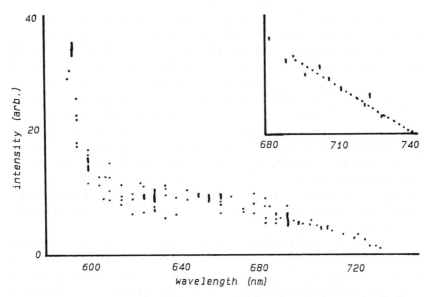

Figure 2.35. Three-beam signal (fluorescence of Na*) versus excitation wavelength. Representative error bars are given at several wavelengths. Large increase near Na D lines may not arise from photoexcitation of the reaction complex. Inset is detailed study near threshold wavelength.

The wavelength dependence of the three-beam signal is shown in Fig. 2.35. The signal is normalized by the product of the molecular beam and laser intensities and corrected for drift during the course of an experiment. It is apparent from Fig. 2.35 that the three-beam signal extends over a very wide wavelength range, is relatively featureless, and anticipates that the apparent threshold at 735 nm is consistent with the process under study. The large increase in three-beam signal to the blue of 600 nm (near the D lines) is probably due to artifact sources such as energy transfer, and has not yet been studied systematically.

The experiments reported by Maguire et al. have established that (1) a positive signal resulting from the production of Na* during the interaction of K, NaCl, and laser light has been observed; (2) the three-beam signal persists over a wavelength range extending at least from 590 to 735 nm; (3) the three-beam signal varies linearly with the laser and molecular beam intensities; (4) the trace amounts of K_2 and Na impurities present in the beams do not appear to be responsible for the three-beam signal on the long-wavelength side of the Na D lines, except perhaps at wavelengths within several nanometers of the D lines. They interpret these results as an observation for the laser-assisted reaction

$$K + NaCl + hv \rightarrow Na^* + KCl \qquad (2.46)$$

They view the process as photoexcitation of the KNaCl complex, followed by decomposition of the reaction complex into Na*. This interpretation is consistent with reactive scattering experiments on the alkali metal–alkali halide reactions and with the results of trajectory calculations of the K + NaCl reaction. Both studies suggest that the ground-state reaction proceeds on a PES that has no activation barrier, and that complexes, with average lifetimes of several rotational periods, are formed in nearly all reactive collisions.

H.   Control of Selectivity of Reactions

Recently Tannor and Rice[114] have proposed a novel approach to the control of selectivity of reaction products. The central idea is that in a two-photon or multiphoton process that is resonant with an excited electronic state, the resonant excited-state PES can be used to assist chemistry on the ground-state PES. By controlling the delay between a pair of ultrashort (femtosecond) laser pulses, it is possible to control the propagation times on the excited-state PES. Different propagation times, in turn, can be used to generate different chemical products. There are many cases for which selectivity of product formation should be possible using this scheme. They illustrate the methodology with numerical application to a variety of model two-degrees-of-freedom systems with two inequivalent exit channels. Branching ratios obtained using a swarm of classical trajectories are in good qualitative agreement with full quantum mechanical calculations.

The cases considered by Tannor and Rice involved reaction on the ground electronic state PES, but mediated by excitation to, evolution on, and stimulated deexcitation from an excited electronic state. In particular, the proposed methodology exploits the coherence properties of exciting and stimulating ultrashort pulses (femtosecond time scale), and the dynamics of wave-packet evolution between exciting and stimulating pulses. The pulse shapes, durations, and separations required to achieve selectivity of product formation depend on the properties of the excited-state PES. In the relevant time domain, which is defined by the shape of the excited-state PES, it is possible to take advantage of the localization in phase space of the time-dependent quantum mechanical amplitude and thereby carry out selective chemistry.

Theoretically it is assumed that the ground electronic state Born–Oppenheimer PES has two or more exit channels corresponding to the formation of two or more distinct chemical species.

It is also assumed that there exists an excited-state PES whose minimum is displaced from that of the ground-state PES and whose normal coordinates are rotated from those of the ground-state PES (Duschinshky effect). This excited-state PES is used to assist the chemistry on the ground-state PES. The time spent on the excited-state PES is used to select the desired chemical species.

At the instant that the first photon is incident the ground-state wave function makes vertical (Franck–Condon) transition to the excited-state PES. The ground-state wave function is not a stationary state on the excited-state PES, so it must evolve as t increases. The duration of the propagation on the excited-state PES can be regulated by the delay of a second pulse leads to a vertical (Franck–Condon) transition down to the ground-state PES. Note that the wave-function amplitude is unchanged in the Franck–Condon transition. If the delay and width of the second pulse are chosen on the basis of the position and width of the windows in the excited-state PES, it is plausible to expect the wave-packet amplitude on the ground-state PES to select one channel over the other.

The Hamiltonian Tannor and Rice adopt is a 2×2 matrix of operators. It represents the ground and excited electronic states within the Born–Oppenheimer approximation, coupled by the radiation field interacting with the transition dipole operators ($\mu = \mu_{ab} = \mu_{ba}$):

$$H = \begin{bmatrix} H_a & \mu E(t) \\ \mu E(t) & H_b \end{bmatrix} \tag{2.47}$$

The time-dependent Schrödinger equation reads

$$i\hbar \frac{\partial}{\partial t} \begin{pmatrix} \psi_a \\ \psi_b \end{pmatrix} = H \begin{pmatrix} \psi_a \\ \psi_b \end{pmatrix} \tag{2.48}$$

At $t = 0$,

$$H_a \psi_0 = E_a \psi_0 \tag{2.49}$$

Also, $\psi_b = 0$.

The two coupled differential equations in Eq. (2.48) can be transformed to two coupled integral equations,

$$\psi_a(t) = e^{(-iH_a t)/\hbar} \psi_a(0) - \frac{i}{\hbar} \int_{-\infty}^{t} e^{-i[H_a(t-t')]/\hbar} \mu E(t') \psi_b(t') \, dt' \tag{2.50a}$$

$$\psi_b = \frac{-i}{\hbar} \int_{-\infty}^{t} e^{-iH_b[(t-t')/\hbar]} \mu E(t') \psi_a(t') \, dt' \tag{2.50b}$$

In the weak field regime, the time-dependent perturbation theory expression for the second-order amplitude on the ground-state surface $\psi_a^{(2)}(t)$ is

$$\psi_a^{(2)}(t) = \frac{-1}{\hbar^2} \int_{-\infty}^{t} \int_{-\infty}^{t_2} e^{-iH_a(t-t_2)/\hbar} \mu E(t_2) e^{-iH_b(t_2-t_1)/\hbar} \mu E(t_1) \psi_a(0) e^{-\omega_a t_1} \, dt_1 \, dt_2$$

$$(2.51)$$

Note that Eq. (2.51) contains the field strength to second order only. Equations (2.50) have the following simple interpretation: $\psi_a(0)$ evolves on the ground-state PES from $t = 0$ until $t = t_1$. At time t_1 it makes a vertical transition to the excited-state PES. The wave function propagates on the excited-state PES from time t_1 until time t_2. At time t_2 the wave function makes a vertical transition back to the ground-state PES. The wave function then evolves on the ground-state PES from time t_2 until time t. In general, the waveforms $E(t_1)$, $E(t_2)$ are extended; therefore one must integrate over t_1 and t_2, all the instants at which the transition up and the transition down may take place. The probability S for exit from channel 1 is

$$S_1 = \lim_{t \to \infty} \langle \psi_a(t) | P_1 | \psi_a(t) \rangle \qquad (2.52)$$

where P_1 is a projection operator in coordinate space corresponding to channel 1 and S_2 is given similarly.

To calculate the time evolution of the wave function, it is first necessary to evaluate $H\psi$. In Eq. (2.48), for example, the diagonal part of the matrix multiplication is given by

$$H_a \hbar^2 = \left[-\frac{\hbar^2}{2m} \left(\frac{\partial^2}{\partial x^2} + \frac{\partial^2}{\partial y^2} \right) + V_a(xy) \right] \psi_a(xy) = [T + V_a] \psi_a \quad (2.53)$$

and similarly for $H_b \psi_b$. $V_a \psi_a$ is evaluated simply by multiplication at each of the discrete grid points. For a detailed discussion the original papers[114] should be consulted.

In concluding the present chapter, we would like to point out that there are other reactions that have been studied for obtaining transition-state spectra, but they were not included here. For example, there are K + $HgBr_2$ + $h\nu \to$ KBr + HgBr* studied by Hering et al.,[51,113] Mg + H_2 + $h\nu \to$ MgH + H studied by Kleiber et al.,[52] Na* + L_i + $h\nu \to$ NaL_i^+ + e^- studied by Polak-Dingels et al.,[53] F + $Na_2 \to$ Na* + NaF studied by Polanyi et al.,[54] O_3 + $h\nu \to O_2$ + O studied by Imre et al.,[55] and so on. Also recent studies of the isomerization of polyatomic molecules[115] and of the photodissociation of molecules[116] in liquids can be included in the category of transition-state spectroscopy. Reviewing recent rapid developments of femtosecond spectroscopy in this chapter, we discuss three types of commonly used femtosecond spectroscopic techniques and their applications. Due to the fact that the dynamics of coherence play an important role in femtosecond spectroscopy, the analysis of experimental data becomes nontrivial. For example, femtosecond time-resolved spectra usually consist of coherent and incoherent contributions, and thus it becomes much more difficult to analyze those spectra in order to obtain structural and dynamic information.

APPENDIX A. SOME DERIVATIONS OF GENERALIZED LINEAR RESPONSE THEORY

Generalized susceptibilities $\chi_{ab}(\omega_1, \omega_2)$, $\chi_{ab}(\omega, t)$, and $\chi_{ab}(\omega, t)$ can be used as characteristics of the absorption of the probe light pulse. According to Landau and Lifshitz,[1] the change in the mean internal energy of the system is found from the partial time derivative of the complete Hamiltonian (including the perturbation energy):

$$Q = \frac{\partial \langle H \rangle}{\partial t} = \left\langle \frac{dH}{dt} \right\rangle \tag{A.1}$$

where

$$H = H_0 + V \tag{A.2}$$

Since only the perturbation V is explicitly time dependent,

$$Q = \frac{\partial \langle V \rangle}{\partial t} = \left\langle \frac{dV}{dt} \right\rangle \tag{A.3}$$

Then since in V only $f_a(t)$ is explicitly time dependent, we get

$$Q = -\sum_a \dot{f}_a \langle x_a \rangle \tag{A.4}$$

The most straightforward connection exists between the absorption rate and the susceptibilities $\tilde{\chi}_{ab}(\omega, t)$. Substituting f_a from Eq. (1.20) and $\langle x_a \rangle$ from Eq. (1.22) into Eq. (A.3), we get

$$
\begin{aligned}
Q = -\sum_{a,b} \Bigg\{ & [\tilde{f}_a(\omega)e^{-i\omega t} + \tilde{f}_a(-\omega)e^{i\omega t}] \frac{\partial L(t)}{\partial t} \\
& - i\omega L(t)[\tilde{f}_a(\omega)e^{-i\omega t} - \tilde{f}_a(-\omega)e^{i\omega t}] \Bigg\} [\tilde{\chi}_{ab}(\omega, t)e^{-i\omega t}\tilde{f}_b(\omega) \\
& + \tilde{\chi}_{ab}^*(\omega, t)\tilde{f}_b(-\omega)e^{i\omega t}]
\end{aligned} \tag{A.5}
$$

Averaging this expression over the period of the external force and assuming that the shape of the probe pulse is smooth enough so that

$$\left|\frac{1}{L}\frac{\partial L}{\partial t}\right| \ll \omega \tag{A.6}$$

we obtain

$$Q = i\omega \sum_{a,b} [\tilde{\chi}_{ab}^*(\omega,t) - \tilde{\chi}_{ba}(\omega,t)]\tilde{f}_a(\omega)\tilde{f}_b^*(\omega)L(t) \tag{A.7}$$

In the isotropic case, when

$$\tilde{\chi}_{ab} = \tilde{\chi}\delta_{ab} \tag{A.8}$$

we obtain from Eq. (A.7)

$$Q = 2\omega\tilde{\chi}''(\omega,t)|\tilde{f}(\omega)|^2 L(t) \tag{A.9}$$

where $$|\tilde{f}(\omega)|^2 = \sum_a |\tilde{f}_a(\omega)|^2 \tag{A.10}$$

Now we show the connection between the spectrum (Fourier transform) of $Q(t)$ and two-frequency susceptibilities $\chi_{ab}(\omega,\omega_1)$. We substitute $\langle x_a \rangle$ from Eq. (1.11) and $f_a(t)$ from Eq. (1.10) into Eq. (A.4) and take a Fourier transform of $Q(t)$,

$$\hat{Q}(\Omega) = \frac{1}{2\pi}\int_{-\infty}^{\infty} Q(t)e^{i\Omega t}\,dt$$

$$= i\sum_{a,b}\int d\omega_1 \int d\omega_2 \omega_1 \chi_{ab}(\Omega - \omega_1, \omega_2)f_a(\omega_1)f_b(\omega_2) \tag{A.11}$$

In the case of a quasi-monochromatic pulse, characterized by Eqs. (1.14) and (1.25), the expression $\hat{Q}(\Omega)$ takes the form

$$\hat{Q}(\Omega) = i\sum_{a,b}\int d\omega_1 \int d\omega_2 \omega_1 \chi_{ab}(\Omega - \omega_1, \omega_2)[\tilde{f}_a(\bar{\omega})\hat{L}(\omega_1 - \bar{\omega})$$

$$+ \tilde{f}_a(-\bar{\omega})\hat{L}(\omega_1 + \bar{\omega})][\tilde{f}_b(\bar{\omega})\tilde{L}(\omega_2 - \bar{\omega}) + \tilde{f}_b(-\bar{\omega})\tilde{L}(\omega_2 + \bar{\omega})] \tag{A.12}$$

To proceed further, we have to obtain some knowledge about properties of two-frequency susceptibility $\chi_{ab}(\omega_1, \omega_2)$. Using Eqs. (1.13) and (1.8), we get

$$\chi_{ab}(\omega_1, \omega_2)$$

$$= \frac{i}{2\pi h} \int_{-\infty}^{\infty} dt \int_0^{t-t_i} d\tau \sum_{mnk} \left[\rho_{imn} x_{ank} x_{bkm} \right.$$

$$\times e^{i(\omega_1 - \omega_2 + \omega_{nm})(t-t_i)} e^{i(\omega_2 - \omega_{km})\tau} \rho_{imn} x_{bnk} x_{akm} e^{i(\omega_1 - \omega_2 + \omega_{nm})(t-t_i)} e^{i(\omega_2 - \omega_{nk})\tau} \right]$$

(A.13)

Thus we come to the expression

$$\chi_{ab}(\omega_1, \omega_2) = \sum_{m,n,k} \frac{1}{h} \zeta_t(\omega_2 - \omega_{nk}) \left[\rho_{imn} x_{bnk} x_{akm} \delta(\omega_1 - \omega_2 + \omega_{nm}) \right.$$

$$- \rho_{ikm} x_{amn} x_{bnk} \delta(\omega_1 - \omega_2 + \omega_{mk}) \right]$$

(A.14)

where
$$\zeta_t(x) = -i \int_0^{t-t_i} e^{itx} d\tau$$

(A.15)

and $\zeta_\infty(x) = \zeta(x) = \lim_{\sigma \to 0} [1/(x + i\sigma)]$. In the stationary case when the density matrix is diagonal, we get the following expression:

$$\chi_{ab}(\omega_1, \omega_2) = \delta(\omega_1 - \omega_2) \chi_{ab}(\omega_1)$$

(A.16)

where
$$\chi_{ab}(\omega) = \frac{1}{h} \sum_{n,k} (\rho_{nn} - \rho_{kk}) x_{akn} x_{bnk} \zeta_t(\omega - \omega_{nk})$$

(A.17)

is the conventional expression for one-frequency stationary susceptibility $(t \to \infty)$ (see, e.g., Fain and Khanim[21]). To have further insight about the two-frequency susceptibility $\chi_{ab}(\omega_1, \omega_2)$, we consider the monochromatic probe (while the initial pulse is still ultrashort). In this case

$$\hat{L}(\omega) = \delta(\omega)$$

(A.18)

and the expression for the Fourier transform of the energy dissipation is of the form

$$\hat{Q}(\Omega) = i\bar{\omega} \sum_{a,b} \left[\chi_{ab}(\Omega - \bar{\omega}, \bar{\omega}) \tilde{f}_a(\bar{\omega}) \tilde{f}_b(\bar{\omega}) - \chi_{ab}(\Omega + \bar{\omega}, -\bar{\omega}) \tilde{f}_a(-\bar{\omega}) \tilde{f}_b(-\bar{\omega}) \right.$$

$$+ \chi_{ab}(\Omega - \bar{\omega}, -\bar{\omega}) \tilde{f}_a(\bar{\omega}) \tilde{f}_b(-\omega) - \chi_{ab}(\Omega + \bar{\omega}, \bar{\omega}) \tilde{f}_a(-\bar{\omega}) f_b(\bar{\omega}) \right]$$

(A.19)

It follows from inspection of expression (A.14) that if there are no resonances on the frequency $2\bar{\omega}$, we can neglect first two terms in the right-hand side of Eq. (A.19). Thus assuming that there are no resonances on $2\bar{\omega}$, we get

$$\hat{Q}(\Omega) = i\bar{\omega} \sum_{a,b} [\chi_{ab}(\Omega - \bar{\omega}, -\bar{\omega}) - \chi_{ba}(\Omega + \bar{\omega}, \bar{\omega})] \tilde{f}_a(\bar{\omega}) \tilde{f}_b(-\bar{\omega}) \quad \text{(A.20)}$$

It follows from Eq. (A.14) that

$$\chi_{ab}(\Omega + \bar{\omega}, \bar{\omega}) = \sum_{m,n,k} \frac{1}{\hbar} \zeta_t(\bar{\omega} - \omega_{nk}) [\rho_{imn} x_{bnk} x_{akm} \delta(\Omega + \omega_{nm})$$
$$- \rho_{ikm} x_{amn} x_{bnk} \delta(\Omega + \omega_{mk})] \quad \text{(A.21)}$$

Particularly, frequency $\bar{\omega}$ is close to the electronic frequencies ω_{nk}, while Ω is close to the vibrational sublevel energy differences ω_{mn} and ω_{km}. When the off-diagonal density matrix elements between these sublevels are zero,

$$\hat{Q}(\Omega) = \delta(\Omega) i\bar{\omega} \sum_{a,b} [\chi_{ab}^*(\bar{\omega}) - \chi_{ba}(\bar{\omega})] \tilde{f}_a(\bar{\omega}) \tilde{f}_b(-\bar{\omega}) \quad \text{(A.22)}$$

The divergence of this expression reflects the fact that in the second-order perturbation theory there exists a time-independent rate of the energy absorption Q. Notice that in the nonstationary case the susceptibility $\chi_{ab}(\Omega + \bar{\omega}, \bar{\omega})$ can be used as a characteristic of two-frequency spectroscopy, $\bar{\omega}$ giving a spectrum of electronic vibrational transitions between various electronic states, while Ω is tuned to vibrational transitions of the same electronic state.

When the probing pulse is monochromatic, [Eq. (A.18)], we can also obtain a convenient expression of the time-dependent absorption rate $Q(t)$ through the time–frequency representation of the susceptibility $\chi_{ab}(\omega, t)$ in Eqs. (A.15) and (A.16). Neglecting second harmonic contribution to $Q(t)$ we obtain, similar to Eq. (A.7), the following expression:

$$Q(t) = i\bar{\omega} \sum_{a,b} [\chi_{ab}^*(\bar{\omega}, t) - \chi_{ba}(\bar{\omega}, t)] \tilde{f}_a(\bar{\omega}) \tilde{f}_b(-\bar{\omega}) \quad \text{(A.23)}$$

and in the isotropic case,

$$Q(t) = 2\bar{\omega}\chi''(\bar{\omega}, t) |\tilde{f}(\bar{\omega})|^2 \quad \text{(A.24)}$$

It is easy to verify, by using relation (1.21), that the Fourier transform of $Q(t)$ in Eq. (A.23) coincides with expression (A.20).

Similar to Eq. (1.38) we can express $\chi_{ab}(\omega, t)$ through the density matrix elements as

$$\chi_{ab}(\omega, t) = \frac{1}{\hbar} \sum_{m,n,k} \rho_{inm} e^{i\omega_{nm}(t-t_i)} [\zeta_t(\omega - \omega_{mk}) x_{bmk} x_{akn} - \zeta_t(\omega - \omega_{kn}) x_{amk} x_{bkn}]$$
$$\text{(A.25)}$$

In the case when the monochromatic probe on frequency $\bar{\omega}$ is close to the energy differences between various electronic states, the time dependence of $Q(t)$ and $\chi_{ab}(\bar{\omega}, t)$ provides information about vibrational (coherent) movements at the same electronic state.[7-15]

APPENDIX B

In the femtosecond range the dynamics of the system can be ignored and we need only to consider the following Liouville equation:

$$\frac{d\rho_t}{dt} = -iL_0\rho_t - \frac{i}{\hbar}[V, \rho_t] \tag{B.1}$$

The generalized master equations are given by

$$\frac{d(\rho_t)_{nn}}{dt} = \frac{2}{\hbar} \sum_m \text{Im}\left[V_{nm}(\rho_t)_{mn}\right] \tag{B.2}$$

and

$$\frac{d(\rho_t)_{mn}}{dt} = -i\omega_{mn}(\rho_t)_{mn} + \frac{i}{\hbar}V_{mn}[(\rho_t)_{mm} - (\rho_t)_{nn}]$$

$$- \frac{i}{\hbar}\sum_l^{l \pm m, n}[V_{ml}(\rho_t)_{ln} - (\rho_t)_{ml}V_{ln}] \tag{B.3}$$

Using the rotating-wave approximation $(\rho_t)_{mn} = \rho_t(\omega)_{mn}e^{-it\omega}$ (for $\omega_{mn} > 0$), Eq. (B.3) becomes

$$\frac{d\rho_t(\omega)_{mn}}{dt} = i(\omega - \omega_{mn})\rho_t(\omega)_{mn} + \frac{i}{\hbar}e^{it\omega}V_{mn}[(\rho_t)_{mm} - (\rho_t)_{nn}]$$

$$- \frac{i}{\hbar}\sum_l'' e^{it\omega}[V_{ml}(\rho_t)_{ln} - (\rho_t)_{ml}V_{ln}] \tag{B.4}$$

Here we rewrite V as [Eqs. (1.2) and (1.14)]

$$V = [D(\omega)e^{-it\omega} + D(-\omega)e^{it\omega}]L(t) \tag{B.5}$$

For $V_{mn} \neq 0$ we obtain approximately

$$(\rho_t)_{mn} = \frac{i}{\hbar}D_{mn}(\omega)[(\rho_0)_{mm} - (\rho_0)_{nn}]\frac{e^{-it\omega - 2t/T} - e^{-it\omega_{mn}}}{i(\omega_{mn} - \omega) - 2/T} \tag{B.6}$$

where $(\rho_0)_{mm}$ and $(\rho_0)_{nn}$ represent the initial populations. In the short-time region, Eq. (B.6) reduces to

$$\rho_t(\omega)_{mn} = \frac{it}{\hbar} D_{mn}(\omega)[(\rho_0)_{mm} - (\rho_0)_{nn}] \qquad \text{(B.7)}$$

Next we apply the results give by Eqs. (B.1)–(B.7) to the photodissociation of NaI and NaBr. In this case, from Eq. (B.3) we find

$$\frac{d(\rho_t)_{av,av'}}{dt} = -i\omega_{av,av'}(\rho_t)_{av,av'} - \frac{i}{\hbar}\sum_l{}'' [V_{av,l}(\rho_t)_{l,av'} - (\rho_t)_{av,l}V_{l,av'}] \qquad \text{(B.8)}$$

If initially only the $\gamma0$ state is populated, then

$$\frac{d(\rho_t)_{av,av'}}{dt} = i\omega_{av,av'}(\rho_t)_{av,av'} - \frac{i}{\hbar}[V_{av,\gamma0}(\rho_t)_{\gamma0,av'} - (\rho_t)_{av,\gamma0}V_{\gamma0,av'}] \qquad \text{(B.9)}$$

By using Eqs. (B.5) and (B.7), Eq. (B.9) can be integrated to yield

$$(\rho_t)_{av,av'} = \frac{t^2}{\hbar^2} D_{av,\gamma0}(\omega)D_{\gamma0,av'}(-\omega)(\rho_0)_{\gamma0,\gamma0} \qquad \text{(B.10)}$$

or

$$(\rho_t)_{av,av'} = \frac{t^2}{2\hbar^2} V_{av,\gamma0}V_{\gamma0,av'}(\rho_0)_{\gamma0,\gamma0} \qquad \text{(B.11)}$$

APPENDIX C

We consider the evolution of the density matrix ρ of the total system which consists of the molecular system and radiation field,[31]

$$\frac{d\rho}{dt} = -\frac{i}{\hbar}[H,\rho] - R\rho = -iL\rho - R\rho \qquad \text{(C.1)}$$

where R denotes the damping operator due to the coupling between system and heat bath, and L represents the Liouville operator corresponding to the Hamiltonian H of the total system. H can be written as a summation of H_s, the Hamiltonian of the molecular system, H_r, the Hamiltonian of the radiation field, and V, the interaction between molecular system and radiation field,

$$H = H_r + H_s + V = H_0 + V \qquad \text{(C.2)}$$

It follows that

$$\frac{d\rho}{dt} = -iL_0\rho - iL'\rho - R\rho \qquad (C.3)$$

where L' corresponds to V. If we let

$$\rho = \exp(-iL_0't)\sigma \qquad (C.4)$$

where $L_0' = L_0 - iR$, we obtain

$$\frac{d\sigma}{dt} = -ie^{itL_0'}L'e^{-itL_0'}\sigma \qquad (C.5)$$

which can be integrated as

$$\sigma(t) = \sigma_i - i\int_{t_i}^{t} e^{i\tau L_0'}L'e^{-i\tau L_0'}\sigma(\tau)\,d\tau = \sigma_i + \Delta\sigma(t) \qquad (C.6)$$

where σ_i denotes σ at $t = t_i$. For time-resolved emission we calculate

$$\langle \dot{H}_r \rangle = \left\langle \frac{dH_r}{dt} \right\rangle = \text{Tr}\left(\rho\frac{dH_r}{dt}\right) = \frac{i}{\hbar}\text{Tr}[\rho[H, H_r]]$$

$$= \frac{i}{\hbar}\text{Tr}\left[e^{-itL_0'}\Delta\sigma(t)[V, H_r]\right] \qquad (C.7)$$

Notice that for a molecular system, V can be expressed as $V = -\vec{X}\cdot\vec{Y}$. Using this relation Eq. (C.7) can be written as

$$\langle \dot{H}_r \rangle = \text{Tr}\left[G(t)\Delta\sigma(t)(\dot{\vec{X}}\cdot\vec{Y})\right] \qquad (C.8)$$

where $\dot{\vec{X}} = d\vec{X}/dt$, and $G(t) = e^{-itL_0'}$. It follows that

$$\langle \dot{H}_r \rangle = \sum_{nm} G(t)_{nm}^{nm}\Delta\sigma(t)_{nm}(\dot{\vec{X}}\cdot\vec{Y})_{mn} \qquad (C.9)$$

and

$$\Delta\sigma(t)_{nm} = -\frac{i}{\hbar}\sum_{n'}\int_0^{t-t_i}[V(t-\tau)_{nn'}\sigma(t-\tau)_{n'm} - \sigma(t-\tau)_{nn'}V(t-\tau)_{n'm}]\,d\tau$$

$$= \frac{i}{\hbar}\int_0^{t-t_i}[\vec{X}(t-\tau)\cdot\vec{Y}(t-\tau), \sigma(t-\tau)]_{nm}\,d\tau \qquad (C.10)$$

where, for example, $V(t - \tau)_{nn'} = \exp[i(t - \tau)(\omega_{nn'} - iR_{nn'})]V_{nn'}$ and $R_{nn'}$ represents the dephasing constant. Substituting Eq. (C.10) into Eq. (C.8) yields

$$\langle \dot{H}_r \rangle = \frac{i}{\hbar} \int_0^{t-t_i} \text{Tr}\{\sigma(t - \tau)[\dot{\vec{X}}(t) \cdot \vec{Y}(t), \vec{X}(t - \tau) \cdot \vec{Y}(t - \tau)]\} \, d\tau \quad (C.11)$$

which can be rewritten as

$$\langle \dot{H}_r \rangle = \frac{i}{\hbar} \int_0^{t-t_i} \{\langle \dot{\vec{X}}(t)\vec{X}(t - \tau)\rangle_0 : \text{Tr}[\sigma^{(s)}(t - \tau)\vec{Y}(t)\vec{Y}(t - \tau)]$$
$$- \langle \vec{X}(t - \tau)\dot{\vec{X}}(t)\rangle_0 : \text{Tr}[\sigma^{(s)}(t - \tau)\vec{Y}(t - \tau)\vec{Y}(t)]\} \, d\tau \quad (C.12)$$

where, for example, $\langle \dot{\vec{X}}(t)\vec{X}(t - \tau)\rangle_0$ denotes the vacuum average of $\dot{\vec{X}}(t)\vec{X}(t - \tau)$ for the radiation field. $\sigma^{(s)}(t)$ represents the density matrix of the molecular system. Notice that

$$\vec{X} = \sum_r \left(\frac{2\pi\hbar}{V\omega_r}\right)^{1/2} (a_r + a_r^*)\vec{e}_r \quad (C.13)$$

Substituting Eq. (C.13) into Eq. (C.12) we obtain

$$\langle \dot{H}_r \rangle = \frac{2\pi}{V} \sum_r \int_0^{t-t_j} \text{Tr}\{\sigma^{(s)}(t - \tau)[(\vec{e}_r \cdot \vec{Y}(t))(\vec{e}_r \cdot \vec{Y}(t - \tau))e^{-it\omega_r}$$
$$+ (\vec{e}_r \cdot \vec{Y}(t - \tau))(\vec{e}_r \cdot \vec{Y}(t))e^{it\omega_r}]\} \, d\tau \quad (C.14)$$

Using Eq. (1.75), the time-resolved emission spectrum measured in terms of the number of emitted quanta per unit time $P(\omega, t)$ is given by

$$P(\omega, t) = \frac{2\omega}{\pi\hbar c^3} \int_0^{t-t_i} \text{Tr}\{\sigma^{(s)}(t - \tau)[(\vec{e} \cdot \vec{Y}(t))(\vec{e} \cdot \vec{Y}(t - \tau))e^{-it\omega}$$
$$+ (\vec{e} \cdot \vec{Y}(t - \tau))(\vec{e} \cdot \vec{Y}(t))e^{it\omega}]\} \, d\tau \quad (C.15)$$

which can be written as

$$P(\omega, t) = \frac{2\omega}{\hbar c^3} \sum_{nmm'} [G^{(s)}(\Delta t)\rho_0^{(s)}]_{nm}(\vec{e} \cdot \vec{Y}_{mm'})(\vec{e} \cdot \vec{Y}_{m'n})$$
$$\times \left[\delta(\omega - \omega_{mm'}) + \delta(\omega + \omega_{m'n}) + \frac{i}{\pi}P\frac{\omega_{mn}}{(\omega + \omega_{m'n})(\omega - \omega_{mm'})}\right]$$
$$(C.16)$$

In obtaining Eq. (C.16), the Markov approximation has been used. A better approximation will be to replace the delta function by the Lorentzian due to the damping R. For a randomly oriented molecular system, Eq. (C.16) becomes

$$P(\omega, t) = \frac{2\omega}{3\hbar c^3} \sum_{nmm'} [G^{(s)}(\Delta t)\rho_0^{(s)}]_{nm}(\vec{Y}_{mm'} \cdot \vec{Y}_{m'n})$$

$$\times \left[\delta(\omega - \omega_{mm'}) + \delta(\omega + \omega_{m'n}) + \frac{i}{\pi} P \frac{\omega_{mn}}{(\omega + \omega_{m'n})(\omega - \omega_{mm'})} \right]$$

$$(C.17)$$

It should be noted that $P(\omega, t)$ consists of two parts; while $[G^{(s)}(\Delta t)\rho_0^{(s)}]_{nm}$ describes the dynamic behavior of the molecular system, the remaining part of $P(\omega, t)$ describes the emission spectroscopic behavior of the molecular system. The Einstein A coefficient (i.e., the spontaneous emission rate constant) can be obtained from Eq. (C.17) by setting $[G^{(s)}(\Delta t)\rho_0^{(s)}]_{nm} = \delta_{nm} (\rho_0^{(s)})_{nn}$, where $(\rho_0^{(s)})_{nn}$ denotes the equilibrium distribution. In this case we have

$$A = \frac{4}{3\hbar c^3} \sum_{nm'} (\rho_0^{(s)})_{nn} |\vec{Y}_{nm'}|^2 \omega_{nm'} = \frac{4}{3\hbar c^3} \sum_{nm'} (\rho_0^{(s)})_{nn} |\vec{\mu}_{nm'}|^2 \omega_{nm'}^3 \quad (C.18)$$

where $\vec{\mu}_{nm'}$ denotes the transition moment. In Appendix D we show how to calculate $[G^{(s)}(\Delta t)\rho_0^{(s)}]_{nm}$.

APPENDIX D

In this appendix we are concerned with the calculation of $[G^{(s)}(t)\rho_0^{(s)}]_{nm}$. To demonstrate the theoretical approach, we consider a two-level (say m and n) system. We first consider the diagonal case,

$$[G^{(s)}(t)\rho_0^{(s)}]_{mm} = G^{(s)}(t)_{mm}^{mm}(\rho_0^{(s)})_{mm} + G^{(s)}(t)_{mm}^{nn}(\rho_0^{(s)})_{nn} \quad (D.1)$$

Here $[G^{(s)}(t)\rho_0^{(s)}]_{mm}$ denotes the population of the m level at time t, while $(\rho_0^{(s)})_{mm}$ and $(\rho_0^{(s)})_{nn}$ represent the initial populations. We assume that the m level is higher than the n level. Notice that

$$G^{(s)}(t) = \exp[-it(L_s - iR)] \quad (D.2)$$

and

$$\frac{d\rho^{(s)}}{dt} = -iL_s\rho^{(s)} - R\rho^{(s)} \quad (D.3)$$

Carrying out the Laplace transformation of Eq. (D.2) yields

$$G^{(s)}(p) = \int_0^\infty e^{-pt} G^{(s)}(t)\, dt = \frac{1}{p + i(L_s - iR)} \tag{D.4}$$

We now calculate $G^{(s)}(p)_{mm}^{mm}$ and $G^{(s)}(p)_{mm}^{nn}$. From Eq. (D.4) we find

$$(p + R_{mm}^{mm})G^{(s)}(p)_{mm}^{mm} + R_{mm}^{nn}G^{(s)}(p)_{nn}^{mm} = 1 \tag{D.5}$$

and
$$(p + R_{nn}^{nn})G^{(s)}(p)_{nn}^{mm} + R_{nn}^{mm}G^{(s)}(p)_{mm}^{mm} = 0 \tag{D.6}$$

where R_{mm}^{mm} and R_{nn}^{nn} denote the relaxation rate constants for $m \to n$ and $n \to m$, respectively, and $-R_{mm}^{nn} = R_{nn}^{nn}$ and $-R_{nn}^{mm} = R_{mm}^{mm}$.

Solving Eqs. (D.5) and (D.6) for $G^{(s)}(p)_{mm}^{mm}$ and $G^{(s)}(p)_{nn}^{mm}$, we obtain

$$G^{(s)}(p)_{mm}^{mm} = \frac{p + R_{nn}^{nn}}{p(p + R_{nn}^{nn} + R_{mm}^{mm})} \tag{D.7}$$

and
$$G^{(s)}(p)_{nn}^{mm} = \frac{R_{mm}^{mm}}{p(p + R_{nn}^{nn} + R_{mm}^{mm})} \tag{D.8}$$

Carrying out the inverse Laplace transformation of Eqs. (D.7) and (D.8) yields

$$G^{(s)}(t)_{mm}^{mm} = \frac{R_{nn}^{nn}}{R_{nn}^{nn} + R_{mm}^{mm}} + \frac{R_{mm}^{mm}}{R_{nn}^{nn} + R_{mm}^{mm}} \exp\left[-t(R_{nn}^{nn} + R_{mm}^{mm})\right] \tag{D.9}$$

and
$$G^{(s)}(t)_{nn}^{mm} = \frac{R_{mm}^{mm}}{R_{nn}^{nn} + R_{mm}^{mm}} \left\{1 - \exp\left[-t(R_{nn}^{nn} + R_{mm}^{mm})\right]\right\} \tag{D.10}$$

Similarly, we obtain

$$G^{(s)}(t)_{nn}^{nn} = \frac{R_{mm}^{mm}}{R_{nn}^{nn} + R_{mm}^{mm}} + \frac{R_{nn}^{nn}}{R_{nn}^{nn} + R_{mm}^{mm}} \exp\left[-t(R_{nn}^{nn} + R_{mm}^{mm})\right] \tag{D.11}$$

$$G^{(s)}(t)_{mm}^{nn} = \frac{R_{nn}^{nn}}{R_{nn}^{nn} + R_{mm}^{mm}} \left\{1 - \exp\left[-t(R_{nn}^{nn} + R_{mm}^{mm})\right]\right\} \tag{D.12}$$

and
$$[G^{(s)}(t)\rho_0^{(s)}]_{nm} = G^{(s)}(t)_{nm}^{nm}(\rho_0^{(s)})_{nm} \tag{D.13}$$

where
$$G^{(s)}(t)_{nm}^{nm} = \exp\left[-t(i\omega_{nm} + R_{nm}^{nm})\right] \tag{D.14}$$

R_{nm}^{nm} denoting the dephasing constant.

It should be noted that to obtain $[G^{(s)}(t)\rho_0^{(s)}]_{nm}$ and $[G^{(s)}(t)\rho_0^{(s)}]_{nn}$, one can also solve the generalized master equations given by Eq. (D.3).

APPENDIX E

Here we consider the calculation of $K_{v_0}(t)$ defined by

$$K_{v_0}(t) = \sum_{w_0} |\langle X_{\beta v_0}|X_{\alpha w_0}\rangle|^2 \exp[-\mu_0'(w_0 + \tfrac{1}{2}) - \lambda_0(v_0 + \tfrac{1}{2})] \quad (E.1)$$

where $\mu_0' = -it\omega_0'$ and $\lambda_0 = it\omega_0$. Using the Mehler formula,[33] we can rewrite Eq. (E.1) as

$$K_{v_0}(t) = \frac{\beta_0' \exp[-\lambda_0(v_0 + \tfrac{1}{2})]}{(2\pi \sinh \mu_0')^{1/2}} \int_{-\infty}^{\infty} \int X_{\beta v_0}(Q) X_{\beta v_0}(\bar{Q}) \, dQ \, d\bar{Q}$$

$$\times \exp\left\{-\frac{\beta_0'^2}{4}\left[(Q' + \bar{Q}')^2 \tanh\frac{\mu_0'}{2} + (Q' - \bar{Q}')^2 \coth\frac{\mu_0'}{2}\right]\right\} \quad (E.2)$$

where $\beta_0' = (\omega_0'/\hbar)^{1/2}$. Using the contour integral representation for the Hermite polynomial and performing the integrations with respect to Q and $\bar{Q}$, we obtain

K_{v_0}

$$= \frac{\beta_0 \beta_0' \exp[-\lambda_0(v_0 + \tfrac{1}{2})]}{\{[\beta_0^2 \sinh(\mu_0'/2) + \beta_0'^2 \cosh(\mu_0'/2)][\beta_0^2 \cosh(\mu_0'/2) + \beta_0'^2 \sinh(\mu_0'/2)]\}^{1/2}}$$

$$\times \left(\frac{v_0!}{2^{v_0}}\right)\left(\frac{1}{2\pi i}\right)^2 \int_c \frac{dz_1}{z_1^{v_0+1}} \int_c \frac{dz_2}{z_2^{v_0+1}} \exp\left[-(z_1^2 + z_2^2) - \frac{\beta_0^2 \beta_0'^2 d_0^2}{\beta_0'^2 + \beta_0^2 \coth(\mu_0'/2)}\right.$$

$$\left. + \frac{\beta_0^2(z_2 - z_1)^2}{\beta_0^2 + \beta_0'^2 \coth(\mu_0'/2)} + \frac{\beta_0^2(z_2 + z_1)^2}{\beta_0^2 + \beta_0'^2 \tanh(\mu_0'/2)} + \frac{2\beta_0'^2 \beta_0 d_0(z_1 + z_2)}{\beta_0'^2 + \beta_0^2 \coth(\mu_0'/2)}\right]$$

$$(E.3)$$

where $Q_0' = Q_0 + d_0$, that is, d_0 denotes the normal coordinate displacement. Carrying out the contour integrals in Eq. (E.3) yields

$$K_{v_0}(t) = K_{0_0}(t)\Delta K_{v_0}(t) \quad (E.4)$$

where

$K_{0_0}(t)$

$$= \frac{\beta_0 \beta_0' \exp\left[-\dfrac{\lambda_0}{2} - \dfrac{\beta_0^2 \beta_0'^2 d_0^2}{\beta_0'^2 + \beta_0^2 \coth(\mu_0'/2)}\right]}{\{[\beta_0^2 \sinh(\mu_0'/2) + \beta_0'^2 \cosh(\mu_0'/2)][\beta_0^2 \cosh(\mu_0'/2) + \beta_0'^2 \sinh(\mu_0'/2)]\}^{1/2}}$$

$$(E.5)$$

and

$$\Delta K_{v_0}(t)$$

$$= \exp(-\lambda_0 v_0) \sum_{n_0=0}^{v_0} \frac{v_0!}{(n_0!)^2(v_0 - n_0)!} \left(\frac{a_2}{2}\right)^{v_0-n_0} \left(\sqrt{\frac{a_1}{2}}\right)^{2n_0} H_{n_0}\left(\frac{-a_3}{\sqrt{4a_1}}\right)^2$$

(E.6)

Here $H_{n_0}(x)$ denotes the Hermite polynomial and

$$a_1 = \frac{-\beta_0^4 + \beta_0'^4}{[\beta_0^2 + \beta_0'^2 \tanh(\mu_0'/2)][\beta_0^2 + \beta_0'^2 \coth(\mu_0'/2)]}$$

(E.7)

$$a_2 = \frac{2\beta_0^2 \beta_0'^2}{[\beta_0^2 \cosh(\mu_0'/2) + \beta_0'^2 \sinh(\mu_0'/2)][\beta_0^2 \sinh(\mu_0'/2) + \beta_0'^2 \cosh(\mu_0'/2)]}$$

(E.8)

$$a_3 = \frac{2\beta_0 \beta_0'^2 d_0}{\beta_0'^2 + \beta_0^2 \coth(\mu_0'/2)}$$

(E.9)

Acknowledgment

This is publication 032 from the Arizona State University Center for the Study of Early Events in Photosynthesis. The Center is funded by the U.S. Department of Energy under grant DE-FG02-88ER13969 as part of the USDA/DOE/NSF Plant Science Center program. This work was supported in part by the National Science Foundation. The authors wish to thank Dr. C. Y. Yeh, C. C. Kuo, W. L. Chang, P. C. Chen, and H. Ma for helping with the preparation of the manuscript.

References

1. L. Landau and E. Lifshitz, *Statistical Physics*, Pergamon, London, 1958; *Electrodynamics of Continuous Media*, Pergamon, Oxford, 1960.
2. V. M. Fain, *Sov. Phys. JETP* **23**, 882 (1966).
3. C. V. Shank, *Science* **233**, 1276 (1986); W. Domcke and H. Köppel, *Chem. Phys. Lett.* **140**, 133 (1987); A. L. Harris, J. K. Brown, and C. B. Harris, *Ann. Rev. Phys. Chem.* **39**, 341 (1988).
4. (a) M. Dantus, M. Rosker, and A. Zewail, *J. Chem. Phys.* **87**, 2395 (1987); (b) T. Rose, M. Rosker, and A. Zewail, *J. Chem. Phys.* **88**, 6672 (1988); (c) M. Rosker, T. Rose, and A. Zewail, *Chem. Phys. Lett.* **146**, 175 (1988); (d) V. Engel, H. Moria, R. Almeida, R. A. Marcus, and A. H. Zewail, *Chem. Phys. Lett.* **152**, 1 (1988).
5. G. R. Fleming, *Chemical Applications of Ultrafast Spectroscopy*, Oxford Univ. Press, London, 1985; M. J. Rosker, F. W. Wise, and C. L. Tang, *Phys. Rev. Lett.* **57**, 321 (1986); I. A. Walmsley, M. Mitsunaga, and C. L. Tang, *Phys. Rev.* **A38**, 4681 (1988).
6. R. Bersohn and A. Zewail, *Ber. Bunsenges. Phys. Chem.* **92**, 373 (1988).
7. I. Riess, *J. Chem. Phys.* **52**, 871 (1960).
8. J. Rosenfeld, B. Voigt, and C. A. Mead, *J. Chem. Phys.* **53**, 1960 (1970).
9. W. M. Gelbart and J. Jortner, *J. Chem. Phys.* **54**, 2070 (1970).
10. B. Fain, *Phys. Rev.* **A37**, 546 (1988).
11. O. K. Rice, *J. Chem. Phys.* **1**, 375 (1933).

12. U. Fano, *Phys. Rev.* **124**, 1866 (1961).
13. L. Landau and E. Lifshitz, *Quantum Mechanics*, Pergamon, Oxford, 1977.
14. B. Fain, *Theory of Rate Processes in Condensed Media*, Springer, Berlin, 1980.
15. A. Boeglin, B. Fain, and S. H. Lin, *J. Chem. Phys.* **84**, 4838 (1986).
16. S. H. Lin, A. Boeglin, and S. M. Lin, *J. Photochem.* **39**, 173 (1987).
17. W. Domcke, A. L. Sobolewski, and S. H. Lin, *J. Chem. Phys.* **89**, 6209 (1988).
18. B. Fain, S. H. Lin, and W. X. Wu, *Phys. Rev.* **A40**, 824 (1989).
19. S. H. Schaefer, D. Bender, and E. Tiermann, *Chem. Phys.* **89**, 65 (1984).
20. R. S. Berry, *J. Chem. Phys.* **27**, 1288 (1957).
21. V. M. (B.) Fain and Y. I. Khanin, *Quantum Electronics*, vol. I, Pergamon, Oxford, 1969; V. M. (B.) Fain, *Photons and Non-Linear Media*, Sov. Radio, Moscow, 1972 (in Russian).
22. R. Kubo, T. Takagahara, and E. Hanamura, *Solid State Commun.* **32**, 1 (1979).
23. Y. R. Shen, *The Principle of Nonlinear Optics*, Wiley-Interscience, New York, 1984.
24. M. A. Kahlow, W. Jarzeba, T. J. Kang, and P. F. Barbara, *J. Chem. Phys.* **90**, 151 (1989); W. Jarzeba, G. C. Walker, A. E. Johnson, M. A. Kahlow, and P. F. Barbara, *J. Phys. Chem.* **92**, 7039 (1988).
25. M. Maroncelli and G. R. Fleming, *J. Chem. Phys.* **86**, 6221 (1989); E. W. Castner, Jr., M. Maroncelli, and G. R. Fleming, *J. Chem. Phys.* **86**, 1090 (1989).
26. J. D. Simon, *Acct. Chem. Res.* **21**, 128 (1988); N. Mataga and Y. Harata, in *Advances in Multiphoton Processes and Spectroscopy*, vol. 5, S. H. Lin, Ed., World Scientific, Singapore, 1989, pp. 175–276.
27. A. Declemy and C. Rulliere, *Chem. Phys. Lett.* **146**, 1 (1988); **145**, 262 (1988).
28. J. M. Hicks, M. T. Vandersalli, E. V. Sitzmann, and K. B. Eisenthol, *Chem. Phys. Lett.* **135**, 413 (1987).
29. E. M. Kosower and D. Hupper, *Ann. Rev. Phys. Chem.* **39**, 127 (1986).
30. W. R. Ware, P. P. Chow, G. J. Braut, and S. K. Lee, *J. Chem. Phys.* **54**, 4729 (1971).
31. S. H. Lin, Y. Fujimura, H. J. Neusser, and E. W. Schlag, *Multiphoton Spectroscopy of Molecules*, Academic Press, New York, 1984, chaps. 1 and 2.
32. W. Louisell, *Quantum Statistical Properties of Radiation*, Wiley-Interscience, New York, 1973, chap. 5.
33. S. H. Lin, *J. Chem. Phys.* **59**, 4458 (1973).
34. S. H. Lin and B. Fain, *Chem. Phys. Lett.* **155**, 216 (1989).
35. B. Fain, S. H. Lin, and N. Hamer, *J. Chem. Phys.*, **91**, 4485 (1989).
36. (a) S. De Silvestri, J. G. Fujimoto, E. P. Ippen, E. B. Gamble Jr., L. R. Williams, and K. A. Nelson, *Chem. Phys. Lett.* **116**, 146 (1985); Xong-xin Yan, E. B. Gamble, Jr., and K. A. Nelson, *J. Chem. Phys.* **83**, 5391 (1985); Yong-xin Yan and K. A. Nelson, *J. Chem. Phys.* **87**, 6240, 6257 (1987); S. Ruhman, A. G. Joly, B. Kohler, L. R. Williams, and K. A. Nelson, *Rev. Phys. Appl.* **22**, 1717 (1987); S. Ruhman, A. G. Joly, and K. A. Nelson, *IEEE J. Quant. Electron.*, **24**, 460, 470 (1988); Lap-Tak Cheng, and K. A. Nelson, *Phys. Rev.* **B39**, 9437 (1989); (b) L. R. Williams, E. B. Gamble, K. A. Nelson, S. De Silvestri, A. M. Weiner, and E. P. Ippen, *Chem. Phys. Lett.* **39**, 244 (1987); A. G. Jolly, S. H. Ruhman, B. Kohler, and K. A. Nelson, in *Ultrafast Phenomena VI*, Eds., T. Yajima, K. Yoshihars, C. B. Harris, and S. Shionoya, vol. 48, Springer Series in Chemical Physics; Berlin, 1988; (c) A. G. Joly and K. A. Nelson, *J. Phys. Chem.* **93**, 2876 (1989).
37. F. W. Wise M. J. Rosker, C. L. Tang, *J. Chem. Phys.* **86**, 2827 (1987); A. Walmsey, F. W. Wise, and C. L. Tang, *Chem. Phys. Lett.* **154**, 315 (1989); Mitsunaga and C. L. Tang, *Phys. Rev.* **A24**, 1099 (1987).
38. N. Bloembergen, *Phys. Rev.* **104**, 324 (1956).
39. B. Fain and S. H. Lin, *Surf. Sci.* **147**, 497 (1984); B. Fain, A. Boeglin, and S. H. Lin, *J. Chem. Phys.* **88**, 7559 (1988).

40. G. Placzek, *Marx Handbuch der Radiologie*, 2nd ed., vol. VI, pt. II, E. Marx, Ed., Verlagsges. Leipzig, Germany, 1934, pp. 209–379.
41. M. Born and K. Huang, *Dynamical Theory of Crystal Lattices*, Clarendon, Oxford, 1963, p. 207.
42. G. M. Genkin, W. M. (B.) Fain, and E. G. Yashchin, *Sov. Phys. JETP* **25**, 592 (1967).
43. R. Kubo, T. Takagahara, and E. Hanamura, *Solid State Commun.* **32**, 1 (1979).
44. S. H. Lin, B. Fain, N. Hamer, and C. X. Yeh, *Chem. Phys. Lett.*, **162**, 73 (1989).
45. W. Bernard, and H. B. Callen, *Rev. Mod. Phys.* **31**, 1017 (1959); S. B. Zhu, J. Lee, and G. W. Robinson, *Phys. Rev.* **A38**, 5810 (1988).
46. M. Dantus, R. M. Bowman, J. S. Baskin, and A. H. Zewail, *Chem Phys. Lett.* **159**, 406 (1989).
47. J. Chesnoy and A. Mokhtary, *Phys. Rev.* **A38**, 3566 (1988).
48. M. Hayashi, Y. Nomura, and Y. Fujimura, *J. Chem. Phys.* **89**, 34 (1988).
49. Y. Ohtsuki and Y. Fujimura, *J. Chem. Phys.*, in press.
50. H. Kono and Y. Fujimura, *J. Chem. Phys.*, in press.
51. (a) P. Hering, P. R. Brooks, R. F. Curl, Jr., R. S. Judson, and R. S. Lowe, *Phys. Rev. Lett.* **44**, 687 (1980); (b) P. R. Brooks, R. F. Curl, and T. C. Maguire, *Ber. Bunsenges. Phys. Chem.* **86**, 401 (1982); (c) T. C. Maguire, P. R. Brooks, and R. F. Curl, *Phys. Rev. Lett.* **50**, 1918 (1983).
52. P. D. Kleiber, A. M. Lyyra, K. M. Sando, S. P. Heneghan, and W. C. Stwalley, *Phys. Rev. Lett.* **54**, 2003 (1985).
53. P. Polak-Dingels, J. F. Delpech, and J. Weiner, *Phys. Rev. Lett.* **44**, 1663 (1980); (b) P. Polak-Dingels, J. Keller, J. Weiner, J. C. Gauthier, and N. Bras, *Phys. Rev.* **A24**, 1107 (1981).
54. (a) P. Arrowsmith, F. E. Bartoszek, S. H. P. Bly, T. Carington, Jr., P. E. Charters, and J. C. Polanyi, *J. Chem. Phys.* **73**, 5895 (1980); (b) P. Arrowsmith, S. H. P. Bly, P. E. Charters, and J. C. Polyani, *J. Chem. Phys.* **79**, 283 (1983); (c) H. J. Foth, J. C. Polanyi, and H. H. Telle, *J. Phys. Chem.* **86**, 5027 (1982).
55. (a) D. G. Imre, J. L. Kinsey, R. W. Field, and D. H. Katayama, *J. Phys. Chem.* **86**, 2564 (1982); (b) D. G. Imre, J. L. Kinsey, A. Sinha, and J. Krenos, *J. Phys. Chem.* **88**, 3956 (1984).
56. W. Kamke, B. Kamke, I. Hertel, and A. Gallagher, *J. Chem. Phys.* **80**, 4879 (1984); W. J. Alford, N. Anderson, K. Burnett, and J. Copper, *Phys. Rev.* **A30**, 2366 (1984).
57. J. Alvarellos and H. Metiu, *J. Chem. Phys.* **88**, 4957 (1988).
58. R. A. Marcus, *Chem. Phys. Lett.* **152**, 8 (1988).
59. V. Engel and H. Metiu, *J. Chem. Phys.* **90**, 6116 (1989).
60. S. E. Choi and J. C. Light, *J. Chem. Phys.* **90**, 2602 (1989).
61. See, for example, E. J. Heller, *Acct. Chem. Res.* **14**, 368 (1981).
62. M. B. Faist and R. D. Levine, *J. Chem. Phys.* **64**, 2953 (1976).
63. S.-Y. Lee, W. T. Pollard, and R. A. Mathies, *J. Chem. Phys.* **90** 6146 (1989).
64. V. Engel and H. Metiu, *J. Chem. Phys.*, **91**, 1596 (1989).
65. R. D. Bower, P. Chevrier, P. Das, H. J. Foth, J. C. Polanyi, M. G. Prisant, and J. P. Visticot, *J. Chem. Phys.* **89**, 4478 (1988).
66. V. Engel and H. Metiu, *Chem. Phys. Lett.* **155**, 77 (1989).
67. N. F. Scherer, J. L. Knee, D. D. Smith, and A. H. Zewail, *J. Phys. Chem.* **89**, 5141 (1985).
68. A. H. Zewail, in *Ultrafast Phenomena* V, vol. 45, Springer Series in Chem. Phys., Berlin, 1986, pp. 356–365, and references therein.
69. M. J. Rosker, M. Dantus, and A. H. Zewail, *J. Chem. Phys.* **89**, 6113 (1988).
70. M. Dantus, M. J. Rosker, and A. H. Zewail, *J. Chem. Phys.* **89**, 6128 (1988).
71. R. B. Bernstein and A. H. Zewail, *J. Chem. Phys.* **90**, 829 (1989).
72. S. O. Williams and D. G. Imre, *J. Phys. Chem.* **92**, 6636 (1988).
73. S. O. Williams and D. G. Imre, *J. Phys. Chem.* **92**, 6648 (1988).
74. D. Kosloff and R. Kosloff, *J. Comput. Phys.* **52**, 35 (1983).

75. R. Kosloff and D. Kosloff, *J. Chem. Phys.* **79**, 1823 (1983).
76. R. Heather and H. Metiu, *Chem. Phys. Lett.* **157**, 505 (1989).
77. H. B. Gray and N. A. Beach, *J. Am. Chem. Soc.* **90**, 5713 (1968).
78. J. D. Simon and X. Xie, *J. Phys. Chem.* **90**, 6751 (1986).
79. L. J. Rothberg, N. J. Cooper, K. S. Peters, and V. Vaida, *J. Am. Chem. Soc.* **10**, 3536 (1982).
80. R. A. Levenson and H. B. Gray, *J. Am. Chem. Soc.* **97**, 6042 (1975).
81. T. Kobayashi, H. Ohtani, H. Noda, S. Teratani, H. Yamazaki, and K. Yasufuku, *Organometallics* **5**, 110 (1986).
82. R. W. Wegman, R. J. Olsen, D. R. Gard, L. R. Faulkner, and T. L. Brown, *J. Am. Chem. Soc.* **103**, 6089 (1981).
83. (*a*) A. F. Hepp and M. S. Wrighton, *J. Am. Chem. Soc.* **105**, 5934 (1983); (*b*) S. P. Church, H. Hermann, F. W. Grevels, and K. Schaffner, *J. Chem. Soc. Chem. Comm.* **785** (1984).
84. A. L. Harris, M. Berg, and C. B. Harris, *J. Chem. Phys.* **84**, 788 (1986).
85. J. R. Hill, E. L. Chronister, T. Chang, H. Kim, J. C. Postlewaite, and D. D. Dlott, *J. Chem. Phys.* **88**, 949 (1988).
86. H. R. Mayne, R. A. Poirier, and J. C. Polanyi, *J. Chem. Phys.* **80**, 4025 (1984); J. C. Polyani and R. J. Wolf, *J. Chem. Phys.* **75**, 5951 (1981).
87. B. Liu, *J. Chem. Phys.* **58**, 1925 (1973).
88. P. Siegbahn and B. Liu, *J. Chem. Phys.* **68**, 2457 (1978).
89. D. G. Truhlar and C. J. Horowitz, *J. Chem. Phys.* **68**, 2457 (1978).
90. V. Engel, Z. Bacic, R. Shinke, and M. Shapiro, *J. Chem. Phys.* **82**, 4844 (1985).
91. N. F. Scherer, L. R. Khundkar, R. B. Bernstein, and A. H. Zewail, *J. Chem. Phys.* **87**, 1451 (1987).
92. R. D. Levine and R. B. Bernstein, *Molecular Reaction Dynamics and Chemical Reactivity*, Oxford Univ. Press, New York, 1987.
93. E. H. Taylor and S. Datz, *J. Chem. Phys.* **23**, 1711 (1955).
94. D. R. Hershcach, *Disc. Faraday Soc.* **33**, 149 (1962); **55**, 233 (1973).
95. W. B. Miller, S. A. Safron, and D. R. Herschbach, *Disc. Faraday Soc.* **44**, 108 (1967).
96. S. Stolte, A. E. Proctor, and R. B. Bernstein, *J. Chem. Phys.* **65**, 4990 (1976).
97. G. A. Oldershaw and D. A. Porter, *Nature* **223**, 490 (1969).
98. C. R. Quick, Jr., and J. J. Tiee, *Chem. Phys. Lett.* **100**, 223 (1983).
99. K. Kleinermanns and J. Wolfrum, *Chem. Phys. Lett.* **104**, 157 (1984).
100. G. Radhakrishnan, S. Buelow, and C. Wittig, *J. Chem. Phys.* **84**, 727 (1986).
101. S. Buelow, M. Noble, G. Radhakrishnan, H. Reisler, C. Wittig, and G. Hancock, *J. Phys. Chem.* **90**, 1015 (1986).
102. I. W. M. Smith and R. Zellner, *J. Chem. Soc., Faraday Trans. 2* **69**, 1617 (1973).
103. I. W. M. Smith, *Chem. Phys. Lett.* **49**, 112 (1977).
104. D. E. Milligan and M. E. Jacox, *J. Chem. Phys.* **54**, 927 (1971).
105. R. M. Bowman, M. Dantus, and A. H. Zewail, *Chem. Phys. Lett.* **156**, 131 (1989).
106. H. Hofman and S. R. Leone, *J. Chem. Phys.* **69**, 3819 (1978); J. A. McGarvey Jr., N. H. Cheung, A. C. Erlandson, and T. A. Cool, *J. Chem. Phys.* **74**, 5133 (1981).
107. P. M. Felker and A. H. Zewail, *Adv. Chem. Phys.* **70**, 265 (1988); E. W. Schlag, W. E. Henke, and S. H. Lin, *Int. Rev. Phys. Chem.* **2**, 1 (1982).
108. J. Maya, *J. Chem. Phys.* **67**, 4976 (1977); K. Wieland, *Helv. Phys. Acta* **2**, 46 (1929); **14**, 420 (1941); *Z. Elektrochem.* **64**, 761 (1960).
109. B. E. Wilcomb and R. B. Bernstein, *J. Mol. Spect.* **62**, 442 (1976).
110. J. Jiang and J. S. Hutchinson, *J. Chem. Phys.* **89**, 6973 (1987).
111. C. Eckart, *Phys. Rev.* **35**, 1303 (1930).
112. A. C. Roach and M. S. Child, *Mol. Phys.* **14**, 1 (1968).

113. T. C. Maguire, P. R. Brooks, R. F. Curl, J. Spence, and S. D. Ulvick, *J. Chem. Phys.* **85**, 844 (1986).

114. D. J. Tannor and S. A. Rice, *J. Chem. Phys.* **83**, 5013 (1985); D. J. Tannor, R. Kosloff, and S. A. Rice, *J. Chem. Phys.* **85**, 5805 (1986).

115. For review, see J. Schröder and J. Troe, *Ann. Rev. Phys. Chem.* **38**, 163 (1987); S. B. Zhu and G. W. Robinson, *Chem. Phys. Lett.* **153**, 539 (1988).

116. A. L. Harris, J. K. Brown, and C. B. Harris, *Ann. Rev. Phys. Chem.* **39**, 341 (1988).

FEEDBACK ANALYSIS OF MECHANISMS FOR CHEMICAL OSCILLATORS*

YIN LUO and IRVING R. EPSTEIN

Department of Chemistry, Brandeis University, Waltham, MA 02254

CONTENTS

* Part 58 in the series *Systematic Design of Chemical Oscillators*.

Advances in Chemical Physics, Volume LXXIX, Edited by I. Prigogine and Stuart A. Rice.
ISBN 0-471-52768-8   © 1990 John Wiley & Sons, Inc.

INTRODUCTION

Probably the most important single factor in transforming chemical oscillation from an amusing lecture demonstration of somewhat questionable relevance to "mainstream" chemistry to a subfield of major importance in physical chemistry and chemical physics has been the ability of chemists to develop mechanistic explanations of oscillating reactions. In particular, the Field–Körös–Noyes mechanism[1] for the Belousov–Zhabotinskii (BZ) reaction demonstrated clearly for the first time that chemical oscillators are governed by the same principles of stoichiometry and kinetics as any other chemical reaction.

The development of the first mechanisms for oscillating chemical reactions was a considerable intellectual achievement. There were few examples or general patterns to serve as guides, and computational capabilities for simulating the resulting rate equations were considerably less powerful than the "user-friendly" packages now available. During the past decade, the number of oscillating reactions has grown by roughly an order of magnitude, thanks in large measure to the introduction of flow reactor (CSTR) techniques[2] and the development of a systematic design procedure.[3] The number of systems for which mechanisms have been proposed has increased apace. In the early days a mechanism might have been considered satisfactory (or even unique!) if it yielded oscillations in some region of parameter space near, but not necessarily overlapping with, the experimentally observed region of oscillations. Agreement, even of a qualitative nature, with such "details" as the measured amplitude, waveform, or period of oscillation was rarely approached. Today, however, with advances in both experimental and numerical methods, one can achieve qualitative and even quantitative agreement in many systems. An example is shown in Fig. 1.

As investigators have developed greater facility in constructing mechanisms of chemical oscillators, it has become increasingly clear that these systems, like any other, are subject to the fundamental dogma of chemical kinetics that no mechanism is unique and that mechanisms may only be disproved, never proved, by comparison with experiment. In two cases, the Briggs–Rauscher[5,6] and the Edblom–Orbán–Epstein (EOE)[7,8] oscillators, different groups, working independently, have arrived at remarkably similar mechanisms that differ only in minor details. On the other hand, for two other systems, the BZ[9,10] and the Jensen[11,12] reactions, rival mechanisms with significant substantive differences have led to lively controversy and ultimately to further experimental work that revealed new details about the reactions in question.

In this chapter, we discuss some general strategies for building, and ways of categorizing, kinetic models that describe chemical oscillation and related dynamic properties of chemical systems. We demonstrate that there are routes

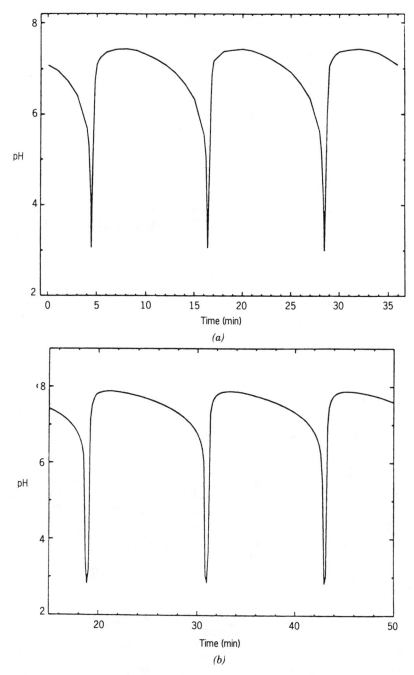

Figure 1. (a) Experimental and (b) calculated pH in the EOE reaction in a stirred tank reactor.[4] Input concentrations: $[IO_3^-]_0 = 0.075M$, $[SO_3^{2-}]_0 = 0.0893M$, $[H_2SO_4]_0 = 0.0045M$, $[Fe(CN)_6^{4-}]_0 = 0.0204M$. Reciprocal residence time $k_0 = 0.0022$ s^{-1}.

that one can follow to "synthesize" an oscillating mechanism with or without knowledge of the detailed chemistry of the system. Using feedback as a unifying principle, we group mechanisms of chemical oscillators in a small number of classes according to the similarities and differences in their cores. The representative model of each class can be viewed as a route that leads to oscillation with the proper combination of parameters, regardless of the extent to which the detailed chemistry has been idealized.

There have been several earlier reviews aimed at systematizing what is known about oscillatory mechanisms. Higgins,[13] drawing exclusively on biochemical systems for his examples, presented a perceptive treatment of many aspects of oscillatory behavior at a time when the mechanistic aspects of chemical oscillators were only beginning to be understood. Tyson,[14] also using primarily examples from biochemical oscillators, showed that all processes that destabilize the steady state of a chemical system, thus making oscillatory behavior possible, fall into one of three classes derivable from consideration only of the signs of the partial derivatives

$$a_{ij} = \left(\frac{\partial v_i}{\partial c_j} \right)_s \tag{1}$$

where the c_j are the concentration variables, $v_i = dc_i/dt$, and the subscript s means evaluated at the steady state. Field and Noyes reviewed the conceptual[15] and experimental[16] aspects of mechanisms of chemical oscillators in 1977, but of necessity devoted nearly all their detailed considerations to the single system that had been well characterized at that time, the BZ reaction. Franck[17] presented a more general treatment of oscillatory mechanisms, emphasizing the role of feedback, but discussed only abstract models rather than actual mechanisms. Over the years, a great deal of terminology has developed (backward activation, end-product inhibition, direct autocatalysis, etc.). We shall employ some of this nomenclature and introduce more of our own, trying at each stage to make very clear the meanings of all terms used.

So as not to complicate the analysis, we limit our consideration to homogeneous isothermal systems. We treat only mono- and bimolecular reactions in models and mechanisms, except when fast protonation–deprotonation equilibria are involved.

I. THE PRINCIPLE OF FEEDBACK COMPETITION

It has long been recognized that an essential requirement for any chemical system to oscillate is that it possess some form of *feedback*, or "self-influence,"[17] that is, a product of some reaction in the mechanism must affect the rate at which one or more species involved in that reaction are produced or consumed.

If the feedback accelerates the reaction whose rate it affects, it is a positive feedback; if it slows that rate, it is a negative feedback.

Franck summarizes the situation as follows[17]:

> In chemical reactions, oscillatory behaviour and instability are primarily the result of self-influence of the reactions involved ... It is an essential fact that all chemical and physicochemical oscillators known so far obviously possess positive and negative feedback properties simultaneously ... As is well-known, positive feedback labilizes the system and may bring about instability. Negative feedback acts antagonistic to the positive feedback. It stabilizes the system and causes recovery and adaptation behavior. Obviously, oscillations can be regarded as a result of an antagonistic interaction or interplay between a labilizing positive feedback releasing a conversion of state and a coupled negative feedback, which recovers that conversion and restores the initial state, so that the closed chain of reactions can start again.

In our view it is this interplay of the positive and negative feedback elements of a mechanism that is most important in generating chemical oscillation. We do not, however, share the view that rigorous separation of negative and positive feedback in a system is always desirable or possible. For example, as we shall see, only explosive autocatalysis is always a positive feedback in an absolute sense, but such behavior is rarely seen in actual systems. A nonexplosive autocatalysis can have both activation and inhibition effects, depending on which species one refers to. In other words, the same autocatalytic reaction can provide positive feedback for one component, but negative feedback for another. Therefore a feedback network should always be analyzed as a whole. This view is implemented throughout the discussions of models and mechanisms that follow.

Most previous efforts to design mechanisms and new oscillating reactions have focused on the positive feedback necessary to destabilize the steady state. Examination of the known oscillators suggests, however, that there is little variation in the nature of the positive feedback in these systems. Where they differ is in the character of the negative feedback. We therefore discuss positive feedback only briefly and focus our attention on previously neglected aspects of negative feedback and on the interactions between the two types of feedback.

We seek to answer several questions. What forms does negative feedback take in realistic chemical systems? While each mechanism may appear to be unique, how different are they from the viewpoint of structure and function? What are the common features? What can we learn about how oscillatory behavior arises from looking at mechanisms and abstract oscillatory models?

Since all of the systems we consider have essentially the same type of positive feedback, our approach will be to classify known models and mechanisms into three groups according to the structural distinctions and similarities in their negative feedbacks. A key result that emerges is the necessity for a time delay between positive and negative feedbacks. Our treatment does not exclude other possibilities, but the examples of real mechanisms show that the classes discussed here are the best known and most frequently encountered to date. Many of the skeleton models, chosen for their simplicity and the clarity with which they illustrate each type of negative feedback, are from examples given by Franck[17] We use no more than three variables in the skeleton models, but, as we see in the detailed mechanisms, both positive and negative feedback networks can be expanded to any number of variables as long as the net result has the same pattern as the models. In all models considered, X, Y, and Z represent species whose concentrations vary, while A designates an input species held at fixed concentration, and P is an inert product.

II. POSITIVE FEEDBACK

In order for oscillations to occur in a chemical system, there must be an unstable steady state, that is, a state of the system in which all reaction rates are zero, but in which an appropriate infinitesimal perturbation will tend to grow rather then relax back toward the steady state. In order for the perturbation to grow, at least one of the following situations must exist. An increase in some concentration must lead to an increase in the rate at which that species is generated (or a decrease in the rate at which it is consumed), or a decrease in a concentration must yield a net decrease in the rate of production of that species. Such behavior is known as positive feedback. There are several schemes for characterizing positive feedback, for example, in terms of whether rates are increased or decreased by increases in concentration and whether the interaction is "forward" or "backward" in the reaction network.[13] Since reaction networks that describe chemical oscillators are often highly convoluted, the notions of forward activation or backward inhibition may be difficult to apply. We prefer the more general characterization suggested by Tyson[14] and based on the elements of the "community matrix," that is, the signs of the partial derivatives in Eq. (1).

A. Direct Autocatalysis

If $a_{ii} > 0$ for some species i, then increasing c_i causes i to be generated more rapidly; we have direct autocatalysis. The simplest example is the prototypical autocatalytic reation

$$A + X \to 2X \tag{2}$$

More complex forms of autocatalysis are of course possible. In the Sel'kov model[18] of glycolytic oscillations, the rate equation for ADP (y) is given by

$$\frac{dy}{dt} = \alpha(xy^\gamma - y) \tag{3}$$

where $x = [\text{ATP}]$. The termolecular autocatalytic reaction

$$A + 2X \to 3X \tag{4}$$

plays a key role in several models of theoretical importance[19-21] and has recently been shown[22] to arise quite straightforwardly from plausible sequences of bimolecular steps.

In discussing autocatalysis of the type shown in Eqs. (2) and (4), it is important to distinguish between two cases. If the concentration of A is held constant in some manner, as is assumed in many models, then we have *explosive* autocatalysis. The concentration of X will grow indefinitely, unless limited by some other reactions in the mechanism. Alternatively, and more realistically, if $[A]$ is a variable, then we have *nonexplosive*, or *self-limiting*, autocatalysis. The growth of $[X]$ is limited by the depletion of the precursor A. Only explosive autocatalysis constitutes unambiguously positive feedback. If the autocatalysis is self-limiting, it may provide positive feedback on one species (X) and negative feedback on another (A).

Tyson[14] points out that direct autocatalysis, defined as a positive a_{ii}, can take a more subtle form than shown in Eqs. (2) and (4). If the rate of a reaction decreases with the concentration of one of the reactants, we have substrate inhibition. Such a case is found in the chlorite–iodide reaction, where the rate law takes the form[23]

$$\text{Rate} = -\frac{d[\text{I}^-]}{dt} = \left\{ k_a[\text{H}^+][\text{I}^-] + \frac{k_b[\text{I}_2]}{[\text{I}^-]} + k_c[\text{I}_2] \right\}[\text{ClO}_2^-] \tag{5}$$

The k_b term with its inverse $[\text{I}^-]$ dependence represents substrate inhibition. Epstein and Kustin[24] demonstrate that such a term can be explained mechanistically by a rate-determining step involving hydrolysis of IClO_2, which is formed together with I^- in a rapid equilibrium,

$$\text{I}_2 + \text{ClO}_2^- \rightleftarrows \text{IClO}_2 + \text{I}^- \tag{6}$$

$$\text{IClO}_2 + \text{H}_2\text{O} \to \text{HIO}_2 + \text{HOCl} \tag{7}$$

B. Indirect Autocatalysis

A second form of positive feedback, which does not require $a_{ii} > 0$ for any individual species i, is referred to by Tyson[14] as indirect autocatalysis. The simplest examples of this type of behavior are competition, in which $a_{ij}, a_{ji} < 0$, that is, i inhibits j and j inhibits i, and symbiosis, where a_{ij}, $a_{ji} > 0$, that is, two species mutually activate one another. Such behavior is often found in population biology and is probably present in the half-center oscillator that can occur when two excitable neurons reciprocally inhibit one another.[25] This type of feedback is often found when simplified models are elaborated into full elementary step mechanisms. For example, the model step (2) could result from the real sequence of elementary steps

$$A + X \rightarrow Y \qquad \text{(slow)} \tag{8}$$

$$Y \rightarrow 2X \qquad \text{(fast)} \tag{9}$$

for which a_{XX}, $a_{YY} < 0$, but a_{XY}, $a_{YX} > 0$.

Still another possibility for positive feedback arises when there is a loop, a group of three or more species $i, j, k, \ldots, q$ such that

$$a_{ij} a_{jk} \cdots a_{qi} > 0 \tag{10}$$

Equation (10) signifies that if c_i is increased, its ultimate effect, mediated through changes in the rates of production or consumption of species $j, k, \ldots, q$, is to further increase c_i. Such behavior appears to occur (or at least to be recognized) more frequently in biochemical than in chemical systems, and we shall not consider it further here.

III.   NEGATIVE FEEDBACK

We now consider three types of negative feedback. In each case we begin with a simple model containing positive feedback in the form of direct autocatalysis. After analyzing the skeleton model, we consider examples of mechanisms in which this type of negative feedback is found.

A.   Coproduct Autocontrol and Oregonator-Type Systems

1.   The Oregonator

This pattern is the most frequently encountered among realistic mechanisms. The best known model is Field and Noyes' Oregonator[9]:

$$A + Y \rightarrow X + P \tag{O1}$$

$$X + Y \rightarrow 2P \tag{O2}$$

$$A + X \rightarrow 2X + Z \tag{O3}$$

$$2X \rightarrow A + P \tag{O4}$$

$$Z \rightarrow fY \qquad 0.25 < f < 1 + 2^{1/2} \tag{O5}$$

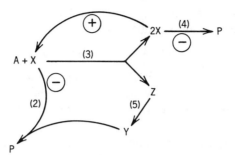

Figure 2. Schematic diagram of reaction network for the Oregonator. Numbers above arrows represent reaction numbers, while + and − signs identify reactions that contribute to the positive and negative feedback, respectively.

where the concentrations of X, Y, and Z are variable and those of A and P are fixed, while f is a stoichiometric coefficient. A schematic diagram of the flow in the reaction network is shown in Fig. 2.

Slow reaction (O1) is the initiation step. The positive feedback, explosive autocatalysis in this case, is provided by (O3). The bimolecular disproportionation (O4) is part of the negative feedback, because it is turned on by the accumulation of X and functions against its further increase. Step (O4) alone, however, is not strong enough to compete with the "explosion." It is incapable of bringing X back to its initial low concentration since the rate of this reaction declines as $[X]^2$.

The crucial feature of the negative feedback in this system is the coproduct autocontrol provided by reactions (O5) and (O2). As X increases autocatalytically in (O3), Z grows simultaneously. Through (O5), Z produces Y, which rapidly consumes X. The transformation of Z to Y provides the essential time delay, which has two essential functions: (1) it allows the positive feedback enough time so that X and Z can accumulate, and (2) after autocatalysis has ceased and the major portion of X has been consumed by (O4), enough Y remains to push $[X]$ down still further. Field and Noyes[9] report that a simulation in which (O5) was omitted and Y was substituted for Z in (O3) gave no oscillations, but only a stable steady state. The delayed formation of Y by (O5) slowly regenerates X through (O1) and reinitiates the cycle. This feature accomplishes the batch oscillation of the model, but it is not necessary in a flow system where the regeneration can be brought about by the inflow. We have been able to confirm this interpretation[26] using the theory of differential delay equations.[27] We omit (O5) and replace Z by Y in (O3). No oscillations are found. We then introduce a time lag by replacing $Y(t)$ in (O1) and/or (O2) by $Y(t - \tau)$. The delay τ is sufficiently long, the steady state is destabilized and oscillations reappear.

We now discuss three examples of real mechanisms in which the negative feedback is of the same autoproduct control type as in the Oregonator.

2. $H_2O_2-KSCN-CuSO_4-NaOH$

A relatively simple example is found in the mechanism of the basic H_2O_2- $KSCN-CuSO_4$ reaction,[28] for which the negative feedback was designed to follow the Oregonator pattern. The core reactions are as follows:

$$^-OS(O)CN + {}^-OOS(O)CN \xrightarrow{H_2O} 2 \cdot OS(O)CN + 2OH^- \tag{F18}$$

$$\cdot OS(O)CN + Cu^+ \longrightarrow {}^-OS(O)CN + Cu^{2+} \tag{F20}$$

$$H_2O_2 + Cu^{2+} + OH^- \longrightarrow HO_2 \cdot Cu(I) + H_2O \tag{F1}$$

$$HO_2 \cdot Cu(I) \longrightarrow Cu^+ + HO_2 \cdot \tag{F3}$$

$$^-OS(O)CN + HO_2 \cdot \longrightarrow SO_3^- \cdot + HOCN \tag{F21}$$

$$SO_3^- \cdot + Cu^+ \longrightarrow SO_3^{2-} + Cu^{2+} \tag{F22}$$

The combinations (F18) + (F20) and (F18) + 2(F20) provide routes for two-step direct autocatalysis, either by $^-OS(O)CN$ or by $\cdot OS(O)CN$. In fact, since the two reactions have quite different rate constants, the species reacting in the rate-determining step of this loop will have a higher concentration at all stages of the reaction. It is therefore considered the major autocatalytic species, that is, it plays the role of X in the Oregonator. Any negative feedback on this species will markedly affect the dynamics of the system. Since Cu^+ is consumed in (F20), the autocatalysis is nonexplosive. The coproduct of the autocatalysis, coresponding to Z in the Oregonator, is Cu^{2+}. The Cu^{2+} produces $HO_2 \cdot$ through reactions (F1) + (F3). The $HO_2 \cdot$ radical then plays the role of Y in the Oregonator, that is, it suppresses $^-OS(O)CN$ (X) through reaction (F21). These steps would seem to reproduce the main structure of the Oregonator negative feedback. However, the simultaneous production of Cu^+ in (F1) + (F3) destroys the time delay that is necessary for $^-OS(O)CN$ to be depleted to a level near its initial value, the same time delay that the slow step (O1) provides in the Oregonator. Even though the recovery steps regenerate Cu^+, a reagent consumed in the autocatalysis, the effect is very similar to that of (O1). If the recovery is too fast, the autocatalytic process cannot be turned down, and cyclic behavior cannot appear. To break the synchrony between the production of $HO_2 \cdot$ and the recovery of Cu^+ we need to invoke (F22). This variation from the simplest coproduct autocontrol model is required by the details of the chemistry in this particular system. Any realistic mechanism can be expected to contain similar departures from the idealized model.

3. Benzaldehyde Oxidation

Two mechanisms have been proposed for the oscillatory O_2 oxidation of benzaldehyde catalyzed by $CoBr_2$ (the Jensen reaction), by Roelofs et al.[11] and by Yuan and Noyes.[12] The two mechanisms differ significantly in some aspects of their chemistry, but are quite similar in their feedback structures; both belong to the Oregonator family. Since the Yuan–Noyes mechanism involves fewer intermediates and steps, we take it as the example. We shall not comment here on the disagreements between the two mechanisms.

The mechanism is given by reactions (S0)–(S14). Unlike the mechanism for the H_2O_2–$KSCN$–$CuSO_4$ system, in which a small group of reactions accounts for all the feedback effects, the majority of the reactions in this mechanism are involved in either the positive or the negative feedback networks. Nevertheless, the structural analysis is not necessarily more difficult, since the Oregonator pattern is followed without much variation.

$$PhCHO + O_2(sol) \rightarrow PhCO\cdot + HO_2\cdot \tag{S0}$$

$$PhCO\cdot + O_2(sol) \rightarrow PhCO_3\cdot \tag{S1}$$

$$PhCHO + PhCO_3\cdot \rightarrow PhCO\cdot + PhCO_3H \tag{S2}$$

$$PhCO_3H + Co^{2+} + H^+ \rightarrow PhCO_2\cdot + Co^{3+} + H_2O \tag{S3}$$

$$PhCHO + PhCO_2\cdot \rightarrow PhCO\cdot + PhCO_2H \tag{S4}$$

$$PhCO_3H + Br^- + H^+ \rightarrow PhCO_2H + HOBr \tag{S5}$$

$$HOBr + Br^- + H^+ \rightarrow Br_2 + H_2O \tag{S6}$$

$$Co^{3+} + PhCHO \rightarrow Co^{2+} + PhCO\cdot + H^+ \tag{S7}$$

$$Co^{3+} + Br^- \rightarrow Co^{2+} + Br\cdot \tag{S8}$$

$$2Br\cdot \rightarrow Br_2 \tag{S9}$$

$$PhCO\cdot + Br_2 \rightarrow PhCOBr + Br\cdot \tag{S10}$$

$$PhCOBr + H_2O \rightarrow PhCO_2H + Br^- + H^+ \tag{S11}$$

$$O_2(g) \rightarrow O_2(sol) \tag{S12}$$

$$2PhCO\cdot \rightarrow \text{stable products} \tag{S13}$$

$$Br_2 \rightarrow 2Br^- + ? \tag{S14}$$

The sequence of reactions (S1), (S2), (S3), (S4), and (S7) gives the following nonexplosive autocatalysis [O_2 is consumed and then is slowly replenished from the air by (S12)]:

$$PhCO\cdot + O_2 + 3PhCHO \xrightarrow{Co^{2+}, H^+} 3PhCO\cdot + PhCO_2H + H_2O \quad (11)$$

If (S7) is not fast enough to participate in the positive feedback process, the autocatalysis will generate only one molecule rather than two of PhCO·, and the (S8) + (S9) pathway (see below) will dominate the negative feedback.

Benzoic acid, $PhCO_2H$, might be a candidate for the role of Z in the Oregonator, because it is produced but not consumed in the positive feedback process. However, in a multistep autocatalysis, any one of the intermediates that appear with a nonnegligible concentration, such as $PhCO_3H$, Co^{3+}, $PhCO_2\cdot$, and $PhCO_3\cdot$ in this example, may be able to fuel the negative feedback. Yuan and Noyes chose two species from this list and built two parallel $Z \rightarrow fY$ paths, where the role of Y is played by Br_2: (S5) + (S6) is the path for $PhCO_3H$, and (S8) + (S9) is the path for Co^{3+}. These steps are followed by (S10) and (S13), which serve, respectively, as the $X + Y \rightarrow P$ and $2X \rightarrow P$ type Oregonator reactions.

4. $BrO_3^- - I^-$

The mechanism proposed by Citri and Epstein[29] for the bromate–iodide oscillator, Eqs. (E1–E13), is complicated, even from the structural point of view. It contains one multistep positive feedback, two parallel multistep negative feedbacks of the coproduct autocontrol type, and three single-step supplementary negative feedbacks, which look quite different from (O4) of the Oregonator, but function very similarly.

$$2H^+ + BrO_3^- + I^- \rightarrow HBrO_2 + HOI \quad (E1)$$
$$HBrO_2 + HOI \rightarrow HIO_2 + HOBr \quad (E2)$$
$$I^- + HOI + H^+ \rightleftarrows I_2 + H_2O \quad (E3)$$
$$BrO_3^- + HOI + H^+ \rightarrow HBrO_2 + HIO_2 \quad (E4)$$
$$BrO_3^- + HIO_2 \rightarrow IO_3^- + HBrO_2 \quad (E5)$$
$$HOBr + I_2 \rightleftarrows HOI + IBr \quad (E6)$$
$$IBr + H_2O \rightleftarrows HOI + Br^- + H^+ \quad (E7)$$
$$HBrO_2 + Br^- + H^+ \rightarrow 2HOBr \quad (E8)$$
$$HOBr + Br^- + H^+ \rightleftarrows Br_2 + H_2O \quad (E9)$$
$$BrO_3^- + Br^- + 2H^+ \rightleftarrows HBrO_2 + HOBr \quad (E10)$$
$$Br^- + HIO_2 + H^+ \rightleftarrows HOI + HOBr \quad (E11)$$
$$HIO_2 + HOBr \rightarrow IO_3^- + Br^- + 2H^+ \quad (E12)$$
$$BrO_3^- + IBr + H_2O \rightarrow IO_3^- + Br^- + HOBr + H^+ \quad (E13)$$

Steps (E1) and (E3) are the initiating processes that generate the essential species $HBrO_2$ and I_2 for the autocatalytic process. Since step (E6) is relatively slow, the sequence (E6) + (E7) + (E8) gives direct autocatalysis in HOBr while consuming I_2 and $HBrO_2$:

$$HOBr + I_2 + HBrO_2 + H_2O \rightarrow 2HOBr + 2HOI \tag{12}$$

The mechanism does not have a step that regenerates I_2 or I^-, so an inflow of I^- is necessary. The autocatalytic coproduct HOI participates in two parallel negative feedback pathways:

$$(E2)+(E12)+(E9): \quad HOI + HOBr + HBrO_2 \rightarrow IO_3^- + Br_2 + H_2O + H^+ \tag{13}$$

$$(E4)+(E12)+(E9): \quad HOI + 2HOBr + BrO_3^- \rightarrow IO_3^- + HBrO_2 + Br_2 + H_2O \tag{14}$$

In each multistep pathway Z, that is, HOI, is converted to the same $Y (Br^-)$ via HIO_2, and Br^- reacts with HOBr (X) to form the product Br_2. The two pathways are not totally independent. If we make the alternative assignment of HIO_2 as Y, the pathways differ only in the transition from Z to Y, but are identical in the processes that follow.

Reactions (E10), (E11), and (E13) provide an independent source of additional negative feedback. They help in controlling the active intermediates in the autocatalysis, Br^- and IBr. These steps are not essential for oscillation, but help to reproduce some of the quantitative details. A revised version of the mechanism[30] yields equally good agreement with experiment by omitting reaction (E13) and making small adjustments in some of the other rate constants.

A new feature of this mechanism is that Br^- participates in both the positive and the negative feedback networks. As a result of these competing processes, $[Br^-]$ differs by only a factor of 4 between the two bistable steady states, while some of the other concentrations change by two or three orders of magnitude.[29] The bromide ion concentration also shows a smaller amplitude of oscillation. Exactly how Br^- functions in the two antagonistic processes is not totally clear and the question of how closely the above mechanism fits into the Oregonator family is still open.

B. Double Autocatalysis

1. The Lotka–Volterra–Franck Model

Probably the first plausible model for chemical oscillation was that proposed by Lotka[31] and later developed by Volterra[32] in an ecological context. The original Lotka–Volterra model suffers from the deficiency that the oscillations it generates are conservative rather than of the limit cycle type.[33] By adding one more variable and one simple bimolecular reaction, Franck[17] was able to construct a simple skeleton model of a limit cycle oscillator based on double autocatalysis. We refer to this model as the LVF model:

$$A + X \rightarrow 2X \tag{L1}$$

$$X + Z \rightarrow 2Z \tag{L2}$$

$$X + Y \rightarrow P \tag{L3}$$

Initially $[X]$ and $[Z]$ are both low. The two autocatalytic steps (L1) and (L2), which are coupled through X, are activated when $[X]$ reaches its threshold, but with a brief time delay between them. Since it is turned on first in each oscillation cycle, and because it is responsible for the explosive nature of the autocatalysis, (L1) is the primary source of instability in the system. The "explosion" is terminated by the autocatalytic consumption of X by Z in (L2). This second autocatalysis is nonexplosive and terminates when X is consumed. The LVF model modifies the Lotka–Volterra model by adding flow terms in X, Y, and Z and by replacing the first-order consumption of Z following (L1) and (L2) with the bimolecular destruction of X by the additional variable Y.[17]

In the Lotka–Volterra version the direct coupling of the two variables and the lack of sufficient time delay make the oscillation sensitive to initial conditions and perturbations. There is an infinite set of oscillatory solutions, and fluctuations will push the system from one to another, resulting in irregular behavior in any real system.[33] In the LVF model the third variable, Y, through reaction (L3), reacts with X and provides the essential phase shift between X and Z. As shown in Fig. 3, Y accumulates sharply as Z consumes X autocatalytically, then holds X at a low level long enough for Z to regain its low initial

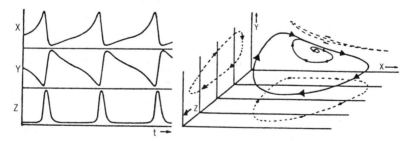

Figure 3. Computer-simulated oscillations and phase portrait of limit cycle in the LVF model.[17]

value. Although step (L3) is a crucial part of the negative feedback in this model, its concrete form in a realistic mechanism, as we shall see in the following examples, varies according to the chemistry of the individual chemical system. The invariant character of this group is the consecutive autocatalysis, with the second process constituting the negative feedback with respect to the autocatalytic species in the first.

2. The Limited Explodator

Also abstracted from the full Field–Körös–Noyes (FKN) mechanism[1] for the Belousov–Zhabotinskii reaction, the Limited Explodator proposed by Noszticzius, Farkas, and Schelly[10] is a simplified, semirealistic model that represents a different part of the FKN mechanism from that in the Oregonator. The two models start from the same positive feedback, that is, the auto-catalysis of $HBrO_2$ (X in both models), but differ in their negative feed-back pathways. We discuss only how the Limited Explodator oscillates from the structural point of view, avoiding comment on how well or badly the two models characterize the chemistry of the actual BZ system.

$$A + X \rightarrow (1 + \alpha)X \qquad \text{(N1)}$$

$$X + Y \rightarrow Z \qquad \text{(N2)}$$

$$Z \rightarrow (1 + \beta)Y \qquad \text{(N3)}$$

$$Y \rightarrow P \qquad \text{(N4)}$$

$$A \rightarrow X \qquad \text{(N5)}$$

When α is unity, the autocatalysis of X in (N1) is the same as that in (O3) of the Oregonator. If β is greater than 0, (N2) + (N3) gives autocatalysis in Y, consuming X at the same time. Here Z emerges as an intermediate in the second autocatalytic loop, breaking the direct coupling of X and Y. With appropriate rate constants, this feature provides a sufficient phase shift between the two autocatalytic processes. Step (N4), the monomolecular disappearance of Y, enables the model to oscillate in batch conditions, as does the BZ reaction. For the same reason, X is replenished through (N5). The calculated limit cycle behavior in the $X - Y$ plane is shown in Fig. 4. Other reactions, such as the disproportionation of X, are claimed to be able to replace (N5) in generating limit cycle behavior[10].

3. The Briggs–Rauscher Reaction

As noted earlier, two groups[5,6] arrived independently at almost identical mechanisms for the Briggs–Rauscher reaction, which involves hydrogen peroxide, manganese and iodate ions, and malonic acid (MA). We analyze here

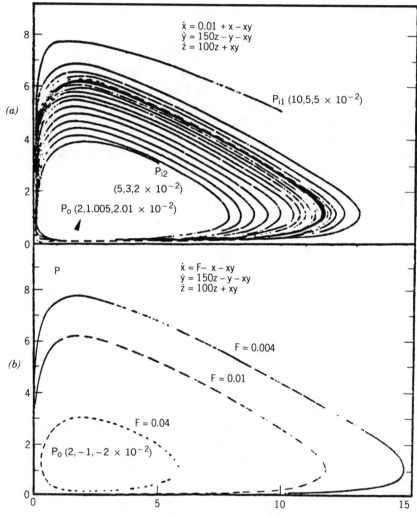

Figure 4. Calculated limit cycle behavior with Limited Explodator.[10] P's denote steady states.

284

the version of De Kepper and Epstein,[6] Eqs. (D1–D10), to show that the mechanism is principally of the double autocatalysis type.

$$I^- + IO_3^- + 2H^+ \rightarrow HOI + HIO_2 \tag{D1}$$

$$HIO_2 + I^- + H^+ \rightarrow 2HOI \tag{D2}$$

$$HOI + I^- + H^+ \rightleftharpoons I_2 + H_2O \tag{D3}$$

$$HIO_2 + IO_3^- + H^+ \rightarrow 2IO_2 \cdot + H_2O \tag{D4}$$

$$2HIO_2 \rightarrow HOI + IO_3^- + H^+ \tag{D5}$$

$$IO_2 \cdot + Mn^{2+} + H_2O \rightarrow HIO_2 + MnOH^{2+} \tag{D6}$$

$$H_2O_2 + MnOH^{2+} \rightarrow HO_2 \cdot + Mn^{2+} + H_2O \tag{D7}$$

$$2HO_2 \cdot \rightarrow H_2O_2 + O_2 \tag{D8}$$

$$I_2 + MA \rightarrow IMA + I^- + H^+ \tag{D9}$$

$$HOI + H_2O_2 \rightarrow I^- + O_2 + H^+ + H_2O \tag{D10}$$

The autocatalysis in HIO_2, composed of (D4) + 2(D6), is initiated, with the help of (D1), when the input reagents are mixed,

$$HIO_2 + IO_3^- + H^+ + 2Mn^{2+} \xrightarrow{H_2O} 2HIO_2 + 2MnOH^{2+} \tag{15}$$

When $[HIO_2]$ is high enough, another autocatalytic process, (D2) + 2(D10), is activated, producing I^- and consuming HIO_2. The stoichiometry of this process is

$$I^- + HIO_2 + 2H_2O_2 \rightarrow 2I^- + 2O_2 + H^+ + 2H_2O \tag{16}$$

The essential time delay, or phase shift, is provided by a kind of "negative feedback's negative feedback" that has not been seen in our previous examples: the autocatalytic species of the negative feedback, I^- in this example, and the intermediate of the same process, HOI, rapidly react in (D3) to form a relatively inert species, I_2. We see in Fig. 5 that [HOI] and $[I^-]$ are almost always out of phase with one another, just the opposite of what is expected from (D2) + 2(D10). While HOI is produced in the fast disproportionation of HIO_2, (D5), at the peak of the positive feedback, the autocatalytic, large-amplitude increase in I^- does not take place until sowewhat later. As the later process is finally turned on, HOI is consumed in (D3) by the large excess of I^-, as is HIO_2 in (D2). The second autocatalysis is therefore shut down. The two autocatalytic processes are so well separated in time that they are clearly

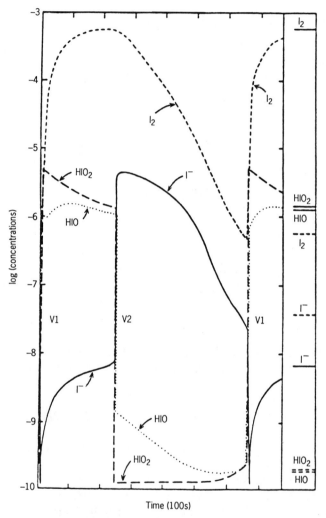

Figure 5. Simulated oscillations[6] in $[\mathrm{I}^-]$, $[\mathrm{I}_2]$, $[\mathrm{HOI}]$, and $[\mathrm{HIO}_2]$ in the Briggs–Rauscher reaction in a flow reactor with residence time 156 s. Input concentrations: $[\mathrm{H}_2\mathrm{O}_2] = 0.33M$, $[\mathrm{MA}]_0 = 0.0015M$, $[\mathrm{H}^+]_0 = 0.056M$, $[\mathrm{Mn}^{2+}]_0 = 0.004M$, $[\mathrm{IO}_3^-]_0 = 0.035M$, $[\mathrm{I}_2]_0 = 1.0 \times 10^{-6}M$. Right-hand column shows steady-state concentrations of the same species at a neighboring bistable composition with all parameters as for oscillation, except $[\mathrm{IO}_3^-]_0 = 0.038M$, $[\mathrm{I}_2]_0 = 2.8 \times 10^{-6}M$.

visible as two vertical lines in Fig. 5, with V_1 as the positive feedback and V_2 as the negative feedback. Other reactions, such as (D7), (D8), and (D9), are for completeness or slow recovery.

C. Flow Control

1. Franck's Model

The last group of oscillatory mechanisms we consider consists of systems that do not have sufficient internal negative feedback, so that the inflow reagents play a crucial role, beyond the usual function of simple replenishment. By internal negative feedback we mean that the chemical reactions, like those we have shown in the previous examples, have the ability both to terminate the positive feedback and to bring the intermediates back near their initial levels.

The simplest three-variable model is found, again, in Franck[17]:

$$X + Z \rightarrow 2X \tag{M1}$$

$$X + Y \rightarrow P_1 \tag{M2}$$

$$A + X \rightarrow P_2 \tag{M3}$$

The autocatalysis with respect to X, (M1), is nonexplosive, that is, it contains a negative feedback (the consumption of Z) to terminate itself. As Fig. 6 shows, Y is first consumed by (M2) when X is increasing. Since the system is open to the flow of X, Y, and Z, if the flow rate and inflow concentrations are such that Y recovers before Z, (M2) can provide the major part of a negative feedback that brings X back toward its initial level. If Z then recovers after X is consumed, the next cycle is ready to begin. If the chosen parameters do not yield the proper delay, (M1) resumes too early, and the cycle is never completed, in which case one obtains at most a few damped wiggles.

Step (M3) adds a linear term to $d[X]/dt$ is addition to the outflow, which apparently serves to break the symmetry among the outflows of the three

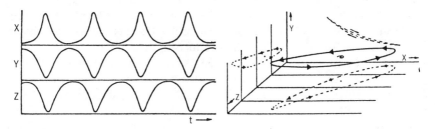

Figure 6. Calculated limit cycle behavior with the simplest skeleton model for flow-control type of negative feedback.[17]

variables. This step is not necessary in systems with more degrees of freedom, as we shall see in the following mechanisms.

2. $ClO_2^- - I^-$

Citri and Epstein's mechanism[34] for the $ClO_2^- - I^-$ system contains only six variables and eight kinetic steps. It is simplified from the original mechanism proposed by Epstein and Kustin[35] by eliminating the intermediates I_2O_2 and $IClO_2$. The simplified mechanism shows significantly better agreement with the experimental results in both batch and flow conditions, and its structural analysis is much easier.

$$Cl(III) + I^- + H^+ \rightarrow HOCl + HOI \tag{C1}$$

$$HOI + I^- + H^+ \rightleftarrows I_2 + H_2O \tag{C2}$$

$$HClO_2 + HOI \rightarrow HOCl + HIO_2 \tag{C3}$$

$$HOCl + I^- \rightarrow HOI + Cl^- \tag{C4}$$

$$HIO_2 + I^- + H^+ \rightleftarrows 2HOI \tag{C5}$$

$$2HIO_2 \rightarrow HOI + IO_3^- + H^+ \tag{C6}$$

$$HIO_2 + HOI \rightarrow IO_3^- + I^- + 2H^+ \tag{C7}$$

$$HIO_2 + HOCl \rightarrow IO_3^- + Cl^- + 2H^+ \tag{C8}$$

The sequence (C3) + (C4) + (C5) gives the following autocatalytic process with respect to HOI:

$$HOI + HClO_2 + 2I^- + H^+ \rightarrow 3HOI + Cl^- \tag{17}$$

Process (17), which consumes both $HClO_2$ and I^-, is the only independent autocatalysis in this mechanism.* To compare Eq. (17) with (M1), we identify Z as $HClO_2 + I^-$. We shall come back to this point later.

The counterpart of Franck's (M2) in the chlorite–iodide mechanism is (C2), with HOI as X and I^- as Y. The I_2 produced may be considered an inert end product in this analysis. The reversibility of (C2) determines the detailed shape of $[I_2]$, but does not alter the negative feedback nature of the reaction. At the end of the autocatalysis, $[I^-]$ has reached its minimum and both [HOI] and $[I_2]$ might be expected to have attained their maximum values. In fact, I_2 actually hydrolyzes ($-$C2) when $[I^-]$ is low, so that $[I_2]$ shows a drop at the time of the fast transition in Fig. 7.

---

* The autocatalysis with respect to HIO_2, Eq. (0) in Ref. 34, involves the same three elementary steps but with different coefficients in the linear combination. The concentrations of HOI and HIO_2 increase almost simultaneously during the autocatalysis.

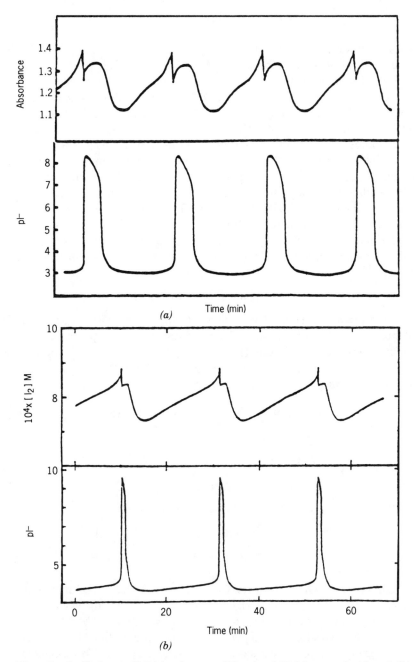

Figure 7. Oscillations in $[I_2]$ (absorbance at 460 nm) and $[I^-]$ (measured with an iodide-selective electrode) in the chlorite–iodide reaction in a CSTR.[34] (a) Experimental. (b) Calculated with simplified mechanism. $[I^-]_0 = 1.8 \times 10^{-3} M$, $[ClO_2^-]_0 = 6.0 \times 10^{-4} M$, pH = 1.5, $k_0 = 1.8 \times 10^{-3}\ s^{-1}$.

As I^- flows back in, (C2) brings about the recovery of $[I_2]$ through the same equilibrium, while suppressing HOI. The level of $[I_2]$ will drop again when HOI is gone, creating the second peak of $[I_2]$ in each oscillation cycle. Full recovery of I^- is not possible until HOI is completely consumed. Since I^- is one part of Z that is essential in the autocatalytic process, the regeneration of ClO_2^- by the inflow is not sufficient to reset the positive feedback. The next pulse cannot be induced before HOI is consumed. It is this linkage of Y and Z, possible only because there are more degrees of freedom, that ensures the proper phase shift. For this reason, no analog of (M3) is necessary here. Instead, there are some additional controlling reactions for the simultaneously generated intermediates (HIO_2 and HOCl) in the multistep autocatalysis, namely, (C6), (C7), and (C8). However, not all of these are essential for oscillation. For example, (C7) "contributes little to the behavior of the system under most conditions and can be omitted without causing significant change."[34]

3. The Minimal Bromate Oscillator

Bar-Eli[36] predicted the oscillation of a BZ type system without any organic substrate in a CSTR on the basis of analytic studies and computer simulations with the seven-step mechanism (I1)–(I7) proposed by Noyes et al.[37] The experimental discovery of such a "minimal bromate" oscillator[38,39] only a short time later was a striking instance of the interaction between theory and experiment.

$$BrO_3^- + Br^- + 2H^+ \rightleftarrows HBrO_2 + HOBr \qquad (I1)$$

$$HBrO_2 + Br^- + H^+ \rightleftarrows 2HOBr \qquad (I2)$$

$$HOBr + Br^- + H^+ \rightleftarrows Br_2 + H_2O \qquad (I3)$$

$$BrO_3^- + HBrO_2 + H^+ \rightleftarrows 2BrO_2 \cdot + H_2O \qquad (I4)$$

$$Ce^{3+} + BrO_2 \cdot + H^+ \rightleftarrows Ce^{4+} + HBrO_2 \qquad (I5)$$

$$Ce^{4+} + BrO_2 \cdot + H_2O \rightleftarrows Ce^{3+} + BrO_3^- + 2H^+ \qquad (I6)$$

$$2HBrO_2 \rightleftarrows BrO_3^- + HOBr + H^+ \qquad (I7)$$

Without the organic substrate, Ce^{4+}, the coproduct of the autocatalysis (I4) + 2(I5) (Z in the Oregonator), no longer produces Br^-, the key species (Y) in the Oregonator or coproduct control type of negative feedback. Nevertheless, bromide ion in step (I2) still plays a role analogous to that of Y in reaction (O2) of the Oregonator. The important difference is that now this species must come from the inflow rather than from a coproduct reaction. This situation resembles that of model (M1)–(M3). Thus by eliminating one reagent and one reaction or group of reactions, the minimal bromate mechanism, also derived from the FKN scheme, becomes a member of the family of flow control oscillators, in which the flow accomplishes both the negative feedback and the replenishment of the reagents.

It is interesting to observe the structural distinctions and the relations among the negative feedbacks of the three simplified mechanisms derived from the same FKN mechanism, that is, the Oregonator, the Limited Explodator, and the minimal bromate oscillator. This single system provides both an overview and new insights into the variety of negative feedback mechanisms found in chemical oscillators.

$$4. \quad BrO_3^- - SO_3^{2-} - Fe(CN)_6^{4-}$$

The nine-reaction mechanism (B1)–(B9) generates both the oscillatory behavior and the bistability found in the bromate–sulfite–ferrocyanide pH oscillator. This mechanism resembles the minimal bromate oscillator or the model (M1)–(M3) only in that the flow finally accomplishes the negative feedback, but it differs greatly from the other two in form. There is one special feature in particular that is found in almost all pH oscillators: fast protonation–deprotonation equilibria serve as important control steps in both positive and negative feedback. As more mechanisms for pH oscillators with very different chemistry are established, they may well form their own class.

$$BrO_3^- + HSO_3^- \rightarrow HBrO_2 + SO_4^{2-} \tag{B1}$$

$$HBrO_2 + Br^- + H^+ \rightarrow 2HOBr \tag{B2}$$

$$HOBr + Br^- + H^+ \rightarrow Br_2 + H_2O \tag{B3}$$

$$Br_2 + H_2O \rightarrow HOBr + Br^- + H^+ \tag{B4}$$

$$2HBrO_2 \rightarrow BrO_3^- + HOBr + H^+ \tag{B5}$$

$$Br_2 + HSO_3^- + H_2O \rightarrow 2Br^- + SO_4^{2-} + 3H^+ \tag{B6}$$

$$H^+ + SO_3^{2-} \rightarrow HSO_3^- \tag{B7}$$

$$HSO_3^- \rightarrow H^+ + SO_3^{2-} \tag{B8}$$

$$BrO_3^- + 2Fe(CN)_6^{4-} + 3H^+ \rightarrow HBrO_2 + 2Fe(CN)_6^{3-} + H_2O \tag{B9}$$

The rapid equilibrium (B7)–(B8) causes the initial solution to be basic. Steps (B1)–(B6) constitute an explicit autocatalysis in H^+ in the sense that H^+ participates directly in the process that increases its concentration.

On the other hand, since HSO_3^- is much more reactive than the unprotonated SO_3^{2-} in all reactions in which sulfite is involved, the equilibrium (B7)–(B8) constitutes an implicit or indirect autocatalysis with respect to H^+. That is, the more free H^+ is produced in (B1)–(B6), the more HSO_3^- is available to facilitate that process. The autocatalysis is terminated by the total consumption of sulfite.

Reaction (B9), which independently consumes large amounts of H^+, is certainly a part of the negative feedback, but it is not powerful enough in itself to generate oscillations. An essential source of negative feedback is the inflow of unprotonated sulfite, which removes H^+ rapidly through, again, the protonation equilibrium. Meanwhile Br_2 hydrolyzes (B3)–(B4) as soon as $[H^+]$ drops, cutting off the reaction route (B6) for the newly formed HSO_3^-, thereby providing the required time delay for the next pulse of autocatalysis. It is clearer in this case than in the other examples, even though it is true in all of them, that sufficient negative feedback for an oscillation requires several reactions or processes having slightly different functions. It is inappropriate in analyzing a system to isolate them from each other, or from the positive feedback.

A simplified model that reflects many of the general features of pH oscillators, as well as some of the details of the mechanism for the $BrO_3^- - SO_3^{2-} - Fe(CN)_6^{4-}$ system, (B1)–(B9), is given by:

$$X' + Y \rightleftarrows X \qquad \text{(Y1)}$$

$$A + X + Y \rightarrow Z \qquad \text{(Y2)}$$

$$X + Z \rightarrow 3Y \qquad \text{(Y3)}$$

$$Z \rightarrow P \qquad \text{(Y4)}$$

$$B + Y \rightarrow Q \qquad \text{(Y5)}$$

The species X, X', and Y correspond to HSO_3^-, SO_3^{2-}, and H^+, respectively, while Z can represent any or all of the important intermediates in the positive feedback, such as $HBrO_2$, $HOBr$, and/or Br_2 in (B1)–(B9). The role of bromine hydrolysis (B4) in removing a key intermediate of the autocatalysis is played by (Y4). Step (Y5) is the analog of (B9) in the full mechanism. The protonation equilibrium (Y1) is rapid, and the autocatalysis terminates when X' is consumed. Only X' and Y have nonzero inflow, to mimic the real system. The excess of X' over Y in the inflow should generate some negative feedback, and the disappearance of Z should provide the time delay.

Both the original[8] and the revised[4] mechanisms for the $IO_3^- - SO_3^{2-} - Fe(CN)_6^{4-}$ system can be represented by the model (Y1)–(Y5). They have different negative feedback pathways only in the sense that different chemical species are used to construct the process (Y5), namely, $I_2 + Fe(CN)_6^{4-}$ in the original and $IO_3^- + Fe(CN)_6^{4-}$ in the revised mechanism. We expect this model to be applicable to most pH oscillators.

IV. COUPLED FEEDBACK NETWORKS

We have described three different types of negative feedback networks and have classified a number of well-established mechanisms in these terms. Several questions now arise. Can one system contain more than one kind of negative feedback? If so, can these feedbacks be independent? Can they cooperate? What can one expect to happen in such cases?

The mechanism for the $BrO_3^- - ClO_2^- - I^-$ oscillator proposed recently by Citri and Epstein[30] gives preliminary answers to some of these questions. That study represents the first attempt to synthesize mechanisms of two subsystems, bromate–iodide[29] and chlorite–iodide,[34] each capable of oscillation, into a description of a chemically coupled oscillator. The combined mechanism is not only able to describe the two different kinds of oscillations originating from the $BrO_3^- - I^-$ and $ClO_2^- - I^-$ suboscillators, but it is also strikingly successful in reproducing the composed[40] and compound[41] oscillations observed in experiments. It is the first and the only realistic mechanism that convincingly demonstrates birhythmicity[41,42] to date.

A composed or compound oscillation observed in a single-state variable time series reflects a complex motion of the trajectory in the multidimensional phase space. If the frequencies or the amplitudes of the fundamental oscillations are distinct, the different kinds of oscillations as well as their various forms of coupling may be distinguishable in a chemically coupled system, as seen in the $BrO_3^- - ClO_2^- - I^-$ reaction. Birhythmicity, that is, bistability between two different modes of oscillations, is another phenomenon that can arise when two oscillations are coupled.

The mechanisms for the oscillatory $BrO_3^- - I^-$ and $ClO_2^- - I^-$ subsystems described in previous sections belong to different negative feedback groups, the coproduct autocontrol and the flow control types, respectively. The two feedback networks remain largely independent when the two mechanisms are combined; each is responsible for one type of oscillation in the region of phase space in which only a single type of oscillation occurs. With certain parameter values the trajectory can encompass both of the previously independent attractors, leading to compound or composed oscillations, or even chaos.

If there exists more than one structurally independent feedback network in a complex chemical system, these may be coupled either by sharing a common positive or negative feedback, as we shall see in the next example, or through a component, such as HOI in the $BrO_3^- - ClO_2^- - I^-$ system, which participates in the negative feedback in one network and in the positive feedback of the other. By structurally independent networks, we mean networks that have different functions and, more importantly, are composed of different sets of chemical processes in their cores. They also have to have different time scales or amplitudes in order to be recognized phenomenologically. Whether or not they must belong to different negative feedback types is an open question. Examples of parallel negative feedback pathways have been seen in the mechanisms for the $BrO_3^- - I^-$ system and the benzaldehyde oxidation. Neither of them shows complex behavior, either because the parallel pathways share important reaction steps, which prevents them from being structurally independent, or because they are essentially the same type of network.

Another example of the generation of complex oscillatory sequences by feedback coupling is found in recent calculations by Richetti et al.,[43] in which a seven-variable, nine-reaction mechanism for the BZ system expanded from the Oregonator was used to simulate the quasi-periodicity, bursting (similar to composed oscillation), and chaos observed in numerous experiments.[44–47] Earlier efforts to reproduce these behaviors with various versions of the Oregonator, which contains only a single source of oscillation, generally resulted in failure. The success of the seven-variable expanded model results from the introduction of a new pseudoindependent negative feedback loop.

$$BrO_3^- + Br^- + 2H^+ \rightarrow HBrO_2 + HOBr \qquad (R1)$$

$$HBrO_2 + Br^- + H^+ \rightarrow 2HOBr \qquad (R2)$$

$$HOBr + Br^- + H^+ \rightarrow Br_2 + H_2O \qquad (R3)$$

$$BrO_3^- + HBrO_2 + H^+ \rightleftarrows 2BrO_2\cdot + H_2O \qquad (R4)$$

$$2HBrO_2 \rightarrow HOBr + BrO_3^- + H^+ \qquad (R5)$$

$$BrO_2\cdot + Ce(III) + H^+ \rightarrow Ce(IV) + HBrO_2 \qquad (R6)$$

$$HOBr + MA \rightarrow BrMA + H_2O \qquad (R7)$$

$$BrMA + Ce(IV) \rightarrow Br^- + R\cdot + Ce(III) + H^+ \qquad (R8)$$

$$R\cdot + Ce(IV) \rightarrow Ce(III) + P \qquad (R9)$$

We first recognize the Oregonator in the above mechanism: (R1) ~ (O1), (R2) ~ (O2), (R4) + 2(R6) ~ (O3), (R5) ~ (O4), and (R8) + (R9) ~ (O5). Only (R3) and (R7) are non-Oregonator ingredients. Evidently, it is these that lead from the simple oscillation of the Oregonator to the toroidal motions found in the new calculations.[43] Upon adding [HOBr] as a variable, the dead end of (O2) is opened up to a process that is autocatalytic in Br^-, consuming $HBrO_2$ more effectively through the process (R2) + 2[(R7) + (R8)]. This is the core of the Explodator (N1)–(N5), with some minor alternation in variable assignment. Since one of the two routes of $HBrO_2$ consumption is autocatalytic (Explodator type) while the other is not (Oregonator type), they may be expected to have different time scales and feedback power. Each is responsible for one of the two dimensions of the torus that results when they are combined with the common positive feedback. Reaction (R3) serves as the limiting step for the second autocatalytic loop—the "negative feedback's negative feedback." As the preceding discussion suggests, it is more appropriate to refer to the mechanism (R1)–(R9) as a modified FKN than as an expanded Oregonator.

As a result of the involvement of Ce(IV) in the common reaction (R8), the two negative feedback loops are not completely independent. The difference in their time scales is limited by the phase of [Ce(IV)]. The inclusion of steps that produce Br^- from HOBr and organic substrate without the metal–ion catalysis (R8) might dissociate the two loops from each other, and thus make it possible to find the double-frequency quasi-periodicity seen in the experiments (Fig. 8). The existence of a large variety of BZ variants that oscillate in the absence of a metal ion (uncatalyzed bromate oscillators[48]) suggests the plausibility of such reactions

In linear systems, behavior like that seen in Fig. 8 results from the addition of two independent waves of different frequencies; the frequency difference determines the envelope frequency of the quasi-periodicity. While chemical systems are nonlinear, some of the same considerations apply. Quasi-periodicity may be expected to arise when there are two well-defined frequencies resulting from independent feedback networks which are then coupled either physically or chemically.[49]

Some "simple" oscillators, certainly simple in terms of their components, have been reported to exhibit complex dynamical behavior. Composed oscillations and chaos, for example, have been found in the $ClO_2^- - S_2O_3^{2-}$ system,[50] while the chlorite–thiourea reaction shows birhythmicity between two distinct modes of oscillations.[51] The analysis presented here implies that there must be more than one independent feedback network in the mechanisms that will be able to simulate the complex dynamics found in these systems. Sharing a common positive feedback would be the simplest form for the coupling.

Coupling through a common positive feedback is not, however, the only way that feedback networks may be joined. Rábai et al.[52] recently discovered oscillatory behavior in the reaction of ferrocyanide ion with hydrogen peroxide in a CSTR. Since this is a subsystem of the newly devised pH oscillator[53] $H_2O_2 - SO_3^{2-} - Fe(CN)_6^{4-}$, additional experiments with the latter were run with much higher $[H_2O_2]_0$. A second mode of oscillation, with much higher frequency than observed in the previous experiments, was found, apparently corresponding to the oscillation of the sulfite-free subsystem. The conventional wisdom has been that in the oscillatory systems based on the reactions of iodate,[54] bromate,[55] and hydrogen peroxide[53] with ferrocyanide and sulfite the SO_3^{2-} participates only in the positive feedback process and $Fe(CN)_6^{4-}$ only in the negative feedback. If the subsystem containing only H_2O_2 and

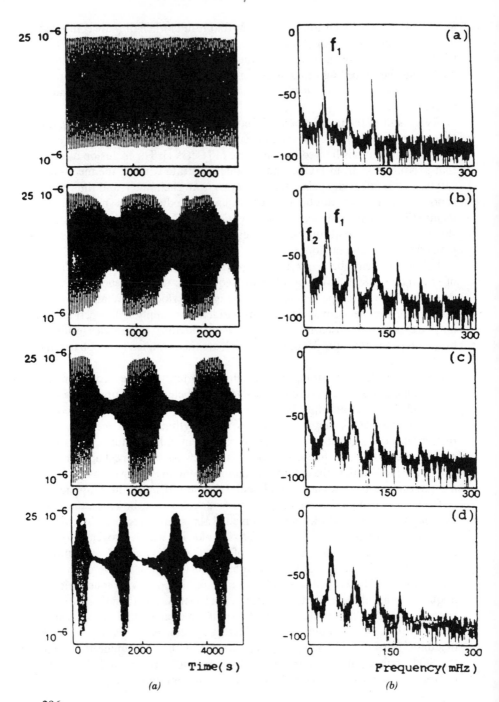

Time(s)

Prequency(mHz)

(a)

(b)

296

$Fe(CN)_6^{4-}$ oscillates, it must have its own instability source, which must be very different from the SO_3^{2-}-generated one. It will be interesting to see how such a system, with independent positive feedbacks but sharing a common negative feedback, behaves when the experiment moves away from the single-oscillation regions.

V. CONCLUSION

We have identified and illustrated with simple models three types of negative feedback. Their characteristics are embodied in the examples of mechanisms of actual chemical oscillators, with variations in form dictated by the chemistry of the individual systems. More than 10 "realistic" mechanisms have been analyzed in varying degrees of detail and found to fit well into the new framework.

The recognition and formalization of different patterns of negative feedback in both abstract models and detailed mechanisms indicate the progress that has been made in understanding the relation between the kinetics and the dynamics of complex chemical reactions. The understanding of kinetic structures and their relation to dynamics may serve as a guide in future mechanistic studies. A feedback pattern can be determined and unnecessary steps avoided if enough information about the dynamics and the chemistry of the system is availble. Even the preliminary analyses presented here have led us to suggestions for improving the "modified FKN" mechanism[43] and to predictions about the structure of mechanisms for the chlorite–thiosulfate and chlorite–thiourea oscillators.

We have embraced and even embellished Franck's view[17] that the interplay of positive and negative feedback is the key to oscillaton. We have also emphasized the crucial role of time lags between the two modes of feedback so that concentrations may repeatedly attain their maximum and minimum values rather than settling into a stable steady state.

While we have limited our treatment here to homogeneous systems, we note that similar ideas emerge from the work of Noyes and coworkers on gas evolution oscillators. In particular, in the Bubbelator model,[56] the oscillations are shown, using the community matrix analysis,[14] to arise from a three-variable negative feedback loop, with no possibility of autocatalysis.[57] Smith and Noyes[56] also point out the importance of delay in the negative feedback and suggest the use of diferential-difference equations to describe this delay.

◁   **Figure 8.**   Experimental quasi–periodicity observed in BZ reaction.[47] Input concentrations: $[BrO_3^-] = 1.2 \times 10^{-2}M$, $[MA]_0 = 0.175M$, $[Ce^{3+}]_0 = 8.3 \times 10^{-5}M$, $[H_2SO_4] = 2N$. Flow rates, from top to bottom: 0.34, 0.355, 0.36, 0.43 mL/min. (a) Amplitude (Ce(IV)). (b) Power spectra.

We emphasize that the three forms of negative feedback discussed in these preliminary studies are not all the possibilities. It will also be of interest to see whether this sort of analysis will be of use in understanding biological oscillators. Enzymatic and neural oscillators, while they are generally modeled at a different level of description, contain feedback elements very similar to those in the chemical systems we have described. Can they be usefully dissected in terms of positive and negative feedback loops with delays, and will the feedback networks found fall into the classes defined here, or will new categories be necessary?

Acknowledgment

We thank Kenneth Kustin, Robert Olsen, and Gyula Rábai for many enlightening discussions. We are grateful to Zoltan Noszticzius, Kenneth Showalter, and, especially, Richard Noyes for their detailed comments on an earlier version of this manuscript. This work was supported by Grant CHE-8800169 from the National Science Foundation.

References

1. R. J. Field, E. Körös, and R. M. Noyes, *J. Am. Chem. Soc.* **94**, 8649 (1972).
2. P. De Kepper and J. Boissonade, in *Oscillations and Traveling Waves in Chemical Systems*, R. J. Field and M. Burger, Eds., Wiley, New Work, 1985, p. 223.
3. I. R. Epstein, K. Kustin, P. De Kepper, and M. Orbán, *Sci. Amer.* **248**(3), 112 (1983).
4. Y. Luo and I. R. Epstein, *J. Phys. Chem.* **93**, 1398 (1989).
5. R. M. Noyes and S. D. Furrow, *J. Am. Chem. Soc.* **104**, 45 (1982).
6. P. De Kepper and I. R. Epstein *J. Am. Chem. Soc.* **104**, 49 (1982).
7. V. Gáspár and K. Showalter, *J. Am. Chem. Soc.* **109**, 4869 (1987).
8. E. C. Edblom, L Györgyi, M. Orbán, and I. R. Epstein, *J. Am. Chem. Soc.* **109**, 4876 (1987).
9. R. J. Field and R. M. Noyes, *J. Chem. Phys.* **60**, 1877 (1974).
10. Z. Noszticzius, H. Farkas, and Z. A. Schelly, *J. Chem. Phys.* **80**, 6062 (1984).
11. M. G. Roelofs, E. Wasserman, and J. H Jensen, *J. Am. Chem. Soc.* **109**, 4207 (1987).
12. Z. Yuan and E. M. Noyes, *J. Am. Chem. Soc.*, in press.
13. J. Higgins, *Ind. Eng. Chem.* **59**, 19 (1976).
14. J. J. Tyson, *J. Chem. Phys.* **62**, 1010 (1975).
15. R. J. Field and R. M. Noyes, *Acc. Chem. Res.* **10**, 214 (1977).
16. R. M. Noyes and R. J. Field, *Acc. Chem. Res.* **10**, 273 (1977).
17. U. F. Franck, in *Temporal Order*, L. Rensing and N. I. Jaeger, Eds., Springer, Berlin, 1985, p. 2.
18. E. E. Sel'kov, *Eur. J. Biochem.* **4**, 79 (1968).
19. I. Prigogine and R. Lefever, *J. Chem. Phys.* **48**, 1695 (1968).
20. F. Schlögl, *Z. Phys.* **253**, 147 (1972).
21. P. Gray and S. K. Scott *Chem. Eng. Sci.* **38**, 29 (1983).
22. G. B. Cook, P. Gray, D. G. Knapp, and S. K. Scott, *J. Phys. Chem.* **93**, 2749 (1989).
23. J. de Meeus and J. Sigalla, *J. Chim. Phys., Phys-Chim. Biol.* **63**, 453 (1966).
24. I. R. Epstein and K. Kustin, *J. Phys. Chem.* **89**, 2275 (1985).
25. D. M. Wilson, and I. Waldron, *Proc. IEEE* **56**, 1058 (1968).
26. I. R. Epstein, *J. Chem. Phys.* **92**, 1702 (1990).
27. R. Bellman and K. L. Cooke, *Differential-Difference Equations*, Academic Press, New York, 1963.

28. Y. Luo, M. Orbán, K. Kustin, and I. R. Epstein, *J. Am. Chem. Soc.* **111**, 4541 (1989).
29. O. Citri and I. R. Epstein, *J. Am. Chem. Soc.* **108**, 357 (1986).
30. O. Citri and I. R. Epstein, *J. Phys. Chem.* **92**, 1865 (1988).
31. A. J. Lotka, *J. Am. Chem. Soc.* **42**, 1595 (1920).
32. V. Volterra, *Leçons sur la Théorie Mathématique de la Lutte pour la Vie*, Gauthier-Villars, Paris, 1931.
33. G. Nicolis and J. Portnow, *Chem. Rev.* **73**, 365 (1973).
34. O. Citri and I. R. Epstein, *J. Phys. Chem.* **91**, 6034 (1987).
35. I. R. Epstein and K. Kustin, *J. Phys. Chem.* **89**, 2275 (1985).
36. K. Bar-Eli, in *Nonlinear Phenomena in Chemical Dynamics*, C. Vidal and A. Pacault, Eds., Springer, Berlin, 1981, p. 228.
37. R. M. Noyes, R. J. Field, and R. C. Thompson, *J. Am. Chem. Soc.* **93**, 7315 (1971).
38. M. Orbán, P. De Kepper, and I.R. Epstein, *J. Am. Chem. Soc.* **104**, 2657 (1982).
39. W. Geiseler, *J. Phys. Chem.* **86**, 4394 (1982).
40. J. Maselko, M, Alamgir, and I. R. Epstein, *Physica* **19D**, 153 (1986).
41. M. Alamgir and I. R. Epstein, *J. Am. Chem. Soc.* **105**, 2500 (1983).
42. O. Decroly and A. Goldbeter, *Proc. Nat. Acad. Sci. USA* **79**, 6971 (1982).
43. P. Richetti, J. C. Roux, F. Argoul, and A. Arneodo, *J. Chem. Phys.* **86**, 3339 (1987).
44. R. A. Schmitz, K. R. Graziani, and J. L. Hudson, *J. Chem. Phys.* **67**, 3040 (1977).
45. J. C. Roux, A. Rossi, S. Bachelart, and C. Vidal, *Phys. Lett. A* **77**, 391 (1980).
46. R. H. Simoyi, A. Wolf, and H. L. Swinney, *Phys. Rev. Lett.* **49**, 245 (1982).
47. F. Argoul, A. Arneodo, P. Richetti, and J. C. Roux, *J. Chem. Phys.* **86**, 3325 (1987).
48. M. Orbán and E. Körös, *J. Phys. Chem.* **82**, 1672 (1978).
49. C. G. Hocker and I. R. Epstein, *J. Chem. Phys.* **90**, 3071 (1989).
50. M. Orbán and I. R. Epstein, *J. Phys. Chem.* **86**, 3907 (1982).
51. M. Alamgir and I. R. Epstein, *Int. J. Chem. Kinet.* **17**, 429 (1985).
52. G. Rábai, K. Kustin, and I. R. Epstein, part 57 in the series Systematic Design of Chemical Oscillators, *J. Am. Chem. Soc.*, **111**, 8271 (1989).
53. G. Rábai, K. Kustin, and I. R. Epstein, *J. Am. Chem. Soc.* **111**, 3870 (1989).
54. E. C. Edblom, M. Orbán, and I. R. Epstein, *J. Am. Chem. Soc.* **108**, 2826 (1986).
55. E. C. Edblom, Y. Luo, M. Orbán, K. Kustin, and I. R. Epstein, *J. Phys. Chem.* **93**, 2722 (1989).
56. K. W. Smith and R. M. Noyes, *J. Phys. Chem.* **7**, 1520 (1983).
57. R. M. Noyes, *J. Phys. Chem.* **88**, 2827 (1984).

AUTHOR INDEX

Numbers in parentheses are reference numbers and indicate that the author's work is referred to although his name is not mentioned in the text. Numbers in *italics* show the pages on which the complete references are listed.

SUBJECT INDEX